MPEG-4 Facial Animation

The Standard, Implementation and Applications

MPEG-4 Facial Animation

The Standard, Implementation and Applications

Edited by

Igor S. Pandzic and Robert Forchheimer

Linköping University, Sweden

JOHN WILEY & SONS, LTD

Telephone (+44) 1243 779777

Email (for orders and customer service enquiries): cs-books@wiley.co.uk
Visit our Home Page on www.wileyeurope.com or www.wiley.com

Other Wiley Editorial Offices

John Wiley & Sons Inc., 111 River Street, Hoboken, NJ 07030, USA

Jossey-Bass, 989 Market Street, San Francisco, CA 94103-1741, USA

Wiley-VCH Verlag GmbH, Pappelallee 3, D-69469 Weinheim, Germany

John Wiley & Sons Australia Ltd, 33 Park Road, Milton, Queensland 4064, Australia

John Wiley & Sons (Asia) Pte Ltd, 2 Clementi Loop #02-01, Jin Xing Distripark, Singapore 129809

John Wiley & Sons Canada Ltd, 22 Worcester Road, Etobicoke, Ontario, Canada M9W 1L1

British Library Cataloguing in Publication Data

A catalogue record for this book is available from the British Library

ISBN 0-470-84465-5

Typeset in 10/12pt Times by Laserwords Private Limited, Chennai, India
Printed and bound in Great Britain by Biddles Ltd, Guildford and King's Lynn
This book is printed on acid-free paper responsibly manufactured from sustainable forestry in which at least two trees are planted for each one used for paper production.

Contents

Appendices

List of Contributors

Jörgen Ahlberg
Department of Electrical Engineering
Linköping University
SE-581 83 Linköping
Sweden
ahlberg@isy.liu.se

Michele Cannella
TAU Tecnologia Automazione Uomo s.c.r.l.
Via XX Settembre 3/6
16121 Genova
Italy
michele.cannella@tetralab.it
michele.cannella@tau-online.it

Roddy Cowie
Queen's University of Belfast
Belfast, N. Ireland
r.cowie@qub.ac.uk

Ellen Douglas-Cowie
Queen's University of Belfast
Belfast, N. Ireland
e.douglas-cowie@qub.ac.uk

Franck Davoine
Université de Technologie de Compiegne
Laboratoire Heudiasyc, BP20529
60205 France
Franck.Davoine@hds.utc.fr

Robert Forchheimer
Department of Electrical Engineering
Linköping University
SE-581 83 Linköping
Sweden
Robert@isy.liu.se

Pengyu Hong
1614 Beckman Institute
Urbana IL61801
hong@ifp.uiuc.edu

Thomas S. Huang
2039 Beckman Institute
Urbana IL61801
huang@ifp.uiuc.edu

Stefanos Kollias
Image, Video and Multimedia Systems Laboratory
National Technical University of Athens
Electrical & Computer Engineering Department
Computer Science Division
ECE Building – 1st Floor – Room 11.23
Athens, Greece
Stefanos@cs.ntua.gr

Fabio Lavagetto
Università degli Studi di Genova
Dipartimento di Informatica, Sistemistica e Telematica
Via all'Opera Pia 13
16145 Genova, Italy
fabio@dist.unige.it

Haibo Li
Digital Media Lab
Umeå University
SE-901 87 Umeå
Haibo.li@tfe.umu.se

Sotiris Malassiotis
ITI, Greece
malasiot@iti.gr

Andrew Marriott
Senior Lecturer
School of Computing
Curtin University of Technology
Hayman Road, Bentley
Western Australia, 6102
raytrace@smtp.cs.curtin.edu.au

Jörn Ostermann
AT&T Labs-Research
Rm A5–4F36
200 Laurel Ave South
Middletown, NJ 07748, USA
Joern.Ostermann@ieee.org

Igor S. Pandzic
Department of Electrical Engineering
Linköping University
SE-581 83 Linköping
Sweden
igor@isy.liu.se

Department of Telecommunications
Faculty of Electrical Engineering and Computing
University of Zagreb
Unska 3
HR-10000 Zagreb
Croatia
Igor.Pandzic@fer.hr

Montse Pardas
Universitat Politècnica de Catalunya
Barcelona, Spain
montse@gps.tsc.upc.es

Catherine Pelachaud
Universita di Roma "La Sapienza"
Dipartimento di Informatica e Sistemistica
via Buonarroti, 12
00185 Roma, Italy
cath@dis.uniroma1.it

Dr. Eric Petajan
Chief Scientist and Founder
face2face animation, Inc.
2 Kent Place Blvd
Summit, NJ 07901
eric@f2f-inc.com

Roberto Pockaj
Eptamedia s.r.l.
c/o Dipartimento di Informatica, Sistemistica e Telematica
Via all'Opera Pia 13
16145 Genova
Italy
roberto.pockaj@eptamedia.com

Amaryllis Raouzaiou
Image, Video and Multimedia Systems Laboratory
National Technical University of Athens
Electrical & Computer Engineering Department
Computer Science Division
ECE Building – 1st Floor – Room 11.23
Athens, Greece
araouz@image.ece.ntua.gr

Nicolas Tsapatsoulis
Image, Video and Multimedia Systems Laboratory
National Technical University of Athens
Electrical & Computer Engineering Department
Computer Science Division
ECE Building – 1st Floor – Room 11.23
Athens, Greece
ntsap@image.ntua.gr

Gael Sannier
W Interactive SARL, France
gael@winteractive.fr

Zhen Wen
1614 Beckman Institute
Urbana IL61801
zhenwen@ifp.uiuc.edu

Liwen You
Linköping University Linköping
Sweden
liwenyou@ieee.org

Author Biographies

Jörgen Ahlberg was born in Karlstad (Sweden) in 1971. He got his M.Sc. in Computer Science and Engineering from Linköping University (Sweden) in 1996, and joined the Image Coding Group, also at Linköping University, as a Ph.D. student the same year. In the Image Coding Group and at Université de Technologie de Compiègne (France, 1999) he has since then undertaken research in different aspects of model-based coding, such as facial parameter compression, texture coding, detection and tracking of faces and facial features and evaluation of face models. He was also an active participant in the development of MPEG-4 Face Animation and is currently working in the European InterFace project. He has seen Andrew Marriott wear shoes. J. Ahlberg can be reached at *ahlberg@isy.liu.se*.

Michele Cannella was born in Genoa on January 14, 1968. He got the 'laurea' degree in Electrical Engineering at DIST, University of Genoa, in 1992. From September 1992 to March 1993 he had a post-degree scholarship from Marconi, working on applications based on Oracle DB. In the period from September 1994 to June 1995, he was a consultant with Modis spa on C applications with Unix-Motif user interface. From April 1996 to May 1999 he worked with Elmer spa, with specific responsibility for the European ACTS project Vidas, with reference to software porting on real-time platform. From January 2000 he has been working with TAU scrl for the European project Interface, with reference to design and implementation of 'real world' applications based on the innovative technologies developed by the partners of the project. In September 2001 he founded Tetralab srl (*www.tetralab.it*), a software development and consulting company based at Genoa. His interests are in the field of software design using object-oriented technologies (UML), the applications of markup languages (XML) and java-based web applications. Michele can be reached at *michele.cannella@tetralab.it*

Roddy Cowie works in the School of Psychology at Queen's University, Belfast. His core interest is in the relationship between the way humans form their impressions of the world and algorithms that it is natural to implement on computers. He has argued that superficially appealing models of perception and cognition tend to be oversimplified, and that systematic attention to the character of human experience can be a key to finding less obvious alternatives. He has used anomalous experiences and illusions to highlight the role that shape plays in human vision and to derive algorithms that are similarly shape-oriented. With Ellen Douglas-Cowie, he has highlighted the way speech conveys impressions of the speaker alongside the overt message, and has developed programs for extracting relevant attributes of the speech signal. He has studied areas where aspects of experience that are profoundly subjective, and difficult to externalize, nevertheless have a major effect on everyday behavior, particularly acquired deafness and more recently religion. Several of these interests link to the area of emotion, and it has been his main focus for the past five years, developing techniques for measuring subjective impressions of emotion and the speech variables that convey them. He can be reached at *r.cowie@qub.ac.uk*

Ellen Douglas-Cowie works in the School of English at Queen's University, Belfast, and she is currently the head of the school. Her research is on the information that speech carries about the speaker, with emphasis on collecting and using natural data. Her Ph.D. research was a widely cited sociolinguistic study, which revealed the complexity of the factors that influence people's choice of speech style. With Roddy Cowie, she carried out seminal research on the way speech is affected when people lose their hearing and the impact of the changes on listeners. This project led them to develop automatic techniques for extracting features of prosody that affect the impression a speaker creates, and it has provided a basis for a broader attack on prosodic speech styles, ranging across clinical varieties, the 'phone voice', skilled and unskilled reading and so on. Recently, their main focus has been on the signs of emotion, vocal and visual, and Ellen has led the assembly of a substantial audio-visual database of emotional speech extracted from real interactions. In 2000, they organized a pioneering workshop on speech and emotion, and they are editing a special issue of Speech Communication arising from it. Ellen can be reached at *e.douglas-cowie@qub.ac.uk*

Franck Davoine received a Ph.D. in Signal, Image and Speech Processing from the Institut National Polytechnique de Grenoble, France, in 1995. He was a visiting researcher at the Division of Image Coding of the University of Linköping, Sweden, from 1996 to 1997 and worked on very low bit rate video representation and coding. He joined the laboratory HEUDIASYC of the University of Technology of Compiègne, France, in 1997 as an assistant professor, and he is currently a CNRS researcher of the same laboratory. His research interests include facial image analysis for human interaction to virtual and augmented environments, and digital image watermarking for content protection, authentication and control. Franck Davoine can be reached at *Franck.Davoine@hds.utc.fr*

Robert Forchheimer received the M.S. degree in electrical engineering from the Royal Institute of Technology, Stockholm (KTH) in 1972 and the Ph.D. degree from Linköping University in 1979. During the academic year 1979 to 1980, he was a visiting research scientist at University of Southern California where he worked in the areas of image coding, computer architectures for image processing and optical computing.

Dr Forchheimer's research areas have involved data security, packet radio communication, smart vision sensors and image coding. He has authored and coauthored papers in all of these areas and also holds several patents. He is the cofounder of several companies within the university science park. Dr Forchheimer is currently in charge of the Image Coding Group at Linköping University. His main work concerns algorithms and systems for image and video communication.

Pengyu Hong received the B. Engr. and M. Engr. degree, both in computer science, from Tsinghua University, Beijing, China, in 1995 and 1997, respectively. He received his doctorate from the Department of Computer Science at the University of Illinois at Urbana-Champaign in December 2001. In 2000, he received the Ray Ozzie fellowship for his research work on face modeling, facial motion analysis and synthesis. He is now a postdoc in the Coordinated Science Laboratory at the University of Illinois at Urbana-Champaign. He is conducting research in the areas of human–computer interaction, multimedia information processing, computer vision and pattern recognition, data mining and machine learning. His home page is *www.ifp.uiuc.edu/~hong*.

Thomas S. Huang received his B.S. Degree in Electrical Engineering from National Taiwan University, Taipei, Taiwan, China and his M.S. and Sc.D. Degrees in Electrical Engineering from the Massachusetts Institute of Technology, Cambridge, Massachusetts. He was on the faculty of the Department of Electrical Engineering at MIT from 1963 to 1973 and on the faculty of the School of Electrical Engineering and director of its Laboratory for Information and Signal Processing at Purdue University from 1973 to 1980. In 1980, he joined the University of Illinois at Urbana-Champaign, where he is now William L. Everitt, distinguished professor of Electrical and Computer Engineering, and research professor at the Coordinated Science Laboratory, and head of the Image Formation and Processing Group at the Beckman Institute for Advanced Science and Technology and cochair of the Institute's major research theme Human Computer Intelligent Interaction.

Dr. Huang's professional interests lie in the broad area of information technology, especially the transmission and processing of multidimensional signals. He has published 12 books and more than 400 papers in Network Theory, Digital Filtering, Image Processing and Computer Vision.

He is a member of the National Academy of Engineering; a foreign member of the Chinese Academy of Engineering and a fellow of the International Association of Pattern Recognition, IEEE, and the Optical Society of America, and has received a Guggenheim Fellowship, an A.V. Humboldt Foundation Senior US Scientist Award, and a fellowship from the Japan Association for the Promotion of Science. He received the IEEE Signal Processing Society's Technical Achievement Award in 1987 and the Society Award in 1991. He was awarded the IEEE Third Millennium Medal in 2000. Also in 2000, he received the Honda Lifetime Achievement Award for 'contributions to motion analysis'. In 2001, he received the IEEE Jack S. Kilby medal.

Stefanos Kollias was born in Athens in 1956. He obtained a Diploma in Electrical Engineering from the National Technical University of Athens (NTUA) in 1979, an M.Sc. in Communication Engineering from the University of Manchester Institute of Science and Technology in England in 1980 and a Ph.D. in Signal Processing from the Computer Science Division of NTUA in 1984. In 1974 he obtained an honorary diploma in the Annual Panhellenic Competition in Mathematics. In 1982 he was given a COMSOC Scholarship from the IEEE Communication Society.

Since 1986 he has served as lecturer, assistant and associate professor of the Department of Electrical and Computer Engineering of NTUA. From 1987 to 1988 he was a visiting research scientist in the Department of Electrical Engineering and the Center for Telecommunications Research of Columbia University in New York, USA, on leave from NTUA. Since 1997 he has been a Professor of NTUA and Director of the Image, Video and Multimedia Systems Lab.

His research interests include image and video processing, analysis, coding, storage, retrieval, multimedia systems, computer graphics and virtual reality, artificial intelligence, neural networks, human–computer interaction and medical imaging. Fifteen graduate students have completed their doctorate under his supervision; another ten are currently doing their Ph.D. He has published more than 140 papers, 60 of them in international journals. In the last few years, he and his team have been leading or participating in forty-five projects, both European and national.

Fabio Lavagetto was born in Genoa, Italy, on August 6, 1962. He received the Masters degree in electrical engineering from the University of Genoa, Genoa, Italy, in March 1987 and the Ph.D. degree from the Department of Communication, Computer and System Sciences (DIST), University of Genoa, in 1992. He was a visiting researcher with AT&T Bell Laboratories, Holmdel, NJ, during 1990 and a contract professor in digital signal processing at the University of Parma, Italy, in 1993. Presently, he is an associate professor with DIST, University of Genoa, where he teaches a course on radio communication systems and is responsible for many national and international research projects. From 1995 to 2000, he coordinated the European ACTS project VIDAS, concerned with the application of MPEG-4 technologies in multimedia telecommunication products. Since January 2000, he has been coordinating the IST European project INTERFACE, which is oriented to speech/image emotional analysis/synthesis. He is the author of more than 70 scientific papers in the area of multimedia data management and coding. He can be reached at *fabio@dist.unige.it.*

Haibo Li is a full professor in Signal Processing in the Department of Applied Physics and Electronics (TFE), Umeå University, Sweden. He received the Technical Doctor degree in Information Theory from Linköping University, Sweden, in 1993. His doctoral thesis dealt with advanced image analysis and synthesis techniques for low bit rate video. Dr Li got the 'Nordic Best Ph.D. Thesis Award' in 1994. In 1997, Dr. Li was awarded the title of 'Docent in Image Coding'. During his period at Linköping University, he developed advanced image and video compression algorithms, including human face image analysis, extremely low bit rate video compression and 3-D video transmission. After joining Umeå University, he is now directing the Digital Media Lab, Umeå Center for Interaction Technology (UCIT), Umeå University, and working on advanced Human, Thing and Information interaction techniques. Prof. Li has been chairing sections at relevant international conferences and has been actively involved in MPEG activities in low bit rate video compression. He has contributed to several EU projects, such as VIDAS, SCALAR, INTERFACE and MUCHI. He has published more than 90 technical papers including chapters in books and holds several patents. Haibo Li can be reached at *haibo.li@tfe.umu.se*.

Sotiris Malassiotis received the B.S. and Ph.D. degrees in Electrical Engineering from the Aristotle University of Thessaloniki in 1993 and 1998, respectively. From 1994 to 1997 he was conducting research in the Information Processing Laboratory of Aristotle University of Thessaloniki. He is currently a senior researcher in the Informatics and Telematics Institute, Thessaloniki. He has participated in several European and National research projects. He is the author of more than ten articles in refereed journals and more than twenty papers in international conferences. His research interests include image analysis, image coding, virtual reality and computer graphics. Dr. Malassiotis may be reached by e-mail at *malasiot@iti.gr* (*http://www.iti.gr/people/malasiot/en/index.html*)

Andrew Marriott is a senior lecturer in the School of Computing at Curtin University of Technology, Perth, Western Australia. His research interests include facial animation, unnatural terrain environments and, of course, pretty pictures. In 1988 he formed the Computer Animation Negus (CAN), a research and development group at Curtin whose aim is to provide a sophisticated environment for animation work at the undergraduate, postgraduate and commercial level. Fax, a facial animation system, was first released into the public domain in 1992 and has been used by many researchers and has 'starred' in a few films. He is developing the Mentor System – a large-scale Java-based graphical mentoring system. He is the principal developer of VHML – the Virtual Human Markup Language (*www.vhml.org*). He is also a full partner in a 5th Framework European Union project called Interface. He has been known to wear shoes. You may find out more about him at *http://www.computing.edu.au/~raytrace* or you may email him via *raytrace@cs.curtin.edu.au.*

Jörn Ostermann studied Electrical Engineering and Communications Engineering at the University of Hannover and Imperial College London, respectively. He received Dr.-Ing. from the University of Hannover in 1994. From 1988 to 1994, he worked as a research assistant at the Institut für Theoretische Nachrichtentechnik, conducting research in low bit rate and object-based analysis–synthesis video coding. In 1994 and 1995 he worked on Visual Communications Research at AT&T Bell Labs. He has been a member AT&T Labs – Research since 1996. He is working on multimodal human–computer interfaces with talking avatars, streaming, video coding and standardization.

From 1993 to 1994, he chaired the European COST 211 sim group coordinating research in low bit rate video coding. Within MPEG-4, he chaired the Adhoc Group on Coding of Arbitrarily shaped Objects in MPEG-4 Video. Jörn was a scholar of the German National Foundation. In 1998, he received the AT&T Standards Recognition Award and the ISO Certificate of Appreciation. He is a senior member of IEEE, the IEEE Technical Committee on Multimedia Signal Processing, chair of the IEEE CAS Visual Signal Processing and Communications (VSPC) Technical Committee and a distinguished lecturer of the IEEE CAS Society. He has contributed to more than 50 papers, book chapters and patents. He is coauthor of the textbook Video Processing and Communications.

Igor S. Pandzic is currently a visiting scientist at the University of Linköping, Sweden, as well as a visiting professor at the University of Zagreb, Croatia, where he obtained an Assistant Professor position at the time when this book was being published. Formerly he worked as a senior assistant at MIRALab, University of Geneva, Switzerland, where he obtained his Ph.D. in 1998. The same year he worked as visiting scientist at AT&T Labs, USA. Igor received his B.Sc. degree in Electrical Engineering from the University of Zagreb in 1993, and M.Sc. degrees from the Swiss Federal Institute of Technology (EPFL) and the University of Geneva in 1994 and 1995, respectively. His current research interests focus on virtual characters for the Internet and mobile platforms, and include Networked Collaborative Virtual Environments, facial analysis and synthesis, computer-generated film production and parallel computing. He has published one book and around 50 papers on these topics. Igor was one of the key contributors to the Facial Animation specification in the MPEG-4 International Standard for which he received an ISO Certificate of Appreciation in 2000. He is involved in undergraduate and postgraduate teaching activities at Linköping and Zagreb Universities. He served in Program Committees of numerous conferences. Igor can be reached at *igor@isy.liu.se* and *Igor.Pandzic@fer.hr*.

Montse Pardàs received the MS degree in telecommunications and the Ph.D. degree from the Polytechnic University of Catalonia, Barcelona, Spain, in July 1990 and January 1995, respectively. Since September 1994 she has been teaching communication systems and digital image processing at this University, where she is currently associate professor. From January 1999 to December 1999 she was a research visitor at Bell Labs, Lucent Technologies, New Jersey. Her main research activity deals with image and sequence analysis, with a special emphasis on segmentation, motion and depth estimation, mathematical morphology and face analysis for synthetic model extraction. Montse Pardàs can be reached at *montse@gps.tsc.upc.es*.

Catherine Pelachaud received a Ph.D. in Computer Graphics at the University of Pennsylvania, Philadelphia, USA, in 1991. Between 1993 and 1994 she was involved in a project funded by an NSF grant, which implemented a system that automatically generates and animates conversations between multiple humanlike agents, with Prof. Badler, Prof. Cassell and Prof. Steedman. In 1993 she was part of the organization of a workshop, sponsored by NSF, on standards for facial animation held at the University of Pennsylvania. Between 1993 and 1996 she was a postdoctorate, with a Human Capital and Mobility grant, in the computer science department at the University of Rome. In 1998 she worked on the EU EAGLES project, which aims to promote standards and distribution of resources in the spoken language field. Since 2000 she has been working in MagiCster, a EU project, whose goal is to build a believable conversational agent. She is also part of the Natural Interaction and Multimodality Working Group of the EU project ISLE. Her research interest includes language standard for agent, conversational agent and human behavior simulation. Since 2002 she has been a research associate at the University of Rome 'La Sapienza'. Catherine Pelachaud can be reached at *cath@dis.uniroma1.it*

Eric Petajan is chief scientist and founder of face2face animation, inc, 2 Kent Place Blvd, Summit, NJ 07901, USA, and chaired the MPEG-4 Face and Body Animation (FBA) group. Prior to forming face2face, Eric was a Bell Labs researcher, where he developed facial motion capture, HD video coding and interactive graphics systems. Starting in 1989, Eric was a leader in the development of HDTV technology and standards leading up to the US HDTV Grand Alliance. He received a Ph.D. in EE in 1984 and an MS in Physics from the University of Illinois, where he built the first automatic lipreading system. Eric is also associate editor of the IEEE Transactions on Circuits and Systems for Video Technology. He can be reached at *eric@f2f-inc.com*.

Roberto Pockaj was born in Genoa, Italy, in 1967. He received the masters degree in Electronic Engineering in 1993 from the University of Genoa, Genoa, Italy, and the Ph.D. degree in computer engineering from the Department of Communication, Computer and System Sciences (DIST), University of Genoa, in 1999. From June 1992 to June 1996 he was with the Marconi Group, Genoa, Italy, working in the field of real-time image and signal processing, for optoelectric applications (active and passive laser sensors). Between 1996 and 2001 he collaborated on the management of the European projects ACTS-VIDAS and IST-INTERFACE, and participated in the definition of the new standard MPEG-4 for the coding of multimedia contents within the Ad Hoc Group on Face and Body Animation. In 2001 he co-founded a start-up company EPTAMEDIA srl, carrying on business in the area of facial animation software. He is currently a contract researcher at DIST. He has authored many papers on image processing and multimedia management. He can be reached at *roberto.pockaj@eptamedia.com* and *pok@dist.unige.it*.

Amaryllis Raouzaiou was born in Athens, Greece, in 1977. She graduated from the Department of Electrical and Computer Engineering, the National Technical University of Athens in 2000 and is currently pursuing her Ph.D. degree at the same University. Her current research interests lie in the areas of synthetic–natural hybrid video coding, human–computer interaction, machine vision and neural networks. She is a member of the Technical Champer of Greece. She is with the team of IST project ERMIS (Emotionally Rich Man-Machine Interaction Systems, IST-2000-29319). Amaryllis Raouzaiou can be reached by e-mail at *araouz@image.ece.ntua.gr*.

Nicolas Tsapatsoulis was born in Limassol, Cyprus, in 1969. He graduated from the Department of Electrical and Computer Engineering, the National Technical University of Athens in 1994 and received his Ph.D. degree in 2000 from the same University. His current research interests lie in the areas of human–computer interaction, machine vision, image and video processing, neural networks and biomedical engineering. He is a member of the Technical Champers of Greece and Cyprus and a member of IEEE Signal Processing and Computer societies. Dr. Tsapatsoulis has published eight papers in international journals and more than 20 in proceedings of international conferences. He served as Technical Program Cochair for the VLBV'01 workshop and as a member of the Technical Program Committee for the ICANN'02 conference. He is a reviewer of the *IEEE Transactions on Neural Networks* and *IEEE Transactions on Circuits and Systems for Video Technology* journals. Since 1995 he has participated in seven research projects at Greek and European level. Dr. Tsapatsoulis can be reached at *ntsap@image.ntua.gr* or at *http://www.image.ntua.gr/~ntsap/*.

Gael Sannier is a computer scientist who published several papers in the field of virtual humans. He worked in MIRALab, University of Geneva, for several years, where his research interests were focused on realistic texture mapping as well as on improving interactions with the virtual humans. He is a cofounder of W Interactive SARL, a French company that provides solutions for creation and animation of virtual characters on the Internet. More information on *http://www.winteractive.fr*, or email *gael@winteractive.fr*

Zhen Wen received the B. Engr degree from Tsinghua University, Beijing, China, and the MS degree from University of Illinois at Urbana Champaign, Urbana, Illinois, USA, both in computer science. Currently he is a Ph.D. student in the Department of Computer Science at University of Illinois at Urbana Champaign. His research interests are face modeling, facial motion analysis and synthesis, image-based modeling and rendering.

Liwen You received her B.Sc. in Electrical Engineering from Nanchang University of China, in 1997 and M.Sc. in Communication and Information Systems from Dalian

University of Technology of China in 2000. Now she is an international master student of Communication and Interactivity Program in Linköping University, Sweden. She is also a student member of IEEE computer society and communication society. She can be reached by e-mail at *liwenyou@ieee.org*.

Foreword

Ever since it became possible to drive the electron beams of cathode ray tubes (CRT) or the coils of loudspeakers with a computer, the number of applications that have been enabled has been staggering. The practical exploitation can be broadly classified as follows: (1) the computer can take the role of directly presenting some audio and visual samples acquired from the natural world and stored on a peripheral device or (2) it can present the result of computations based on some internal computer program that produces audio or visual information perceivable by human senses.

The former case has been driven by the telecommunication and audiovisual entertainment industries – even though these industries used to shy away from the idea of using a computer as the device driving the presentation – while the latter case has been driven by the Information Technology (IT) industry and by the hybrid mixture of the IT and entertainment industries called *Video games*.

Each of the industries involved used to have their own paradigms. The telecommunication and broadcast industries were obsessed with the idea of representing high-quality information with the smallest number of bits. This implied that the information – and it is a lot of bits for a two-hour movie – had to have a binary form. Further, processing could be virtually anything, provided that it could be implemented in special-purpose VLSI operating in real time and providing the given audio and video quality. On the other hand, the IT industry gave little importance to information representation. It sufficed that it be 'human readable' and understood by the specific computing device at hand. Real-time transmission was not an issue and the need to provide real-time pictures and audio was mitigated by the fact that information was local.

In July 1993, MPEG started its third standard project: MPEG-4. The title of the project 'Very low bit rate audiovisual coding' betrayed the original goal of serving the new unstructured world of ubiquitous fixed and mobile digital networks (not to mention portable storage devices). In a sense it was the continuation of its work done in its preceding two standards that largely targeted at the telecommunication and entertainment industries. Two years later, however, MPEG made the bold decision to try and integrate synthetic audio and visual information as well. This decision signaled the start of the Synthetic–Natural Hybrid Coding and was a major reason that prompted the change of the title of the MPEG-4 project six months later to 'Coding of audiovisual objects'.

This decision posed considerable new technical challenges. To provide what it claimed, MPEG-4 needed technologies to compose heterogeneous (natural and synthetic) audiovisual objects in a 3-D space. The starting point was provided by the technology developed by the Virtual Reality Modeling Language (VRML) Consortium (now Web3D) with which a fruitful collaboration was started in 1997. VRML 97 provided a purely textual representation, but the BInary Format for Scene description (BIFS) provided a bit-efficient representation. In addition, MPEG added the ability to extend the static VRML composition technology with real-time updates. Lastly, MPEG provided a technology to compose objects in a 2-D space as a special case.

This book is particularly concerned with a special part of the MPEG-4 SNHC world: the capability to animate humanlike faces and bodies. MPEG-4 SNHC could draw from a wealth of research results that had been waiting for a conversion into commercial exploitation opportunities and succeeded in striking a balance between accuracy and efficient coded representation of the information needed to animate faces and bodies.

This book, authored by some of the major contributors to MPEG-4 Face and Body Animation (FBA), is recommended to those who want to have an in-depth knowledge of standard face and body animation technologies and to open a window on the wide world of applications enabled by MPEG-4 FBA.

Leonardo Chiariglione
Telecom Italia Lab

Preface

In the recently released MPEG-4 International Standard, the Moving Pictures Expert Group extended its focus from traditional audio and video coding to a much broader multimedia context including images, text, graphics, 3-D scenes, animation and synthetic audio. One of the more revolutionary parts of the new standard is the Face and Body Animation (FBA) – the specification for efficient coding of shape and animation of human faces and bodies. This specification is a result of collaboration of experts with different backgrounds ranging from image coding/compression to video analysis, computer graphics as well as speech analysis and synthesis, all sharing a common interest in computer simulation of humans. The result is a specification that is suitable for a wide range of applications not only in telecommunications and multimedia but also in fields like computer animation and human–computer interfaces. This is the first standard for animation of human faces and bodies, and it has already gained a wide acceptance within the research community. As practical applications emerge, this acceptance is spreading into the commercial areas as well.

In this book we concentrate on the animation of faces. The authors are some of the leading practitioners in this field, and include most of the important contributors to the MPEG-4 Facial Animation (FA) specification. In part one we put the MPEG-4 FA specification against the historical background of research on facial animation and model-based coding and provide a brief history of the development of the standard itself. In Part 2 we provide a comprehensive overview of the FA specification with the goal of helping the reader understand how the standard works, what is the thinking behind it and how it is intended to be used. Part 3, forming the bulk of the book, covers the implementations of the standard on both the encoding and decoding side. We present several face animation techniques for MPEG-4 FA decoders, as well as architectures for building applications based on the standard. While the standard itself actually specifies only the decoder, for applications it is interesting to look at technologies for producing FA content, and we treat a wide range of such technologies including speech analysis/synthesis and video analysis. The last part of the book brings a collection of applications using the MPEG-4 FA specification.

In the appendix, we propose a standard benchmark method, with publicly available data, for assessing the quality of MPEG-4 FA decoders.

We hope that this book will be a valuable companion for practitioners implementing applications based on the MPEG-4 FA specification or for those who simply wish to understand the standard and its implications.

Igor S. Pandzic, Robert Forchheimer
Editors

PART 1
BACKGROUND

To understand the reasons behind the development of MPEG-4 Facial Animation (FA) specification, it is necessary to look at the field of facial animation in a broad sense, including animation techniques, sources of animation parameters, applications and people developing and using such applications. From such a broad look we derive the need for a facial parameterization and its ideal properties. Having established that, we measure the existing standard against these ideal properties and provide a brief historical overview of the development of the standard.

This part provides a high-level understanding of why and how MPEG-4 FA specification was born.

1

The Origins of the MPEG-4 Facial Animation Standard

Igor S. Pandzic and Robert Forchheimer

Linkoping University, Linkoping, Sweden

ABSTRACT

The MPEG-4 Facial Animation (FA) specification [1] is the first facial control parameterization to be standardized. In this chapter we place this specification within the context of a broad and detailed FA taxonomy in order to point out its meaning and importance. We discuss the properties of an ideal facial parameterization and measure MPEG-4 against this ideal. Finally, we present a brief history of facial control parameterizations and of the development of the MPEG-4 FA standard.

1.1 INTRODUCTION

Broadly speaking, the world of computer FA is inhabited by two kinds of people: the researchers and the artists.

The researchers are mostly interested in the more technical aspects of the problem. Many have studied methods for modeling and animating faces based on approaches ranging from ones purely based on observation to pseudomuscle models and various degrees of physical simulation of bone, muscle and tissue dynamics. Others have been busy trying to track facial features in real time in unconstrained video without facial markers, a Holy Grail still to be reached. Yet others have studied synthesis of talking faces from speech, including issues such as coarticulation and simulation of natural facial gestures accompanying speech.

The artists have been busy with rather more immediate and practical tasks of producing high quality FA for film, TV, games, live performances or the Web. For most of them, FA is just a part of whatever project they are currently producing. Obviously, they use the best methods available. However, 'available' means integrated or at least compatible with the major 3-D animation software and procedures they already use.

This often meant that FA was produced using *ad hoc* solutions, which were then forgotten until the next project, and yet another *ad hoc* solution.

The communication between the two communities was not very good. How many of the FA methods described in scientific articles over the past three decades have made their way into a major film or TV production? Fairly few. Success stories such as Pixar's are rare. The majority is left, to be demonstrated by more-or-less adequate efforts of research institutions to produce final results themselves and thus validate their research.

Even within the research community, communication between different branches can be improved. True, in the past decade it became very fashionable for the computer animation and the computer vision people to work together, forming joint projects or coordinating within the same research group. Yet, how many times have we witnessed a demonstration of a face tracking system driving an FA system that was not developed within the same group? Not many. The systems just do not speak the same language.

So is it possible to bridge these gaps, to have a standard language that would enable an artist to control any FA system through the same interface, reuse the FA sequences or to let any face tracker or other source of animation data drive any FA system, on any platform, potentially in a networked or even mobile environment? This is what MPEG-4 FA specification is about. It is already happening in the research community in which MPEG-4 FA has taken a firm foothold and it is spreading out into the commercial world through the efforts of companies participating in the writing of this book and many others.

But what does FA have to do with MPEG (Moving Picture Experts Group) and how did it get there? Isn't MPEG all about video compression? Well, since MPEG-4, this is only partly true. The MPEG-4 standardization effort grew out of the wish to create a video-coding standard more capable of extending into lower bit rates than those of MPEG-1 and MPEG-2. Work groups were set up to focus on the many aspects of the new standard. While the main efforts were concentrated on improving the basic video encoder, it was soon realized that the standard should also reflect part of the model-based image-coding concepts that had been developed since the early 1980s [2, 3]. As the main philosophy of the standardization group was to define only the decoder, the difficult problem of how to do model-based encoding was avoided. A separate *ad hoc* group, the Face and Body Animation (FBA) group was set up as a subgroup of the Synthetic Natural Hybrid Coding (SNHC) activity. SNHC was primarily concerned with how to integrate general synthetic objects such as 2-D and 3-D graphics and synthetic audio into the standard, while the FBA group started working on the definition of a decoder for face images.

It was soon recognized that many of the participating partners had numerous other applications in mind apart from model-based coding. Faces to be generated by the system did not necessarily need to mimic real ones. In fact they could even be cartoonlike, both with regard to the appearance and the involved motion. This fitted well with the now established concept that MPEG-4 should be a general standard for multimedia contents. A 'golden rule' of earlier MPEG standards that was sacrificed, however, concerned the decoded output. Since no particular face model was part of the standard, different decoders would produce different-looking outputs. Although this shift in philosophy may seem fairly undramatic considering that HyperText Markup Language (HTML)-based web browsers had already been around for some years, it was nevertheless a substantial break from earlier developments in the video coding community. This issue was also vigorously debated all through the standardization work. On one hand, people with a background in Internet solutions considered the policy as quite normal, while some of the video people, particularly those involved with content production feared the loss of control over the look of the images on the receiving screen.

Thus, through the FBA part, MPEG-4 started to incorporate a new philosophy that marked the end of a basic principle, that is, that the decoder is completely defined. With MPEG-4 FBA, only the format (and meaning) of the transmitted data is standardized while neither the encoding nor decoding parts are fully defined.

We present the need for the parameterization of FA in the next section by examining the full taxonomy of FA and exposing the parameterization as the central enabling element for many applications. We then discuss the properties of an ideal parameterization and match the MPEG-4 specification against this ideal in the following two sections. Finally, we give a historical overview of the facial parameterizations and a more detailed history of the development of the FA specification in MPEG-4.

1.2 THE NEED FOR PARAMETERIZATION

The complete taxonomy of FA is presented in Figure 1.1. It becomes apparent that FA can be viewed as two separate problems: (1) low-level facial animation or parameterized facial motion implementation and (2) high-level facial animation, creating streams of FA parameters to produce final animation sequences.

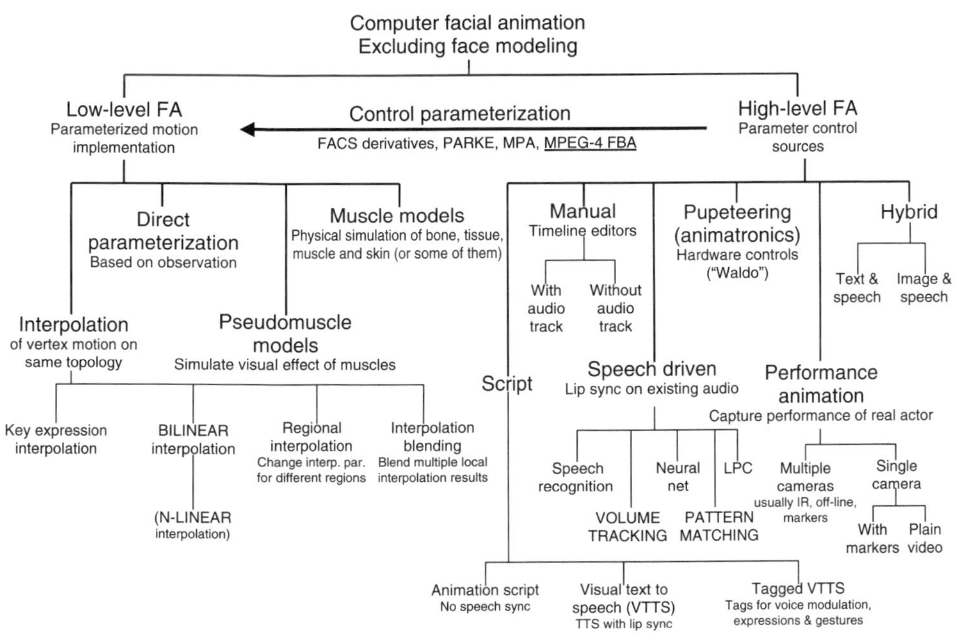

Figure 1.1 Taxonomy of computer facial animation

Parameterized facial motion implementation, or low-level FA, deals with methods to change the facial geometry and/or appearance in time, based on a set of parameters. The approaches include direct parameterization [4, 5], pseudomuscle models [6–8], muscle models [9–11] or, the simplest approach that has recently regained popularity, interpolation [12, 13]. Several implementations of MPEG-4 parameterized facial motion are presented in Part 3 of the book.

High-level facial animation means using the parameters exposed by the low-level model in order to produce final animation sequences. Here we find a multitude of approaches. The most time consuming, but also giving the most precise control, is manual editing, usually with a graphical interface involving a timeline on which key frames are manipulated. Animatronics or puppetry involves hardware controls of all varieties, ranging from knobs and buttons to gloves and skeletons, allowing one or more operators to control the face. In performance animation, the facial motion of a real actor is captured. A high-end optical tracking system is often used on the basis of multiple calibrated infrared (IR) cameras and reflective markers glued on the face. Precise 3-D trajectory of tracked points is achieved, usually off-line and involving a complicated and expensive setup. Single camera systems based on markers glued to the face can give good results, but the Holy Grail is the tracking of facial features in plain video without markers and in real time. This problem is explored in more detail by Ahlberg in Chapter 6. Speech-driven FA aims to synchronize lip motion to existing speech signal, and is examined in detail by Peng *et al.* in Chapter 7. Finally, animation can be produced on the basis of scripts. Such scripts were extensively used in FA [7], and presently this approach is very popular in the form of Visual Text-to-Speech (VTTS) systems explored by Pelachaud in Chapter 8. A step above plain VTTS is the addition of standardized tags in the text in order to increase the naturalness of produced facial motion by introducing nods, eye/eyebrow motion, gaze and so on. This is explored by Marriott in Chapter 12.

It may be noted that the aforementioned methods cover the complete spectrum of FA application classes envisaged today: Film/TV, games, communications and multimodal user interfaces. The central piece holding it all together, and illustrated in Figure 1.2, is the control parameterization: a set of parameters letting different systems talk to each other and exchange information about the animation and appearance of the animated faces.

This parameterization means different things to different people. The researchers working on animation need to decide which parameters should be implemented and how to translate them into actual facial motion. The computer vision researchers' main concern is how the parameters can be obtained from video and whether it is at all possible. To the artists, the parameters are the puppeteer's handle to move the face. They let them express their creativity; everything else is a black box.

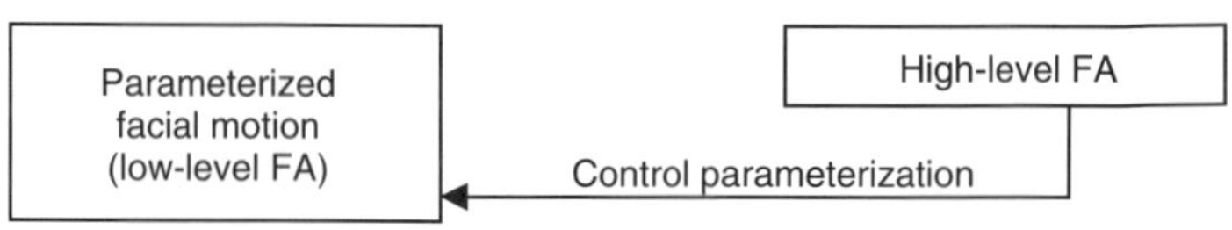

Figure 1.2 Importance of control parameterization for FA

1.3 THE IDEAL PARAMETERIZATION

Satisfying all these demands is not easy, but a first step is certainly to make a list of requirements that an ideal parameterization should satisfy. Parke made such a list [14], and we extend it here with a few more demands resulting from more recent application areas involving video analysis and networking.

Range of possible faces and expressions

Ideally, it should be possible to express any face with any expression. This is a tremendous requirement even if we just consider realistic human faces and expressions and is further extended to cartoon characters with exaggerated features and facial motion, as they often appear in computer animation.

Ease of use

The ease of use of parameterization is determined by the number of parameters, their complexity and intuitivity. It is usually inversely proportional with the previous requirement on the range of faces and expressions – the more we want to extend that range, the more complex the parameters become.

Subtlety

Humans are sensitive to extremely subtle changes of facial expression. The slightest twisting of a mouth corner may indicate mild discontent or skepticism; very small narrowing of the eyelids can produce a threatening look and so on. The parameters should be able to express and precisely control even the slightest motion.

Orthogonality

One parameter should not affect any other parameter, so that they can be adjusted independently.

Ability to be basis for higher-level abstraction

It is desirable to be able to control the face on a high level of abstraction, for example, using a tagged text that is automatically converted into speech with facial expressions. For this purpose, the parameterization should allow aggregation of parameters into higher-level controls such as visemes or expressions.

Predictability

The result of any parameter combination should be predictable.

Portability

It is desirable that the same parameter controls give the same result in terms of high-level expression when applied to different face models. This enables the easy switching of face models in applications and the reuse of facial animation parameter sets or sequences.

Possibility to measure the parameters

In many applications it is interesting to derive the facial parameters by measuring a real face and its movement and this measurement process should preferably be automatic. The parameters should therefore be measurable. This means that they have to be expressed in physically defined units that can be measured visually on a face.

Efficiency (bandwidth)

For applications involving communication of FA through a network, the representation of parameters should be efficient and allow encoding and streaming.

1.4 IS MPEG-4 FA UP TO THE IDEAL?

To answer this question, we check the standard against the list of requirements proposed in the previous section.

Range of possible faces and expressions

Concerning the range of possible faces, the Facial Definition Parameters (FDPs) are scalable and allow modification of face shape by sending a number of feature points (FPs), additional texture or, at the highest level, the whole facial polygon mesh and texture, both of arbitrary complexity. This makes the range of faces that can be represented practically unlimited.

As for the range of expressions, the Facial Animation Parameters (FAPs) cover all main facial articulations in detail and even the possibility of exaggerated motion for cartoon animation.

However, by their nature (being defined on the basis of point movements) FAPs do not allow explicit specification of more subtle effects such as skin creases and wrinkles. FAPs can be interpreted to include such effects – this is left to each particular implementation of a FAP decoder. For example, an implementation of an old person's face may include automatic wrinkling in the interpretation of FAPs. Still, this information is in no way represented in the parameters themselves.

In order to explicitly specify these effects, it is necessary to use Facial Animation Tables (FATs) that allow precise specification of vertex movements corresponding to each FAP. This mechanism extends the definition of each FAP to any desired level of detail.

Ease of use

In general, the MPEG-4 FA specification is fairly low level and not intended for direct use, that is, controlling the face through a set of sliders corresponding to each FAP would be a cumbersome process. This low-level approach was necessary in order to allow these parameters to be, to the highest possible extent, the universal link between various systems and processes involved in FA.

However, as the parameters are defined geometrically, they are very intuitive and easy to grasp. The numerical issues such as normalization can be hidden from the user by an appropriate user interface. It is not difficult to imagine such an interface in which parameters can be controlled simply by moving the points on the face. Higher-level abstractions can (and should) be included. For example, an interface may allow all lip points to be moved at once by high-level commands such as open, stretch and so on.

At the same time, for most important functionalities, that is, the visemes and basic expressions, high-level parameters are directly provided and can be used for production of visual speech in an extremely straightforward way.

Subtlety

Simulating very subtle facial expressions is one of the toughest challenges in FA. MPEG-4 FAPs cover most articulations of the face and they can be controlled in sufficiently small units to allow barely perceptible movements. 'Barely perceptible' is of course not the same as 'subtle', but it is at least a basis for it. As previously discussed, there is a lack of control over effects outside the main articulatory regions, such as wrinkling on the cheeks and forehead. This can be overcome by using the FAT mechanism.

Orthogonality

Because of their definition as movements of points of the face, FAPs are naturally orthogonal. However, each implementation interprets the movement of neighboring points in the face as well. For example, although the open-jaw parameter is defined as the movement of the chin point only, it is obvious that any reasonable implementation has to move the whole jaw.

For this reason, some of the parameters are in practice not totally independent. This mainly goes for the lip region, defined with inner and outer contour that are clearly not independent.

Ability to be basis for higher-level abstraction

The FAPs already provide high-level parameters (viseme and expression), and low-level FAPs can be used to express other expressions. The high-level FAPs provide the blending parameters that can be used for transitions and simple coarticulation. This makes it very easy to use MPEG-4 FAPs as a basis for automatic visual speech production.

Predictability

It is very difficult to objectively predict the result of a complex set of parameters on a system as complex as a human face. MPEG-4 makes a compromise by making a subset of it predictable – the movements of FPs. The rest is deliberately left open to implementations.

However, full predictability is possible using the FAT mechanism, as already mentioned in the paragraph discussing the range of possible expressions in the beginning of this section. This mechanism allows exact and complete definition of the movement caused by each FAP.

Portability

FAPs are normalized with respect to the face they are applied to, and therefore are portable from one face to another. Experiments during the development of the standard have shown reasonable reproduction of the same high-level expressions on different face models.

Possibility to measure the parameters

The use of point movements as the basis of parameter definition was largely driven by this demand. The parameters are therefore measurable in an exact manner.

There are a few exceptions where this is slightly less exact, like the points on the cheeks whose exact placement is difficult because of the lack of prominent features on the cheek.

Efficiency (bandwidth)

MPEG-4 defines the encoding of the parameters in a highly efficient bit stream that can be used on its own or as a part of an MPEG-4 system. The coding system optimizes the bit stream by allowing the transmission of only the parameters that are actually used. The typical bit rates range from 3 to 6 kbit s^{-1} for full FA at 25 or 30 frames per second. By using only high-level FAPs, the bit rate is reduced to approximately 0.3 kbit s^{-1}. These results are appropriate for any kind of platform or environment in use today.

1.4.1 Conclusion

After discussing each of the criteria separately, can we answer the question from the title of this section? Is MPEG-4 FA specification the ideal parameterization for FA? Obviously, it is not. But it is closer to the ideal than any other available parameterization, matches the state of the art in FA and related technologies and is appropriate to support applications being envisaged in the foreseeable future.

1.5 BRIEF HISTORY OF FACIAL CONTROL PARAMETERIZATION

Although the understanding of the importance of parameterization has been around for quite a long time [14, 15], until recently the parameterizations were developed in a fairly *ad hoc* manner, with practically as many different parameter sets as there were different animation systems. The real effort was usually concentrated on the implementation of the actual facial motion caused by the parameters, rather than a systematic effort to choose an optimal set of parameters.

Perhaps the strongest influence was that of FACS, the Facial Action Coding System [16]. FACS was not meant for animation, it is rather a system for evaluating (scoring) the facial expressions on the faces of real humans in order to code them in a standardized way comprising a set of 46 Action Units (AUs) or basic facial movements. However, as a result of its systematic approach, foundation on muscle activity and detailed description of visual effects of each AU, FACS had obvious attractions to the researchers implementing FA systems. Several systems, in particular muscle-based ones, used subsets of FACS as control parameters.

Another parameterization that had a big influence on the FA community is the one developed by Parke [4]. It was originally developed on the basis of careful observation and analysis of the face and its motion. Parke divided the parameters into expression parameters and conformation parameters, the later ones serving to change the basic appearance of the face, the former to give it expressions.

The expression parameters included: eyelid opening, eyebrow arch, eyebrow separation, jaw rotation, mouth width, mouth expression, upper lip position, mouth corner position and eye gaze.

The conformation parameters were jaw width, forehead shape, nose length/width, cheek shape, chin shape, neck shape, eye size and separation, face proportions.

Even with this relatively modest number of parameters, a wide range of faces and expressions could be obtained. The Parke facial model and parameterization was used by many researchers and had several extensions [17–19].

In 1988, Magnenat-Thalmann *et al.* propose the Abstract Muscle Actions (AMA) [6]. They are pseudomuscle procedures controlling the face. AMA has been inspired by the FACS system, but typically they are on a somewhat higher level of abstraction then the FACS AUs. Each AMA procedure simulates an action of one or more muscles. They are not independent, and the order in which they are executed is important. There are two high-level controls, phoneme and expression, composed of several low-level AMA. Examples of AMA procedures include Close Upper Lip, Right Eyelid, Right Zygomatic, Move Right Eyebrow, Left Lip Raiser, Move Right Eye Horizontal and so on.

In 1991, Kalra *et al.* introduce an approach to FA based on abstraction layers [20]. The central part of this parameterization is the Minimal Perceptible Actions (MPAs). MPAs describe visible effects of one or more muscle actions and are combined into expressions on higher levels of abstraction. Examples of MPAs include raise_eyebrow, squeeze_eyebrow, move_horizontal_eye, move_vertical_eye, mouth_beak, nod_head, turn_head, roll_head etc. The intensity of MPEs is expressed by a normalized value between 0 and 1 for unidirectional MPAs and between −1 and 1 for bidirectional MPAs. The actual minimum and maximum of each movement is dependent on the implementation and characteristics of each particular face model. MPAs are therefore neither exactly predictable, nor portable from one face model to another, that is, the same set of MPA values may produce very different expressions on different face models.

1.6 THE BIRTH OF THE STANDARD

The idea of FA specification arose within the SNHC group of MPEG-4, which began its life in Tokyo in the summer of 1995. Compared with the Audio, Systems and Visual activities, the SNHC was a more recent addition to the MPEG-4 work plan. The SNHC Call for Proposals (CFP) was issued in March 1996. CFP Amendments followed in July

1996 at the MPEG meeting in Tampere. These amendments [21] specifically included a CFP on Face Animation with the intent to standardize a set of parameters that can be used to shape a face model and animate it. The initial evaluation of proposals was performed in September 1996 in Chicago. The proposal adopted as the basis for the standard was the one based on Kalra's MPAs proposed by the University of Geneva. In Chicago, the basis of FAP normalization was already established and a first draft specification of face and body definition and animation parameters was produced [22]. Also, the Face and Body Animation Ad Hoc Group (FBA) was formally initiated within MPEG [23]. At the next MPEG meeting, in November 1996 in Maceio, Brazil, the FA specification became part of a larger MPEG document, the SNHC Verification Model [24], beginning its path through the formal stages of the standardization process. At the same meeting, the predictive coding scheme and bit stream syntax are introduced, and Core Experiments initiated in order to test and refine the scheme. A major change to the parameter set that happened in Maceio is the introduction of separate inner and outer lip contour animation parameters, which we consider the worst feature of the MPEG-4 FA specification, but never quite managed to bring the rest of the FBA standardization group to that opinion. The Sevilla meeting, February 1997, starts the transition of the FA specification from the SNHC documents into the 'main' documents of the future standard: MPEG-4 Systems and MPEG-4 Visual. This transition is temporarily halted at the Bristol meeting, April 1997, amidst a serious soul-searching within the FBA group over concerns about the quality of the produced specification and the lack of thorough testing and verification. Additional Core Experiments [25] are defined and undertaken in the coming months in order to overcome these drawbacks. At the same time, the Bristol meeting saw the introduction into the FA specification of the Facial Animation Table mechanism, proposed by AT&T. Another major addition was the introduction of an alternative discrete cosine transform (DCT) coding scheme, providing even higher compression for the cases when a delay is acceptable. The DCT scheme was first proposed in the meeting in Stockholm, July 1997. Over the next MPEG meetings in Fribourg, San Jose, Tokyo, Dublin, Atlantic City and Rome, the specification was refined on the basis of further experiments and discussions; profiles and levels were defined. Another major task was to try and keep the FA specification as clear as possible, made difficult by the fact that it was spread between MPEG-4 Systems and Visual documents.

In 1999, Version 1 of MPEG-4, including the FA specification, became an International Standard.

ACKNOWLEDGMENTS

This work is partly supported by the VISIT programme of the Swedish Foundation for Strategic Research (SSF).

REFERENCES

[1] Moving Picture Experts Group, ISO/IEC 14496 – MPEG-4 International Standard, *www.cselt.it/mpeg*.

[2] R. Forchheimer, O. Fahlander, Low bit rate coding through animation, *Proceedings Picture Coding Symposium 83*, 1983.

[3] R. Forchheimer, O. Fahlander, T. Kronander, A semantic approach to the transmission of face images, *Proceedings Picture Coding Symposium 84*, 1984.
[4] F. I. Parke, A Parametric Model for Human Faces, Ph.D. Thesis, University of Utah, Salt Lake City, Utah, 1974, UTEC-CSc-75-047.
[5] F. I. Parke, "Parametrized models for facial animation", *IEEE Computer Graphics and Applications*, **2**(9), 61–68 (1982).
[6] N. Magnenat-Thalmann, N. E. Primeau, D. Thalmann, "Abstract muscle actions procedures for human face animation", *Visual Computer*, **3**(5), 290–297 (1988).
[7] P. Kalra, A. Mangili, N. Magnenat-Thalmann, D. Thalmann, "Simulation of facial muscle actions based on rational free-form deformation", *Proceedings of the Annual Conference of the European Association for Computer Graphics-Eurographics '92*, Cambridge, UK, September, 1992, pp. 65–69.
[8] P. Kalra, An Interactive Multimodal Facial Animation System, Ph.D. Thesis nr. 1183, EPFL, Lausanne, 1993.
[9] S. M. Platt, N. I. Badler, "Animating facial expressions", *Computer Graphics*, **15**(3), 245–252 (1981).
[10] K. Waters, "A muscle model for animating three-dimensional facial expressions", *Computer Graphics (SIGGRAPH'87)*, **21**(4), 17–24 (1987).
[11] D. Terzopoulos, K. Waters, "Physically-based facial modeling, analysis and animation", *Journal of Visualization and Computer Animation*, **1**(4), 73–80 (1990).
[12] K. Arai, T. Kurihara, K. Anjyo, "Bilinear interpolation for facial expressions and metamorphosis in real-time animation", *The Visual Computer*, **12**, 105–116 (1996).
[13] I. S. Pandzic, A web-based MPEG-4 facial animation system, Euroimage ICAV3D 2001, Mykonos, Greece, June, 2001; demonstration at *www.icg.isy.liu.se/~igor/MpegWeb*.
[14] N. Magnenat-Thalmann, D. Thalmann, eds., "Control parametrization for facial animation", *Computer Animation '91*, Springer-Verlag, Geneva, 1991, pp. 3–14.
[15] F. I. Parke, K. Waters, Computer Facial Animation, ISBN 1-56881-014-8, A K Peters, 1996.
[16] P. Ekman, W. V. Friesen, *Manual for the Facial Action Coding System*, Consulting Psychologists Press, Palo Alto, Calif., 1978.
[17] A. Pearce, B. Wyvill, G. Wywill, D. Hill, "Speech and expression: a computer solution to face animation", *Proceedings of the Graphics Interface 86*, Canadian Information Processing Society, Calgary, 1986, pp. 136–140.
[18] M. Cohen, D. Massaro, "Synthesis of visible speech", *Behavioral Research Methods and Instrumentation*, **22**(2), 260–263 (1990).
[19] S. DiPaola, "Extending the range of facial types", *Journal of Visualization and Computer Animation*, **2**(4), 129–131 (1991).
[20] P. Kalra, A. Mangili, N. Magnenat-Thalmann, D. Thalmann, SMILE: A Multi Layered Facial Animation System, *IFIP WG 5.10*, Tokyo, 1991, pp. 189–198.
[21] MPEG-4 SNHC Group, ISO/IEC JTC1/SC29/WG11 N1315, "Amendments to the Call for Proposals and PPD on SNHC".
[22] E. Petajan, I. Pandzic, T. Capin, P. Ho, R. Pockaj, H. Tao, H. Chen, J. Shen, P.-E. Chaut, J. Osterman, ISO/IEC JTC1/SC29/WG11 N1365, "Face and Body Definition and Animation Parameters".
[23] MPEG, ISO/IEC JTC1/SC29/WG11 N1360, "Ad Hoc Group on Face and Body Animation".
[24] MPEG-4 SNHC Group, ISO/IEC JTC1/SC29/WG11 N1454, "SNHC Verification Model 2.0".
[25] MPEG-4 SNHC Group, ISO/IEC JTC1/SC29/WG11 N1672, "FBA Core Experiments".

PART 2
THE STANDARD

This part explains the functional details of the MPEG-4 Facial Animation specification. In the actual MPEG-4 standard, the FA specification is somewhat scattered around in the documentation, and even in different parts of the standard (Visual and Systems). In Chapter 2, Joern Ostermann presents the FA specification in a streamlined, easy-to-follow fashion. In Chapter 3, Eric Petajan explains the details of profiles and levels involving FA.

While we do not present the low-level details of coding, this part is an invaluable companion to the MPEG-4 documentation for anyone attempting to implement MPEG-4 FA specification. On its own, it provides a thorough understanding of the specification and its intended use.

2

Face Animation in MPEG-4

Jörn Ostermann

AT&T Labs-Research, Middletown, New Jersey

ABSTRACT

This chapter presents an overview of face animation (FA) as supported by MPEG-4. Using facial animation parameters (FAPs), MPEG-4 enables the animation of a proprietary face model or a face model downloaded to the MPEG-4 client. FAPs are compressed using either differential pulse code modulation (DPCM) or discrete cosine transform (DCT). For the face model, MPEG-4 defines feature points (FPs) that are animated by the FAPs. An MPEG-4 client would use the motion of these FPs to animate its face model. An MPEG-4 client may animate its proprietary face model or animate a face model that is downloaded to the client using MPEG-4 Binary Format for Scenes (BIFS). Synchronization of animated faces with other media streams like audio is achieved using time stamps in the FAP stream. We also address integration of the FA with the text-to-speech interface (TTSI) as defined by MPEG-4, such that FA can be driven by text input.

2.1 INTRODUCTION

MPEG-4 is an object-based multimedia compression standard that allows for encoding of different audiovisual objects (AVO) in the scene independently. The visual objects may have natural or synthetic content, including arbitrary shape *video objects*, special synthetic objects such as human face and body and generic 2-D/3-D objects composed of primitives such as rectangles, spheres or indexed face sets, which define an object surface by means of vertices and surface patches.

The representation of synthetic visual objects in MPEG-4 is based on the prior Virtual Reality Modeling Language (VRML) standard [1–3] using nodes such as *Transform*, which defines rotation, scale or translation of an object, and *IndexedFaceSet* describing the 3-D shape of an object by an indexed face set. However, MPEG-4 is the first international standard that specifies a compressed binary representation of animated synthetic AVO and scenes. It is important to note that MPEG-4 only specifies the decoding of compliant bit streams in an MPEG-4 terminal. The encoders do enjoy a large degree of freedom in how to generate MPEG-4 compliant bit streams. As specified in MPEG-4 Systems, decoded AVO can be composed into 2-D and 3-D scenes using the BIFS [3], which also allows for implementation of generic animation of objects and their properties. Very efficient special purpose animation techniques are defined for face/body and 2-D mesh animation in MPEG-4 Visual [4]. MPEG-4 also provides synthetic audio tools such as structured audio tools for sound synthesis and a TTSI for speech synthesis [5]. Body animation and 3-D mesh compression are supported in MPEG-4 to complement FA; however, they are not covered here [3, 4, 6, 7].

This chapter presents a detailed overview of the MPEG-4 support for FA as specified in MPEG-4 Audio, Visual and Systems. We describe how to specify and animate 3-D face models, compress FAPs and integrate FA with text-to-speech TTS in MPEG-4, such that FA can be driven by text input. The MPEG-4 standard allows using proprietary 3-D face models that are resident at the decoder as well as transmission of face models such that the encoder can predict the quality of the presentation at the decoder. In Section 2.2, we explain how MPEG-4 specifies a 3-D face model and its animation using face definition parameters (FDP) and FAPs, respectively. Section 2.3 provides details on how to efficiently encode FAPs. The integration of FA into an MPEG-4 terminal with TTS capabilities is shown in Section 2.4. In Section 2.5, we describe briefly the integration of FA with MPEG-4 systems. An overview of MPEG-4 profiles related to FA is given in Section 2.6. Chapter 3 discusses some MPEG-4 profiles and conformance testing in detail.

2.2 SPECIFICATION AND ANIMATION OF FACES

MPEG-4 specifies a face model in its neutral state, a number of FPs on this neutral face as reference points and a set of FAPs, each corresponding to a particular facial action deforming a face model in its neutral state. Deforming a neutral face model according to some specified FAP values at each time instant generates a facial animation sequence. The FAP value for a particular FAP indicates the magnitude of the corresponding action, for example, a big versus a small smile or deformation of a mouth corner. For an MPEG-4 terminal to interpret the FAP values using its face model, it has to have predefined model specific animation rules to produce the facial action corresponding to each FAP. The terminal can either use its own animation rules or download a face model and the associated face animation tables (FAT) to have a customized animation behavior. Since the FAPs are required to animate faces of different sizes and proportions, the FAP values are defined in face animation parameter units (FAPU). The FAPU are computed from spatial distances between major facial features on the model in its neutral state.

In the following text, we first describe what MPEG-4 considers to be a generic face model in its neutral state and the associated FPs. Then, we explain the FAPs for this generic model. Finally, we show how to define MPEG-4 compliant face models that can be transmitted from the encoder to the decoder for animation.

2.2.1 MPEG-4 Face Model in Neutral State

As the first step, MPEG-4 defines a generic face model in its neutral state by the following properties (Figure 2.1):

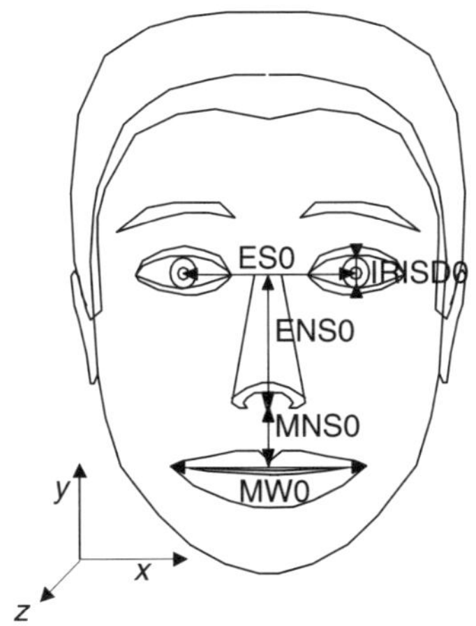

Figure 2.1 A face model in its neutral state and the feature points used to define FAP units (FAPU). Fractions of distances between the marked key features are used to define FAPU (from ISO/IEC IS 14496-2 Visual, 1999 [4])

- Gaze is in the direction of z-axis
- All face muscles are relaxed
- Eyelids are tangent to the iris
- The pupil is one third the diameter of the iris
- Lips are in contact; the line of the lips is horizontal and at the same height at lip corners
- The mouth is closed and the upper teeth touch the lower ones
- The tongue is flat, horizontal with the tip of the tongue touching the boundary between upper and lower teeth.

An FAPU and the FPs used to derive the FAPU are defined next with respect to the face in its neutral state.

Face animation parameter units

In order to define FAPs for arbitrary face models, MPEG-4 defines FAPUs that serve to scale FAPs for any face model. FAPUs are defined as fractions of distances between key facial features (Figure 2.1). These features, such as eye separation are defined on a face model that is in the neutral state. The FAPU allows interpretation of the FAPs on

Table 2.1 Facial animation parameter units and their definitions

IRISD0	Iris diameter (by definition it is equal to the distance between upper and lower eyelid) in neutral face	IRISD = IRISD0/1024
ES0	Eye separation	ES = ES0/1024
ENS0	Eye–nose separation	ENS = ENS0/1024
MNS0	Mouth–nose separation	MNS = MNS0/1024
MW0	Mouth width	MW = MW0/1024
AU	Angle unit	10E-5 rad

Note: AU: Access Unit.

any facial model in a consistent way producing reasonable results in terms of expression and speech pronunciation. The measurement units are shown in Table 2.1.

Feature points

MPEG-4 specifies 84 FPs on the neutral face (Figure 2.2). The main purpose of these FPs is to provide spatial references for defining FAPs. Some FPs such as the ones along the hairline are not affected by FAPs. However, they are required for defining the shape of a proprietary face model using FPs (Section 2.2.3). FPs are arranged in groups such as cheeks, eyes and mouth. The location of these FPs has to be known for any MPEG-4-compliant face model. The FPs on the model should be located according to Figure 2.2 and the hints given in Table 2.6 in the Annex of this chapter.

2.2.2 Face Animation Parameters

The FAPs are based on the study of minimal perceptible actions (MPA) and are closely related to muscle actions [8–11]. The 68 parameters are categorized into 10 groups related to parts of the face (Table 2.2). FAPs represent a complete set of basic facial actions including head motion, tongue, eye and mouth control. They allow representation of natural facial expressions (see Table 2.7 in the Annex). For each FAP, the standard defines the appropriate FAPU, FAP group, direction of positive motion and

Table 2.2 FAP groups

Group	Number of FAPs
1. Visemes and expressions	2
2. Jaw, chin, inner lowerlip, cornerlips, midlip	16
3. Eyeballs, pupils, eyelids	12
4. Eyebrow	8
5. Cheeks	4
6. Tongue	5
7. Head rotation	3
8. Outer-lip positions	10
9. Nose	4
10. Ears	4

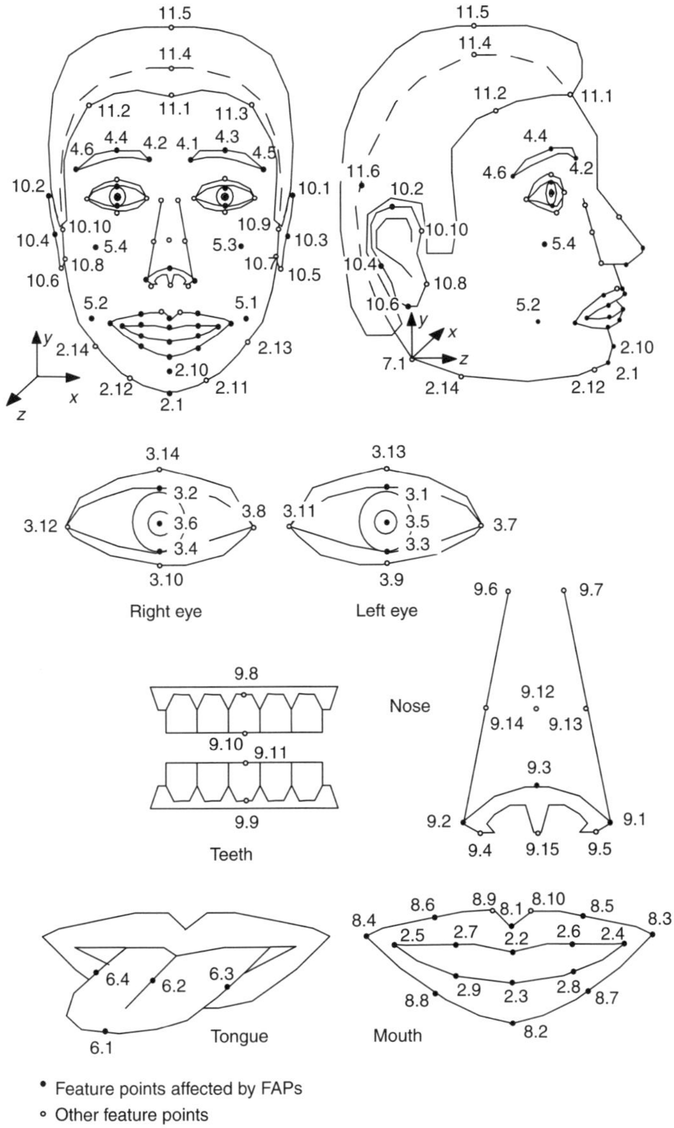

Figure 2.2 Feature points may be used to define the shape of a proprietary face model. The FAPs are defined by motion of some of these FPs (from ISO/IEC IS 14496-2 Visual, 1999 [4])

whether the motion of the FP is unidirectional (see FAP 3, open jaw) or bidirectional (see FAP 48, head pitch). FAPs in groups 2 to 10 are considered low-level parameters. They specify precisely how much a FP of a face has to be moved for a given amplitude. FAPs can also be used to define facial AU [12]. Exaggerated amplitudes permit the definition of actions that are normally not possible for humans but are desirable for cartoonlike characters.

Table 2.3 Visemes and related phonemes

Viseme #	Phonemes	Example
0	None	na
1	p, b, m	put, bed, mill
2	f, v	far, voice
3	T,D	think, that
4	t, d	Tip, doll
5	k, g	call, gas
6	tS, dZ, S	chair, join, she
7	s, z	Sir, zeal
8	n, l	Lot, not
9	r	red
10	A:	car
11	e	bed
12	I	Tip
13	Q	top
14	U	book

The FAP set contains two high-level parameters, visemes and expressions (FAP group 1). A viseme (FAP 1) is a visual correlate to a phoneme. Only 14 static visemes that are clearly distinguished are included in the standard set (Table 2.3). In order to allow for coarticulation of speech and mouth movement [13], the shape of the mouth of a speaking human is not only influenced by the current phoneme, but also the previous and the next phoneme [14]. In MPEG-4, transitions from one viseme to the next are defined by blending only two visemes with a weighting factor. So far, it is not clear how this can be used for high quality visual speech animation.

The expression parameter FAP 2 defines the six primary facial expressions (Table 2.4, Figure 2.3). In contrast to visemes, facial expressions are animated by a value defining the excitation of the expression. Two facial expressions can be animated

Table 2.4 Primary facial expressions as defined for FAP 2

#	Expression name	Textual description
1	Joy	The eyebrows are relaxed. The mouth is open and the mouth corners pulled back toward the ears.
2	Sadness	The inner eyebrows are bent upward. The eyes are slightly closed. The mouth is relaxed.
3	Anger	The inner eyebrows are pulled downward and together. The eyes are wide open. The lips are pressed against each other or opened to expose the teeth.
4	Fear	The eyebrows are raised and pulled together. The inner eyebrows are bent upward. The eyes are tense and alert.
5	Disgust	The eyebrows and eyelids are relaxed. The upper lip is raised and curled, often asymmetrically.
6	Surprise	The eyebrows are raised. The upper eyelids are wide open, the lower relaxed. The jaw is opened.

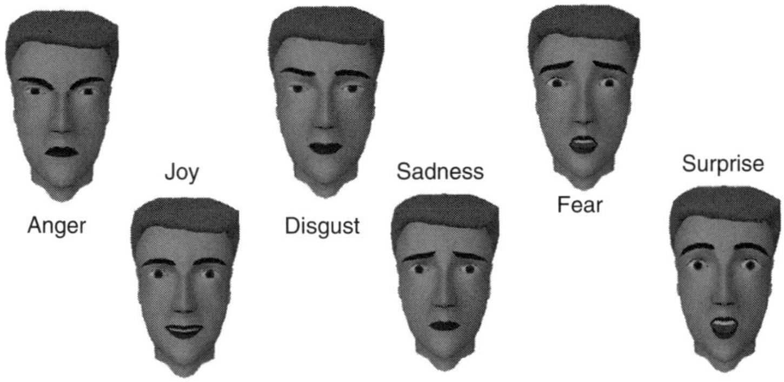

Figure 2.3 Facial expressions

simultaneously with the amplitude in the range of [0–63] defined for each expression. The facial expression parameter values are defined by textual descriptions. The expression parameter allows for an efficient means of animating faces. They are high-level animation parameters. A face model designer creates them for each face model. Since they are designed as a complete expression, they allow animating unknown models with high subjective quality [15, 16]. Another benefit of using FAP 2 is that each face model preserves its 'personality' such that each face in a group of different face models would show its own version of the facial expression.

Using FAP 1 and FAP 2 together with low-level FAPs 3 to 68 that affect the same areas as FAP 1 and 2 may result in unexpected visual representations of the face. Generally, the lower level FAPs have priority over deformations caused by FAP 1 or 2. When specifying an expression with FAP 2, the encoder may send an init_face bit that deforms the neutral face of the model with the expression prior to superimposing FAPs 3 to 68. This deformation is applied with the neutral face constraints of mouth closure, eye opening, gaze direction and head orientation. Since the encoder does not know how FAP 1 and 2 are implemented, we recommend using by and large those low-level FAPs that will not interfere with FAP 1 and 2 in case they are used.

2.2.3 Face Model Specification

Every MPEG-4 terminal that is able to decode FAP streams has to provide an MPEG-4 compliant face model that it animates (section on proprietary face model). Usually, this is a model proprietary to the decoder. The encoder does not know about the look of the face model. Using a FDP node, MPEG-4 allows the encoder to completely specify the face model to animate. This involves defining the static geometry of the face model in its neutral state using a scene graph (section on neutral face model using a scene graph), defining the surface properties and defining the animation rules using FAT that specify how this model gets deformed by the FAPs (section on definition of animation rules using FAT). Alternatively, the FDP node can be used to 'calibrate' the proprietary face model of the decoder (section on face model adaptation). However, MPEG-4 does not specify how to 'calibrate' or adapt a proprietary face model.

Proprietary face model

For a face model to be MPEG-4 compliant, it has to be able to execute all FAPs according to Sections 2.2.1 and 2.2.2. Therefore, the face model has to have at least as many vertices as there are FPs that can be animated. Thus, an MPEG-4-compliant face model may have as little as 50 vertices. Such a model would not generate a pleasing impression. We expect to require at least 500 vertices for pleasant and reasonable face models (Figure 2.3).

A proprietary face model can be built-in 4 steps:

1. We build the shape of the face model and define the location of the feature points on the face model according to Section 2.2.1 and Figure 2.2.
2. For each FAP, we define how the feature point has to move. For most feature points, MPEG-4 defines only the motion in one dimension. As an example, we consider FAP 54, which displaces the outer right lip corner horizontally. Human faces usually move the right corner of the lip backward as they move it to the right. It is left to the face model designer to define a subjectively appealing face deformation for each FAP.
3. After the motion of the feature points is defined for each FAP, we define how the motion of a feature point affects its neighboring vertices. This mapping of feature point motion onto vertex motion can be done using lookup tables such as FAT (section on definition of animation rules using FAT) [17], muscle-based deformation [8, 10, 11], distance transforms [18] or cloning from existing models [19, 20] (see Chapter 4).
4. For expressions, MPEG-4 provides only qualitative hints on how they should be designed (Table 2.4). Similarly, visemes are defined by giving sounds that correspond to the required lip shapes (Table 2.3). FAP 1 and 2 should be designed with care because they will generally be used for visually appealing animations.

Following the above steps, our face model is ready to be animated with MPEG-4 FAPs. Whenever a face model is animated, gender information is provided to the terminal. MPEG-4 does not require using a different face model for male or female gender. We recommend that the decoder reads the gender information and, as a minimum requirement, deforms its model to be male or female. This avoids the presentation of a female face with a male voice and vice versa.

Face model adaptation

An encoder may choose to specify the location of all or some FPs. Then, the decoder is supposed to adapt its own proprietary face model such that the model conforms to the FP positions. Since MPEG-4 does not specify any algorithm for adapting the surface of the proprietary model to the new FP locations, we cannot specify the subjective quality of a face model after its adaptation. Face model adaptation allows also for downloading of texture maps for the face. In order to specify the mapping of the texture map onto the proprietary face model, the encoder sends texture coordinates for each FP. Each texture coordinate defines the location of one FP on the texture map. This does not allow for precise texture mapping at important features such as eyelids or lips. Within the standard, this process of adapting the FP locations of a proprietary face model according to encoder specifications is commonly referred to as 'face model calibration'. As stated above, MPEG-4 does not specify any minimum quality of the adapted face model. Therefore, we prefer to name this process 'face model adaptation'.

In Reference 18, a method for face model adaptation is proposed using an iterative approach based on radial basis functions (RBFs) for scattered data interpolation. For each FP of the proprietary model, a region of interest is defined. When a FP moves, it deforms the model within this region of interest. In order to achieve smooth surfaces, an iterative algorithm was developed.

MPEG-4 allows for a second method of face adaptation by sending an arbitrary mesh to the decoder in addition to FPs. Whereas a possible implementation of this approach is described in Reference 21, MPEG-4 will not mandate a specific implementation in the decoder nor will MPEG-4 define any conformance points for this approach to face model calibration. Therefore, we expect most MPEG-4 terminals not to provide this feature.

The advantage of face model adaptation over downloading a face model from the encoder to the decoder is that the decoder can adapt its potentially very sophisticated model to the desired shape. Since MPEG-4 does not define minimum qualities for proprietary face models and a good adaptation algorithm is fairly difficult to implement, we expect mostly disappointing results as also pointed out in Reference [15]. In order to some what limit the shortcomings, we recommended that the encoder always sends the entire set of FPs for face model adaptation. Sending of partial data may result in completely unpredictable face representations. For applications that specify exactly how the contents are presented at the decoder, downloading a face model using a scene graph seems to be the preferred method (as can be seen in the two following sections).

Neutral face model using a scene graph

In order to download a face model to the decoder, the encoder specifies the static geometry of the head model with a scene graph using MPEG-4 BIFS. For this purpose, BIFS provides the same nodes as VRML. VRML and BIFS describe scenes as a collection of nodes arranged in a scene graph. Three types of nodes are of particular interest for the definition of a static head model. A *Group* node is a container for collecting child objects; it allows for building hierarchical models. For objects to move together as a group, they need to be in the same *Transform* group. The *Transform* node defines geometric affine 3-D transformations such as scaling, rotation and translation, which are performed on its children. When *Transform* nodes contain other *Transforms*, their transformation settings have a cumulative effect. Nested *Transform* nodes can be used to build a transformation hierarchy. An *IndexedFaceSet* node defines the geometry (3-D mesh) and surface attributes (color, texture) of a polygonal object. Texture maps are coded with the wavelet coder of the MPEG still image coder [4].

Figure 2.4 shows the simplified scene graph for a face model. Nested *Transforms* are used to apply rotations about the x-, y- and z-axis one after another. Embedded into these global head movements are the rotations for the left and right eye. Separate IndexedFaceSets define the shape and the surface of the face, hair, tongue, teeth, left

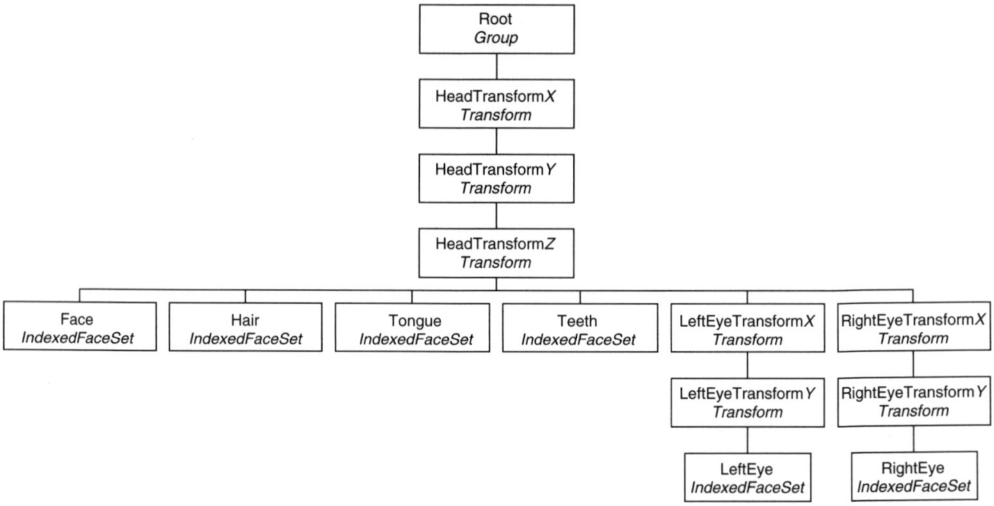

Figure 2.4 Simplified scene graph for a head model. The names of BIFS nodes are given in italic

eye and right eye, thus allowing for separate texture maps. Because the face model is specified with a scene graph, this face model can be easily extended to a head and shoulder model. The surface properties of the face can be specified using colors or still images to define texture mapped models.

The shape of the face models may be generated using interactive modelers, scanners or image analysis software [21, 22]

Definition of animation rules using FAT

FATs define how a model is spatially deformed as a function of the amplitude of the FAPs. Three BIFS nodes provide this functionality: *FaceDefTable*, *FaceDefTransform* and *FaceDefMesh*. These nodes are considered to be part of the face model. Using FaceDefTransform nodes and FaceDefMesh nodes, the FaceDefTable specifies, for an FAP, which nodes of the scene graph are animated by it and how [17].

Animation definition for a transform node

If an FAP causes a transformation such as rotation, translation or scale, a Transform node can describe this animation. The FaceDefTable specifies a FaceDefTransform node that defines the type of transformation and a scaling factor for the chosen transformation. During animation, the received value for the FAP, the FAPU and the scaling factor determine the actual value by which the model is transformed.

Animation definition for an IndexedFaceSet node

If an FAP like joy causes flexible deformation of the face model, the FaceDefTable node uses a FaceDefMesh node to define the deformation of IndexedFaceSet nodes. The animation results in updating vertex positions of the affected IndexedFaceSet nodes. Moving the affected vertices as a piecewise linear function of FAP amplitude values

Table 2.5 Simplified example of a FaceDefMesh and a FaceDefTransform

#FaceDefMesh
FAP 6 (stretch left corner lip)
IndexedFaceSet: *Face*
Interval borders: −1000, 0, 500, 1000
Displacements:
Vertex 50 1 0 0, 0.9 0 0, 1.5 0 4
Vertex 51 0.8 0 0, 0.7 0 0, 2 0 0
#FaceDefTransform
FAP 23 (yaw left eye ball)
Transform: *LeftEyeX*
Rotation scale factor: 0−1 0 (axis) 1 (angle)

approximates flexible deformations of an IndexedFaceSet. The FaceDefMesh defines for each affected vertex its own piecewise linear function by specifying intervals of the FAP amplitude and 3-D displacements for each interval (see Table 2.5 for an example). The VRML community started to define a *Displacer* Node that provides a similar functionality. However, the motion of a vertex is limited to a straight line.

If P_m is the position of the mth vertex of the IndexedFaceSet in neutral state (FAP = 0) and D_{mk} is the 3-D displacement that defines the piecewise linear function in the kth interval, then the following algorithm is used to determine the new position P'_m of the same vertex after animation with the given FAP value (Figure 2.5):

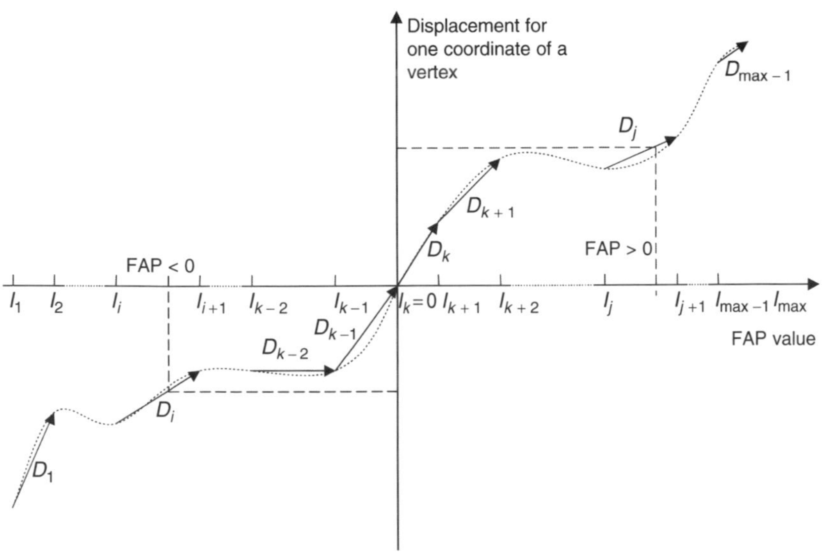

Figure 2.5 Piecewise linear approximation of vertex motion as a function of the FAP Value

1. Determine the interval listed in the FaceDefMesh in which the received FAP value is lying.

2. If the received FAP is in the jth interval $[I_j, I_{j+1}]$ and $0 = I_k \leq I_j$, the new position P'_m of the mth vertex is given by

$$P'_m = P_m + \text{FAPU} * [(I_{k+1} - 0) * D_{m,k} + (I_{k+2} - I_{k+1}) * D_{m,k+1} + \cdots (I_j - I_{j-1}) * D_{m,j-1} + (\text{FAP} - I_j) * D_{m,j}]$$

3. If FAP $> I_{\max}$, then P'_m is calculated by using the equation given in 2 and setting the index $j = \max - 1$.
4. If the received FAP is in the jth interval $[I_j, I_{j+1}]$ and $I_{j+1} \leq I_k = 0$, the new position P'_m of the mth vertex is given by

$$P'_m = P_m + \text{FAPU} * [(I_{j+1} - \text{FAP}) * D_{m,j} + (I_{j+2} - I_{j+1}) * D_{m,j+1} + \cdots (I_{k-1} - I_{k-2}) * D_{m,k-2} + (0 - I_{k-1}) * D_{m,k-1}]$$

5. If FAP $< I_1$, then P'_m is calculated by using the equation in 4 and setting the index $j = 1$.
6. If for a given FAP and 'IndexedFaceSet' the table contains only one interval, the motion is strictly linear:

$$P'_m = P_m + \text{FAPU} * \text{FAP} * D_{m1}$$

Strictly speaking, these animation rules are not limited to faces. Using this technology, MPEG-4 allows for a very efficient mechanism of animating *IndexedFaceSet* and *Transform* nodes of arbitrary objects with up to 68 FAPs. In Figure 2.6, we see a head and shoulder model that can be animated using 68 FAPs. Obviously, the interpretation of the FAPs by the model is partially, not according to the standard because the standard does not define a means for moving an arm. Therefore, this model should only be animated by an encoder that knows the interpretation of FAPs by this model.

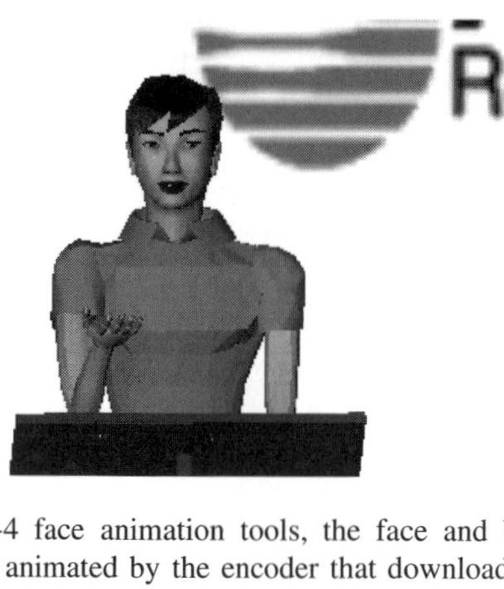

Figure 2.6 Using MPEG-4 face animation tools, the face and body of this model can be downloaded and efficiently animated by the encoder that downloads the model to the decoder. (Reprinted from Tekalp and Ostermann, "Face and 2-D mesh animation in MPEG-4", *Signal Processing: Image Communication*, Vol. 15, 2000, pp. 387–421, Copyright 2000, with permission from Elsevier Science)

Example for a FaceDefTable

In Table 2.5, two FAPs are defined by children of a FaceDefTable, namely, the FaceDefMesh and the FaceDefTransform: FAP 6 that stretches the left corner lip and FAP 23 that manipulates the horizontal orientation of the left eyeball.

FAP 6 deforms the IndexedFaceSet named *Face*. For the piecewise linear motion function three intervals are defined: [−1000, 0], [0, 500] and [500, 1000]. Displacements are given for the vertices with indices 50 and 51. The displacements for vertex 50 are: (1 0 0), (0.9 0 0) and (1.5 0 4), the displacements for vertex 51 are (0.8 0 0), (0.7 0 0) and (2 0 0). Given an FAP amplitude of 600, the resulting displacement for vertex 50 would be:

$$P'_{50} = P_{50} + 500 * (0.9\ 0\ 0)^{\mathrm{T}} + 100 * (1.5\ 0\ 4)^{\mathrm{T}} = P_{50} + (600\ 0\ 400)^{\mathrm{T}}$$

FAP 23 updates the rotation field of the Transform node *LeftEyeX*. The rotation axis is (0, −1, 0), and the neutral angle is 0 radians. The FAP value determines the rotation angle.

Figure 2.7 shows 2 phases of a left eye blink (plus the neutral phase), which have been generated using simple animation architecture [17].

The creation of the FaceDefMesh nodes for large models can be time consuming. However, the process depicted in Figure 2.8 uses a FaceDefTable generator that computes these tables from a set of face models. The face model is described as a VRML file and read into the modeler. In order to design the behavior of the model for one animation parameter, the model is deformed using the tools of the modeler. The modeler exports the deformed model as a VRML file without changing the topology of the model [2].

The FaceDefMesh generator compares the output of the modeler with its input, the face model in its neutral state. By comparing vertex positions of the two models, the

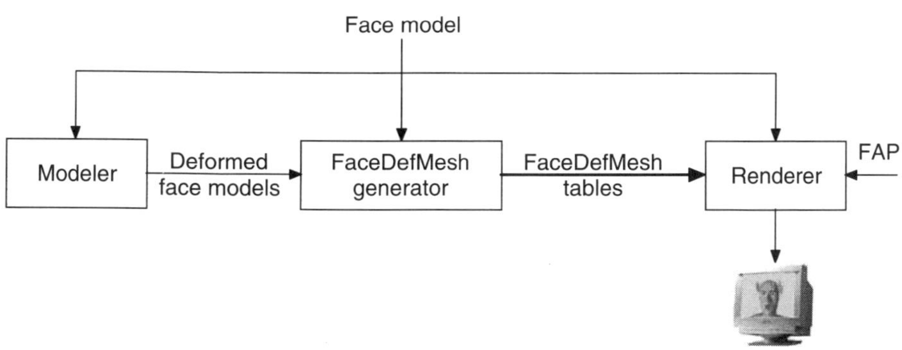

Figure 2.7 Neutral state of the left eye (left) and two deformed animation phases for the eye blink (FAP 19). The FAP definition defines the motion of the eyelid in negative y-direction; the FaceDefTable defines the motion in one of the vertices of the eyelid in x- and z-direction. Note that positive FAP values move the vertices downwards (Table 2.7)

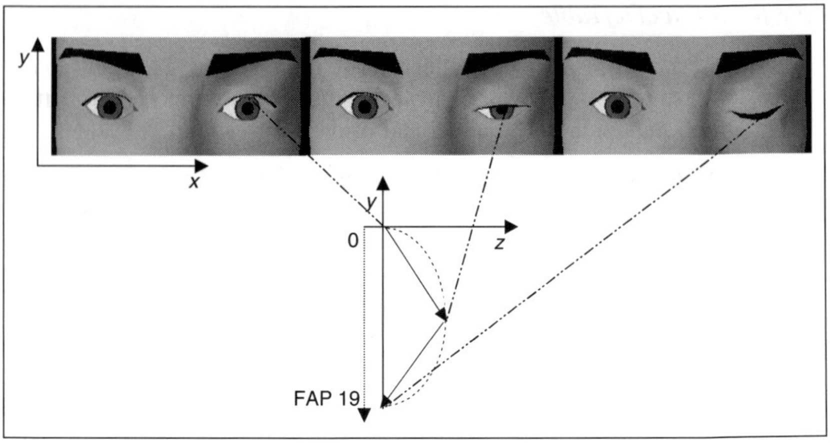

Figure 2.8 FaceDefMesh Interface – the modeler is used to generate VMRL files with the object in different animated positions. The generator computes one FaceDefMesh for each animation parameter

vertices affected by the newly designed animation parameter are identified. The generator computes a 3-D-displacement vector defining the deformation for each affected vertex and exports this information in a FaceDefMesh table. The renderer reads the VRML file of the model and the table in order to learn the definition of the new animation parameter. Now, the renderer can use the newly defined animation as required by the animation parameters.

2.3 CODING OF FACE ANIMATION PARAMETERS

MPEG-4 provides two tools for coding of FAPs. Coding of quantized and temporally predicted FAPs using an arithmetic coder allows for low-delay FAP coding (Section 2.3.1). Alternatively, DCT coding of a sequence of FAPs introduces a larger delay but achieves higher coding efficiency (Section 2.3.2).

MPEG-4 provides a special mode (def_bit) that allows downloading definitions of expressions and visemes (FAP 1 and 2) in terms of low-level FAPs. Although the syntax for this capability is defined, MPEG-4 does not require the decoder to store a minimum number of these definitions. Therefore we recommend not using this tool until MPEG-4 provides clarifications on this tool in a potential future revision of the standard. Instead, we recommend the use of the FAP Interpolation Table (FIT) as described in Section 2.3.3.

2.3.1 Arithmetic Coding of FAPs

Figure 2.9 shows the block diagram for low-delay encoding of FAPs. The first set of FAP values FAP_0 at time instant 0 is coded without prediction (intracoding). The value of an FAP at time instant k FAP_k is predicted using the previously

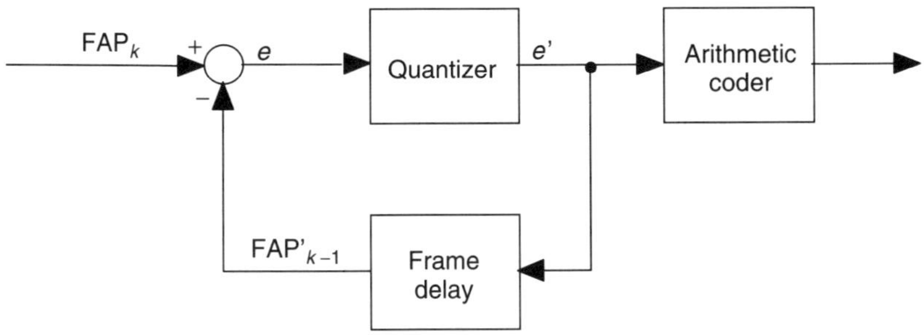

Figure 2.9 Block diagram of the low-delay encoder for FAPs. (Reprinted from Tekalp and Ostermann, "Face and 2-D mesh animation in MPEG-4", *Signal Processing: Image Communication*, Vol. 15, 2000, pp. 387–421, Copyright 2000, with permission from Elsevier Science)

decoded value FAP'_{k-1}. The prediction error e is quantized using a quantization step size QP*FAP_QUANT. QP is specified in Table 2.7. The quantization parameter FAP_QUANT is used to adjust the bit rate with $0 < \mathrm{FAP\_QUANT} < 31$. FAP_QUANT is identical for all FAP values at one-time instant k. Using the FAP dependent quantization step size QP, FAP_QUANT assures that quantization errors are subjectively evenly distributed between different FAPs. The quantized prediction error e' is arithmetically encoded using a separate adaptive probability model for each FAP. Since the encoding of the current FAP value depends only on one previously coded FAP value, this coding scheme allows for low-delay communications. At the decoder, the received data is arithmetically decoded, dequantized and added to the previously decoded value in order to recover the encoded FAP value. When using $\mathrm{FAP\_QUANT} > 15$, the subjective quality of the animation deteriorates significantly such that we recommend not to increase FAP_QUANT above 15 [15].

In order to avoid transmitting all FAPs for every frame, the encoder can transmit a mask indicating for which groups (Table 2.2) FAP values are transmitted. The encoder can also specify for which FAPs within a group values will be transmitted. This allows the encoder to send incomplete sets of FAPs to the decoder. FAP values that have been initialized in an intracoded FAP set are assumed to retain those values if subsequently no update is transmitted. However, the encoder can also signal to the decoder that a previously transmitted FAP value is not valid anymore.

The decoder can extrapolate values of FAPs that have been invalidated or have never been specified in order to create a more complete set of FAPs. The standard provides only limited specifications on how the decoder is supposed to extrapolate FAP values. Examples are that if only FAPs for the left half of a face are transmitted, the corresponding FAPs of the right side have to be set such that the face moves symmetrically. If the encoder only specifies motion of the inner lip (FAP group 2), the motion of the outer lip (FAP group 8) has to be extrapolated in an unspecified way. Letting the decoder extrapolate FAP values may create unexpected results depending on the particular decoder. However, the encoder can always prevent the decoder from using extrapolation by defining all FAP values or defining FITs (Section 2.3.3).

2.3.2 DCT Coding of FAPs

The second tool that is provided for coding FAPs is the DCT applied to 16 consecutive FAP values. This introduces a significant delay into the coding and decoding process. Hence, this coding method is mainly useful for applications in which animation parameter streams are retrieved from a database. After computing the DCT of 16 consecutive values of one FAP, DC and AC coefficients are coded differently (Figure 2.10). While the DC value is coded predictively using the previous DC coefficient as prediction, the AC coefficients are directly coded. The AC coefficient and the prediction error of the DC coefficient are linearly quantized. Whereas the quantizer step size can be controlled, the ratio between the quantizer step size of the DC coefficients and the AC coefficients is set to 1/4. The quantized AC coefficients are encoded with one variable length code word (VLC) defining the number of zero-coefficients prior to the next nonzero coefficient and one VLC for the amplitude of this nonzero coefficient. The handling of the decoded FAPs with respect to masking and interpolation is not changed (see Section 2.3.1).

Figure 2.11 compares the coding performance of the DCT FAP coder and the arithmetic FAP coder. The Peak Signal to Noise Ratio (PSNR) is measured by comparing the amplitude of the original and coded FAP averaging over all FAPs. This PSNR does not relate to picture quality but to the smoothness of temporal animation. In contrast to the arithmetic coder, the DCT coder is not able to code FAPs with near lossless quality. At low data rates, the DCT coder requires up to 50% less data rate than the arithmetic coder at the price of an increased coding delay. This advantage in coding efficiency disappears with increasing fidelity of the coded parameters.

2.3.3 FAP Interpolation Tables

As mentioned in Section 2.3.1, the encoder may allow the decoder to extrapolate the values of some FAPs from the transmitted FAPs [23]. Alternatively, the decoder can specify the interpolation rules using FITs. A FIT allows a smaller set of FAPs to be

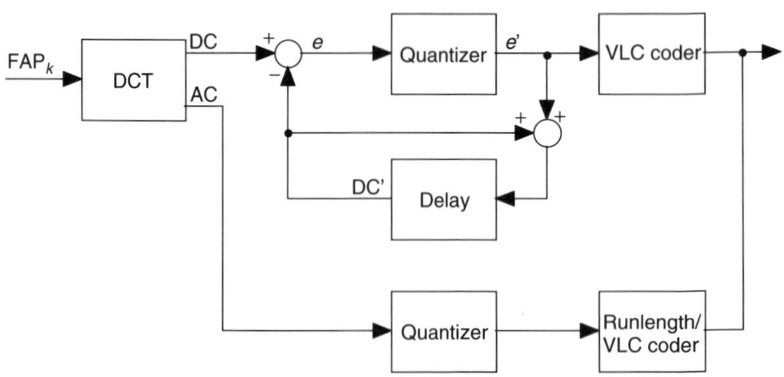

Figure 2.10 Block diagram of the FAP encoder using DCT. DC coefficients are predictively coded. AC coefficients are directly coded. (Reprinted from Tekalp and Ostermann, "Face and 2-D mesh animation in MPEG-4", *Signal Processing: Image Communication*, Vol. 15, 2000, pp. 387–421, Copyright 2000, with permission from Elsevier Science)

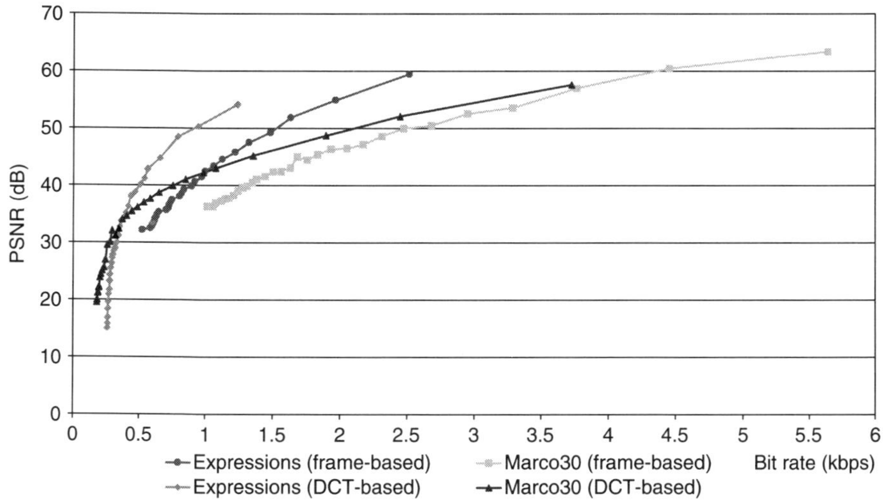

Figure 2.11 Rate distortion performance of the arithmetic and DCT coding mode of FAPs for the sequences Marco30 (30 Hz) and Expressions (25 Hz) (from G. Abrantes, F. Pereira, "MPEG-4 facial animation technology: survey, implementation and results", *IEEE CSVT*, **9**(2), 290–305 (1999)

sent for a facial animation. This small set can then be used to determine the values of other FAPs, using a rational polynomial mapping between parameters. For example, the top inner-lip FAPs can be sent and then used to determine the top outer-lip FAPs. The inner-lip FAPs would be mapped to the outer-lip FAPs using a rational polynomial function that is specified in the FIT.

An FAP interpolation graph (FIG) is used to specify which FAPs are interpolated from other FAPs. The FIG is a graph with nodes and directed links. Each node contains a set of FAPs. Each link from a parent node to a child node indicates that the FAPs in a child node can be interpolated from those of the parent node. In a FIG, an FAP may appear in several nodes, and a node may have multiple parents. For a node that has multiple parent nodes, the parent nodes are ordered as 1st parent node, 2nd parent node and so on. During the interpolation process, if this child node needs to be interpolated, it is first interpolated from the 1st parent node if all FAPs in that parent node are available. Otherwise, it is interpolated from the 2nd parent node, and so on. An example of FIG is shown in Figure 2.12. Each node has an ID. The numerical label on each incoming link indicates the order of these links.

Each directed link in an FIG is a set of interpolation functions. Suppose $F_1, F_2, \ldots, F_n$ are the FAPs in a parent set and $f_1, f_2, \ldots, f_m$ are the FAPs in a child set. Then, there are m interpolation functions denoted as: $f_1 = I_1(F_1, F_2, \ldots, F_n)$, $f_2 = I_2(F_1, F_2, \ldots, F_n)$, $f_m = I_m(F_1, F_2, \ldots, F_n)$. Each interpolation function I_k () is in the form of a rational polynomial

$$I(F_1, F_2, \ldots, F_n) = \sum_{i=0}^{K-1} \left(c_i \prod_{j=1}^{n} F_j^{l_{ij}} \right) \Bigg/ \sum_{i=0}^{P-1} \left(b_i \prod_{j=1}^{n} F_j^{m_{ij}} \right) \qquad (2.1)$$

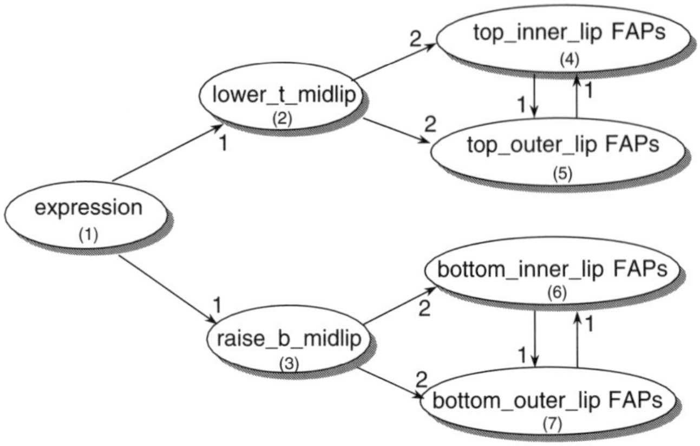

Figure 2.12 A FIG example for interpolating unspecified FAP values of the lip. If only the expression is defined, the FAPs get interpolated from the expression. If all inner-lip FAPs are specified, they are used to interpolate the outer-lip FAPs. (Reprinted from Tekalp and Ostermann, "Face and 2-D mesh animation in MPEG-4", *Signal Processing: Image Communication*, Vol. 15, 2000, pp. 387–421, Copyright 2000, with permission from Elsevier Science)

where K and P are the numbers of polynomial products, c_i and b_i are the coefficients of the ith product. l_{ij} and m_{ij} are the power of F_j in the ith product. The encoder should send an interpolation function table that contains all $K, P, c_i, b_i, l_{ij}, m_{ij}$ to the decoder for each link in the FIG.

Here, we provide some simple examples in which the use of FIT can be useful to reduce the bit rate for transmitting FAPs:

1. Precise specification of the extrapolation of FAPs from their counterparts on the other side of the face. If desired, this mechanism allows even for unsymmetrical face animation.
2. Extrapolation of outer-lip FAPs from inner-lip FAPs.
3. Extrapolation of eyebrow motion from FAP 34 (raise right middle eyebrow). This can be done with linear polynomials.
4. Definition of facial expression (FAP 1 and 2) using low-level FAPs instead of using the def_bit.

In order to specify the FITs for the examples, linear polynomials are usually sufficient. If it is desired to simulate the varying elasticity the of skin for large FAP amplitudes, nonlinear mappings might be useful. Following example 3, we might want the inner and outer eyebrows to follow the middle eyebrow roughly at first, linearly and then to a lesser extent. This gives eyebrows with increasing curvature as the FAP amplitude increases.

2.4 INTEGRATION OF FACE ANIMATION AND TEXT-TO-SPEECH SYNTHESIS

MPEG-4 provides interfaces for a proprietary TTS synthesizer [5] that allows driving a talking head from text (see Figure 2.13) [8, 13, 24–27]. This section discusses the

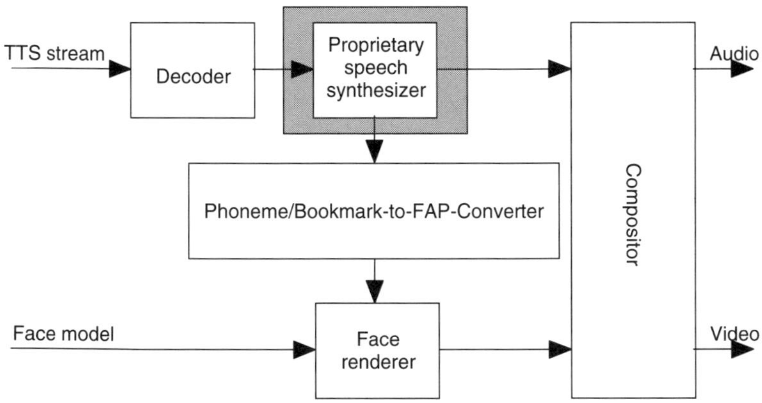

Figure 2.13 Block diagram showing the integration of a proprietary text-to-speech synthesizer into an MPEG-4 face animation system

integration of FA and TTS [28] allowing for animation of a talking face using a TTS synthesizer [16]. A key issue here is the synchronization of the speech stream with the FAP stream. Synchronization of an FAP stream with TTS synthesizers using the TTS Interfaces (TTSI) is only possible, if the encoder sends prosody and timing information. This is due to the fact that a conventional TTS system driven by text only behaves as an asynchronous source. Animation with text only allows for animation of talking heads at a data rate of less than 200 bits s^{-1}.

Given a TTS stream that contains text or prosody in binary form, the MPEG-4 TTSI decoder decodes the gender, text and prosody information according to the interface defined for the TTS synthesizer. The synthesizer creates speech samples that are handed to the compositor. The compositor presents audio and if required video to the user. The second output interface of the synthesizer sends the phonemes of the synthesized speech as well as start time and duration information for each phoneme to the Phoneme/Bookmark-to-FAP-Converter [16]. The converter translates the phonemes and timing information into face animation parameters that the face renderer uses in order to animate the face model. The precise method of how the converter derives visemes from phonemes is not specified by MPEG and left to the implementation of the decoder. This also allows using a coarticulation model at the decoder that uses the current, previous and next phoneme in order to derive the correct mouth shape.

Most speech synthesizers do not have a synchronous behavior. This means that the time they require to speak a sentence is not predictable. Therefore, synchronizing the output of a TTS with facial expressions defined in an FAP stream is not possible. Bookmarks in the text of the TTS are used to animate facial expressions and non-speech-related parts of the face [16]. The start time of a bookmark is derived from its position in the text. When the TTS finds a bookmark in the text it sends this bookmark to the Phoneme/Bookmark-to-FAP-Converter at the same time as it sends the first phoneme of the following word. The bookmark defines the *start* point and *duration* of the transition to a new FAP amplitude. Consequence: No additional delay, no look ahead in the bit stream but no precise timing control on when the amplitude will be reached relative to the spoken text.

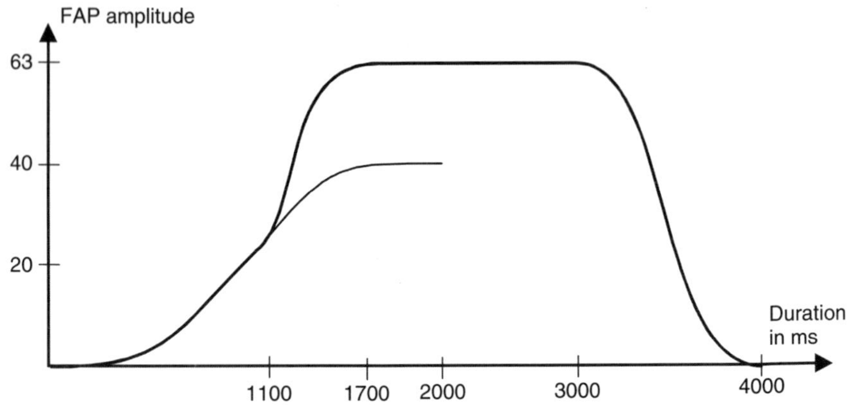

<FAP 2 1 40 1 40 2000 3>.text.<FAP 2 1 63 1 63 600 3>...........text.........<FAP 2 1 0 1 0 1000 3>...text.....

Figure 2.14 Example for text with bookmarks for one facial expression (joy) and the related amplitude of the animated FAP. The syntax of a bookmark is: ⟨FAP 2 (expression) 1 (joy) amplitude 1 (joy) amplitude duration 3 (Hermite time curve)⟩. The amplitude of joy over time is computed according to the bookmarks (see Section 2.2.2). (Reprinted from Tekalp and Ostermann, "Face and 2-D mesh animation in MPEG-4", *Signal Processing: Image Communication*, Vol. 15, 2000, pp. 387–421, Copyright 2000, with permission from Elsevier Science)

An example of a TTS stream with bookmarks is given in Figure 2.14 [16]. The renderer will generate the visemes associated with each word, following the timing information derived by the speech synthesizer. It will also start to deform the model to generate a smile with an amplitude of 40. To simulate a more natural expression, which typically goes through three phases (onset, climax and relax), a desired temporal behavior [29] for a prescribed FAP can be specified in the bookmark. Three functions are defined: A linear interpolation function and a Hermite function can be used to specify the transition of an FAP from its current value to the target value. A triangular function can be specified to linearly increase the amplitude of an FAP to the target value and to decrease it back to its starting amplitude. The bookmark also specifies the desired duration to reach the FAP value in the bookmark. If another bookmark appears before this duration, the renderer starts to deform the face according to the newly specified FAP information from the current position. This is illustrated in Figure 2.14 using Hermite functions.

MPEG-4 assumes that the TTS is located at the MPEG-4 terminal. However, the IETF [30] develops a real-time transport protocol (RTP) payload format for the phoneme and bookmark information that the TTS sends to the Phoneme/Bookmark to FAP Converter. This enables network-based TTS servers with a large variety of high-quality voices such that not only can the look of a face model be easily customized but the voice as well.

2.5 INTEGRATION WITH MPEG-4 SYSTEMS

In order to use FA in the context of MPEG-4 systems, a BIFS scene graph has to be transmitted to the MPEG-4 terminal. The minimum scene graph contains a Face node and an FAP node. An FAP stream is streamed as a BifsAnim stream from the server to the terminal. The FAP decoder writes the amplitude of the FAPs into fields of the FAP node. The FAP node might have the children Viseme and Expression, which are FAPs requiring a special syntax (see Section 2.2.2). This scene graph would enable an encoder to animate the proprietary face model of the decoder. If a face model is to be controlled from a TTS system, an AudioSource node is to be attached to the face node.

A face model may be downloaded to the decoder; the face node requires an FDP node as one of its children. This FDP node contains the position of the FPs in the downloaded model, the scene graph of the model and the FaceDefTable, FaceDefMesh and FaceDefTransform nodes required to define the action caused by FAPs. Figure 2.15 shows how these nodes relate to each other.

Usually, the FA has to be synchronized with other parts of a multimedia presentation. Therefore, MPEG-4 systems require to add time stamps to each data stream. A talking face reporting in front of a video would require the video, audio and FAP streams to carry time stamps. Whereas MPEG-4 provides transport mechanisms for streaming of its contents, the Audiovisual Transport group of IETF also develops real-time streaming protocols for MPEG-4 streams [30].

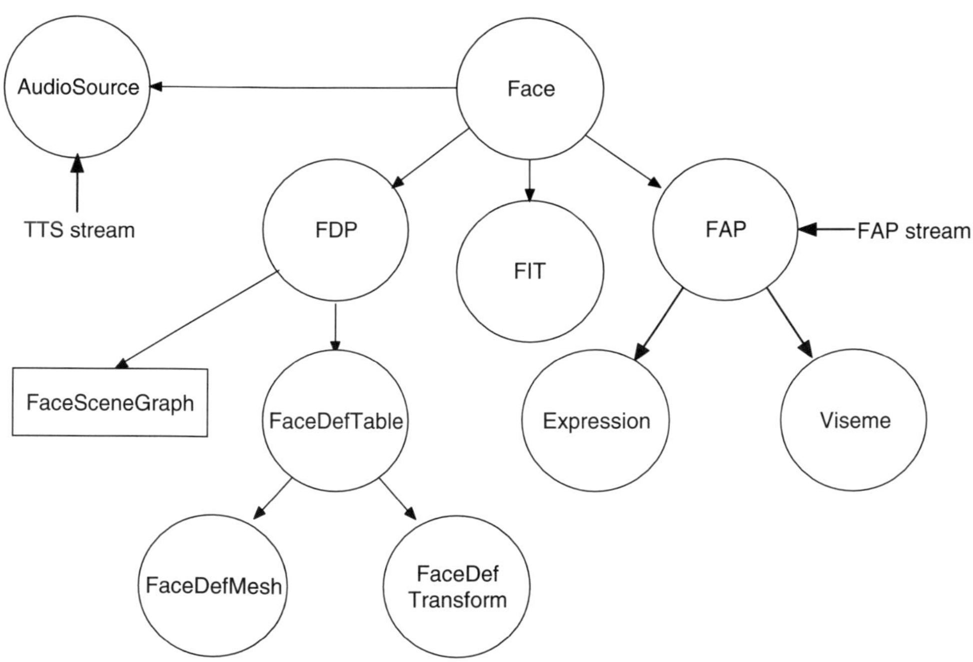

Figure 2.15 Nodes of a BIFS scene graph that are used to describe and animate a face. The FaceSceneGraph contains the scene graph of the static face. Here, it is assumed that the streams are already decoded. (Reprinted from Tekalp and Ostermann, "Face and 2-D mesh animation in MPEG-4", *Signal Processing: Image Communication*, Vol. 15, 2000, pp. 387–421, Copyright 2000, with permission from Elsevier Science)

2.6 MPEG-4 PROFILES FOR FACE ANIMATION

MPEG-4 defines profiles to which decoders have to conform. A profile consists of *objects* defining the tools of the profile. Levels of a profile and object put performance and parameter limits on the tools. MPEG-4 Audio, Visual and Systems define parts of FA. A detailed overview of MPEG-4 profiles related to FA is given in Chapter 3.

In Visual, the neutral face with its FPs and FAPs, the coding of FAPs as well as the Phoneme/Bookmark-to-FAP-Converter with its interface to TTSI are defined. The corresponding object type is called *Simple Face*. The *Simple Face* object allows animating a proprietary face model using an FAP stream or from a TTSI provided that the terminal supports MPEG-4 audio. Two levels are defined for this object: At level 1, it requires to animate one face model with a maximum bit rate of 16 kbit s^{-1} and a render frame rate of at least 15 Hz. At level 2, up to 4 faces can be animated with a total bit rate not to exceed 32 kbit s^{-1} and a render frame rate of 60 Hz shareable between faces. This *Simple Face* object is included in the following visual profiles: *Hybrid*, *Basic Animated Texture* and *Simple FA*. Whereas the *Simple FA* requires only face animation capabilities, *Basic Animated Texture* adds scalable texture coding and mesh-based animation of these textures. The *Hybrid* profile adds video decoding according to the *Core* profile (video and binary shape) [4].

In MPEG-4 Audio, the TTSI with the bookmark identifiers for FA and the interface to the Phoneme/Bookmark-to-FAP-converter is defined. It is part of all Audio profiles. Using a TTS, any Audio profile and a Visual profile containing the *Face* object allows to define interactive services with FA at extremely low data rates. Without using a TTS, any Audio profile and a Visual profile containing the *Face* object allows to play speech and animate the proprietary face model.

In order to enable the specification of the face, the BIFS node FDP and its children have to be transmitted. This is possible for terminals that support the *Complete* Scene Graph profile and the *Complete* Graphics profile.

2.7 CONCLUSION

MPEG-4 integrates synthetic and natural content in multimedia communications and documents. For animated faces, MPEG-4 defines a complete set of FAPs tailored toward animation of the human face. These FAPs include high-level FAPs like facial expressions and low-level FAPs that define the deformation of individual parts of a face as well as head and eye rotation. In order to enable animation of a face model over low bit rate communication channels, for point-to-point and multipoint connections, MPEG-4 encodes the FAPs using temporal prediction. For applications requiring low latency, an arithmetic coder should be used to code the prediction error. For more efficient coding of FAP streams, a DCT coder is provided. Depending on the selected quantizer, face models can be animated with a data rate of 300 to 2000 bits s^{-1}. MPEG-4 also defines an application program interface for TTS synthesizer. Using this interface, the synthesizer can be used to provide phonemes and related timing information to the face model. Adding facial expressions to the talking head is achieved by using bookmarks in the text. This allows for animated talking heads driven just by one text stream at a data rate of less than 200 bits s^{-1}.

MPEG-4 FA requires an MPEG-4 terminal to provide a proprietary face model. MPEG-4 also provides the capability to download a face model from a server to the terminal such that a service provider or electronic store may have his own customized agent interacting with the viewer. Subjective tests reported in Reference [31] show that an Electronic Commerce Web site with talking faces gets higher ratings than the same Web site without talking faces.

REFERENCES

[1] J. Hartman, J. Wernecke, *The VRML Handbook*, Addison-Wesley, Reading, Mass., 1996.
[2] ISO/IEC 14772-1:1997, "Information Technology – Computer Graphics and Image Processing – The Virtual Reality Modeling Language – Part 1: Functional Specification and UTF-8 Encoding", 1997.
[3] ISO/IEC IS 14496-1 Systems, 1999.
[4] ISO/IEC IS 14496-2 Visual, 1999.
[5] ISO/IEC IS 14496-3 Audio, 1999.
[6] T. K. Capin, E. Petajan, J. Ostermann, "Efficient modeling of virtual humans in MPEG-4", *ICME 2000*, New York, 2000, p. TPS9.1.
[7] T. K. Capin, E. Petajan, J. Ostermann, "Very low bit rate coding of virtual human animation in MPEG-4", *ICME 2000*, New York, 2000, p. TPS9.2.
[8] P. Kalra, A. Mangili, N. Magnenat-Thalmann, D. Thalmann, "Simulation of facial muscle actions based on rational free form deformations", *Proceedings of the Eurographics 92*, 1992, pp. 59–69.
[9] F. I. Parke, "Parameterized models for facial Animation", *IEEE Computer Graphics and Applications*, **2**, 61–68 (1982).
[10] D. Terzopolous, K. Waters, "Physically-based facial modeling, analysis and animation", *Journal of Visualization and Computer Animation*, **1**, 73–80 (1990).
[11] K. Waters, "A muscle model of animating three dimensional facial expression", *Computer Graphics*, **22**(4), 17–24 (1987).
[12] P. Ekman, W. V. Friesen, *Manual for the Facial Action Coding System*, Consulting Psychologist Press, Palo Alto, Calif., 1978.
[13] M. M. Cohen, D. W. Massaro, "Modeling coarticulation in synthetic visual speech", in M. Thalmann, D. Thalmann, eds., *Computer Animation '93*, Springer-Verlag, Tokyo.
[14] C. Pelachaud, "Visual text-to-speech", Chapter 8, this book.
[15] G. Abrantes, F. Pereira, "MPEG-4 facial animation technology: survey, implementation and results", *IEEE CSVT*, **9**(2), 290–305 (1999).
[16] J. Ostermann, M. Beutnagel, A. Fischer, Y. Wang, Integration of talking heads and text-to-speech synthesizers for visual TTS, *ICSLP 99*, Australia, December, 1999.
[17] J. Ostermann, E. Haratsch, "An animation definition interface: rapid design of MPEG-4 compliant animated faces and bodies", *International Workshop on Synthetic – Natural Hybrid Coding and Three Dimensional Imaging*, Rhodes, Greece, September, 1997, pp. 216–219.
[18] F. Lavagetto, R. Pockaj, "The facial animation engine: toward a high-level interface for the design of MPEG-4 compliant animated faces", *IEEE CSVT*, **9**(2), 277–289 (1999).
[19] M. V. Mani, J. Ostermann, Cloning of MPEG-4 face models, *International Workshop on Very Low Bit rate Video Coding (VLBV 01)*, Athens, October, 2001.
[20] J. Noh, U. Neumann, Expression cloning, *ACM SIGGRAPH 2001*, Los Angeles, August, 2001.
[21] M. Escher, N. M. Thalmann, Automatic 3D cloning and real-time animation of a human face, *Proceedings of Computer Animation '97*, Geneva, June, 1997.

[22] L. Chen, J. Ostermann, T. Huang, "Adaptation of a generic 3D human face model to 3D range data", in Y. Wang, S. Panwar, S. Kim, H. Bertoni, eds., *First IEEE Workshop on Multimedia Signal Processing*, IEEE, Princeton, N.J., June, 1997, pp. 274–279.

[23] H. Tao, H. H. Chen, W. Wu, T. S. Huang, "Compression of facial animation parameters for transmission of talking heads", *IEEE CSVT*, **9**(2), 264–276 (1999).

[24] S. Morishima, H. Harashima, "A media conversion from speech to facial image for intelligent man-machine interface", *IEEE Journal of Selected Areas in Communications*, **9**(4), 594–600 (1991).

[25] K. Waters, T. Levergood, "An automatic lip-synchronization algorithm for synthetic faces", *Proceedings of the Multimedia Conference*, ACM, San Francisco, Calif., September, 1994, pp. 149–156.

[26] C. Bregler, M. Covell, M. Slaney, Video rewrite: driving visual speech with audio, *Proceedings of ACM SIGGRAPH 97*, in Computer Graphics Proceedings, Annual Conference Series, Los Angeles, August, 1997.

[27] S. Morishima, "Modeling of facial expression and emotion for human communication system", *Display*, **17**, 15–25 (1996).

[28] R. Sproat, J. Olive, "An approach to text-to-speech synthesis", in W. B. Kleijn, K. K. Paliwal, eds., *Speech Coding and Synthesis*, Elsevier Science, Amsterdam, 1995.

[29] S. Murakami, H. Tadokoro, "Generation of facial expressions based on time functions", *Proceedings of IWSNHC3DI*, Rhodes, Greece, September, 1997, pp. 212–215.

[30] The Internet Engineering Task Force, *http://www.ietf.org*.

[31] I. Pandzic, J. Ostermann, D. Millen, "User evaluation: synthetic talking faces for interactive services", accepted for publication in *The Visual Computer*, Special Issue on Real-time Virtual Worlds, **15**(7/8), 330–340 (1999).

ANNEX

Table 2.6 Location of feature points on a face model (Figure 2.2). Recommended location constraints define for some or all coordinates of a feature point the value as a function of other feature points, that is, feature points 2.1, 2.2, 2.3 and 7.1 have the same x-coordinate thus locating them in the same yz-plane

Feature points		Recommended location constraints		
#	Text description	x	y	z
2.1	Bottom of the chin	7.1.x		
2.2	Middle point of inner upper-lip contour	7.1.x		
2.3	Middle point of inner lower-lip contour	7.1.x		
2.4	Left corner of inner-lip contour			
2.5	Right corner of inner-lip contour			
2.6	Midpoint between FP 2.2 and 2.4 in the inner upper-lip contour	$(2.2.x + 2.4.x)/2$		
2.7	Midpoint between FP 2.2 and 2.5 in the inner upper-lip contour	$(2.2.x + 2.5.x)/2$		
2.8	Midpoint between FP 2.3 and 2.4 in the inner lower-lip contour	$(2.3.x + 2.4.x)/2$		
2.9	Midpoint between FP 2.3 and 2.5 in the inner lower-lip contour	$(2.3.x + 2.5.x)/2$		
2.10	Chin boss	7.1.x		

(*continued overleaf*)

Table 2.6 *(continued)*

Feature points		Recommended location constraints		
#	Text description	x	y	z
2.11	Chin left corner	$>8.7.x$ and $<8.3.x$		
2.12	Chin right corner	$>8.4.x$ and $<8.8.x$		
2.13	Left corner of jawbone			
2.14	Right corner of jawbone			
3.1	Center of upper inner-left eyelid	$(3.7.x + 3.11.x)/2$		
3.2	Center of upper inner-right eyelid	$(3.8.x + 3.12.x)/2$		
3.3	Center of lower inner-left eyelid	$(3.7.x + 3.11.x)/2$		
3.4	Center of lower inner-right eyelid	$(3.8.x + 3.12.x)/2$		
3.5	Center of the pupil of left eye			
3.6	Center of the pupil of right eye			
3.7	Left corner of left eye			
3.8	Left corner of right eye			
3.9	Center of lower outer-left eyelid	$(3.7.x + 3.11.x)/2$		
3.10	Center of lower outer-right eyelid	$(3.7.x + 3.11.x)/2$		
3.11	Right corner of left eye			
3.12	Right corner of right eye			

3.13	Center of upper outer-left eyelid	$(3.8.x + 3.12.x)/2$		
3.14	Center of upper outer-right eyelid	$(3.8.x + 3.12.x)/2$		
4.1	Right corner of left eyebrow			
4.2	Left corner of right eyebrow			
4.3	Uppermost point of the left eyebrow	$(4.1.x + 4.5.x)/2$ or x coord of the uppermost point of the contour		
4.4	Uppermost point of the right eyebrow	$(4.2.x + 4.6.x)/2$ or x coord of the uppermost point of the contour		
4.5	Left corner of left eyebrow			
4.6	Right corner of right eyebrow			
5.1	Center of the left cheek		$8.3.y$	
5.2	Center of the right cheek		$8.4.y$	
5.3	Left cheek bone	$>3.5.x$ and $<3.7.x$	$>9.15.y$ and $<9.12.y$	
5.4	Right cheek bone	$>3.6.x$ and $<3.12.x$	$>9.15.y$ and $<9.12.y$	
6.1	Tip of the tongue	$7.1.x$		
6.2	Center of the tongue body	$7.1.x$		
6.3	Left border of the tongue			$6.2.z$
6.4	Right border of the tongue			$6.2.z$
7.1	Top of spine (center of head rotation)			

(*continued overleaf*)

Table 2.6 *(continued)*

Feature points		Recommended location constraints		
#	Text description	x	y	z
8.1	Middle point of outer upper-lip contour	$7.1.x$		
8.2	Middle point of outer lower-lip contour	$7.1.x$		
8.3	Left corner of outer-lip contour			
8.4	Right corner of outer-lip contour			
8.5	Midpoint between FP 8.3 and 8.1 in outer upper-lip contour	$(8.3.x + 8.1.x)/2$		
8.6	Midpoint between FP 8.4 and 8.1 in outer upper-lip contour	$(8.4.x + 8.1.x)/2$		
8.7	Midpoint between FP 8.3 and 8.2 in outer lower-lip contour	$(8.3.x + 8.2.x)/2$		
8.8	Midpoint between FP 8.4 and 8.2 in outer lower-lip contour	$(8.4.x + 8.2.x)/2$		
8.9	Right high point of Cupid's bow			
8.10	Left high point of Cupid's bow			

9.1	Left nostril border		
9.2	Right nostril border		
9.3	Nose tip	7.1.x	
9.4	Bottom right edge of nose		
9.5	Bottom left edge of nose		
9.6	Right upper edge of nose bone		
9.7	Left upper edge of nose bone		
9.8	Top of the upper teeth	7.1.x	
9.9	Bottom of the lower teeth	7.1.x	
9.10	Bottom of the upper teeth	7.1.x	
9.11	Top of the lower teeth	7.1.x	
9.12	Middle lower edge of nose bone (or nose bump)	7.1.x	(9.6.y + 9.3.y)/2 or nose bump
9.13	Left lower edge of nose bone		(9.6.y + 9.3.y)/2
9.14	Right lower edge of nose bone		(9.6.y + 9.3.y)/2
9.15	Bottom middle edge of nose	7.1.x	
10.1	Top of left ear		
10.2	Top of right ear		
10.3	Back of left ear		(10.1.y + 10.5.y)/2
10.4	Back of right ear		(10.2.y + 10.6.y)/2

(continued overleaf)

Table 2.6 *(continued)*

Feature points		Recommended location constraints		
#	Text description	*x*	*y*	*z*
10.5	Bottom of left ear lobe			
10.6	Bottom of right ear lobe			
10.7	Lower contact point between left lobe and face			
10.8	Lower contact point between right lobe and face			
10.9	Upper contact point between left ear and face			
10.10	Upper contact point between right ear and face			
11.1	Middle border between hair and forehead	7.1.*x*		
11.2	Right border between hair and forehead	<4.4.*x*		
11.3	Left border between hair and forehead	>4.3.*x*		
11.4	Top of skull	7.1.*x*		>10.4.*z* and <10.2.*z*
11.5	Hair thickness over FP 11.4	11.4.*x*		11.4.z
11.6	Back of skull	7.1.*x*	3.5.*y*	

Table 2.7 FAP definitions, group assignments, and step sizes. FAP names may contain letters with the following meaning: l = left, r = right, t = top, b = bottom, i = inner, o = outer, m = middle. The quantizer step size is a scaling factor for coding as described in Section 2.3

#	FAP name	FAP description	Units	Uni-orBidir	Pos motion	G r p	FDP subgrp num	Quant step size QP	Min/Max I-Frame quantized values	Min/Max P-Frame quantized values
1	Viseme	Set of values determining the mixture of two visemes for this frame (e.g. pbm, fv, th)	na	na	na	1	na	1	Viseme_blend: +63	Viseme_blend: ±63
2	Expression	A set of values determining the mixture of two facial expressions	na	na	na	1	na	1	Expression_intensity1, expression_intensity2: +63	Expression_intensity1, expression_intensity2: ±63
3	Open_jaw	Vertical jaw displacement (does not affect mouth opening)	MNS	U	Down	2	1	4	+1080	+ 360
4	Lower_t_midlip	Vertical top middle inner-lip displacement	MNS	B	Down	2	2	2	±600	±180
5	Raise_b_midlip	Vertical bottom middle inner-lip displacement	MNS	B	Up	2	3	2	±1860	±600

(*continued overleaf*)

Table 2.7 *(continued)*

#	FAP name	FAP description	Units	Uni-orBidir	Pos motion	G r p	FDP subgrp num	Quant step size QP	Min/Max I-Frame quantized values	Min/Max P-Frame quantized values
6	Stretch_l_cornerlip	Horizontal displacement of left inner-lip corner	MW	B	Left	2	4	2	±600	±180
7	Stretch_r_cornerlip	Horizontal displacement of right inner-lip corner	MW	B	Right	2	5	2	±600	±180
8	Lower_t_lip_lm	Vertical displacement of midpoint between left corner and middle of top inner lip	MNS	B	Down	2	6	2	±600	±180
9	Lower_t_lip_rm	Vertical displacement of midpoint between right corner and middle of top inner lip	MNS	B	Down	2	7	2	±600	±180
10	Raise_b_lip_lm	Vertical displacement of midpoint between left corner and middle of bottom inner lip	MNS	B	Up	2	8	2	±1860	±600

11	Raise_b_lip_rm	Vertical displacement of midpoint between right corner and middle of bottom inner lip	MNS	B	Up	2	9	2	±1860	±600
12	Raise_l_cornerlip	Vertical displacement of left inner-lip corner	MNS	B	Up	2	4	2	±600	±180
13	Raise_r_cornerlip	Vertical displacement of right inner-lip corner	MNS	B	Up	2	5	2	±600	±180
14	Thrust_jaw	Depth displacement of jaw	MNS	U	Forward	2	1	1	+600	+180
15	Shift_jaw	Side to side displacement of jaw	MW	B	Right	2	1	1	±1080	±360
16	Push_b_lip	Depth displacement of bottom middle lip	MNS	B	Forward	2	3	1	±1080	±360
17	Push_t_lip	Depth displacement of top middle lip	MNS	B	Forward	2	2	1	±1080	±360
18	Depress_chin	Upward and compressing movement of the chin (like in sadness)	MNS	B	Up	2	10	1	±420	±180
19	Close_t_l_eyelid	Vertical displacement of top left eyelid	IRISD	B	Down	3	1	1	±1080	±600

(continued overleaf)

Table 2.7 *(continued)*

#	FAP name	FAP description	Units	Uni- orBidir	Pos motion	G r p	FDP subgrp num	Quant step size QP	Min/Max I-Frame quantized values	Min/Max P-Frame quantized values
20	Close_t_r_eyelid	Vertical displacement of top right eyelid	IRISD	B	Down	3	2	1	±1080	±600
21	Close_b_l_eyelid	Vertical displacement of bottom left eyelid	IRISD	B	Up	3	3	1	±600	±240
22	Close_b_r_eyelid	Vertical displacement of bottom right eyelid	IRISD	B	Up	3	4	1	±600	±240
23	Yaw_l_eyeball	Horizontal orientation of left eyeball	AU	B	Left	3	na	128	±1200	±420
24	Yaw_r_eyeball	Horizontal orientation of right eyeball	AU	B	Left	3	na	128	±1200	±420
25	Pitch_l_eyeball	Vertical orientation of left eyeball	AU	B	Down	3	na	128	±900	±300
26	Pitch_r_eyeball	Vertical orientation of right eyeball	AU	B	Down	3	na	128	±900	±300
27	Thrust_l_eyeball	Depth displacement of left eyeball	ES	B	Forward	3	na	1	±600	±180
28	Thrust_r_eyeball	Depth displacement of right eyeball	ES	B	Forward	3	na	1	±600	±180
29	Dilate_l_pupil	Dilation of left pupil	IRISD	B	Growing	3	5	1	±420	±120
30	Dilate_r_pupil	Dilation of right pupil	IRISD	B	Growing	3	6	1	±420	±120

31	Raise_l_i_eyebrow	Vertical displacement of left inner eyebrow	ENS	B	Up	4	1	2	±900	±360
32	Raise_r_i_eyebrow	Vertical displacement of right inner eyebrow	ENS	B	Up	4	2	2	±900	±360
33	Raise_l_m_eyebrow	Vertical displacement of left middle eyebrow	ENS	B	Up	4	3	2	±900	±360
34	Raise_r_m_eyebrow	Vertical displacement of right middle eyebrow	ENS	B	Up	4	4	2	±900	±360
35	Raise_l_o_eyebrow	Vertical displacement of left outer eyebrow	ENS	B	Up	4	5	2	±900	±360
36	Raise_r_o_eyebrow	Vertical displacement of right outer eyebrow	ENS	B	Up	4	6	2	±900	±360
37	Squeeze_l_eyebrow	Horizontal displacement of left eyebrow	ES	B	Right	4	1	1	±900	±300
38	Squeeze_r_eyebrow	Horizontal displacement of right eyebrow	ES	B	Left	4	2	1	±900	±300

(*continued overleaf*)

Table 2.7 *(continued)*

#	FAP name	FAP description	Units	Uni- orBidir	Pos motion	G r p	FDP subgrp num	Quant step size QP	Min/Max I-Frame quantized values	Min/Max P-Frame quantized values
39	Puff_l_cheek	Horizontal displacement of left cheek	ES	B	Left	5	1	2	±900	±300
40	Puff_r_cheek	Horizontal displacement of right cheek	ES	B	Right	5	2	2	±900	±300
41	Lift_l_cheek	Vertical displacement of left cheek	ENS	U	Up	5	3	2	±600	±180
42	Lift_r_cheek	Vertical displacement of right cheek	ENS	U	Up	5	4	2	±600	±180
43	Shift_tongue_tip	Horizontal displacement of tongue tip	MW	B	Right	6	1	1	±1080	±420
44	Raise_tongue_tip	Vertical displacement of tongue tip	MNS	B	Up	6	1	1	±1080	±420
45	Thrust_tongue_tip	Depth displacement of tongue tip	MW	B	Forward	6	1	1	±1080	±420
46	Raise_tongue	Vertical displacement of tongue	MNS	B	Up	6	2	1	±1080	±420
47	Tongue_roll	Rolling of the tongue into U shape	AU	U	Concave upward	6	3,4	512	+300	+60
48	Head_pitch	Head pitch angle from top of spine	AU	B	Down	7	na	170	±1860	±600

49	Head_yaw	Head yaw angle from top of spine	AU	B	Left	7	na	170	±1860	±600
50	Head_roll	Head roll angle from top of spine	AU	B	Right	7	na	170	±1860	±600
51	Lower_t_midlip_o	Vertical top middle outer-lip displacement	MNS	B	Down	8	1	2	±600	±180
52	Raise_b_midlip_o	Vertical bottom middle outer-lip displacement	MNS	B	Up	8	2	2	±1860	±600
53	Stretch_l_cornerlip_o	Horizontal displacement of left outer-lip corner	MW	B	Left	8	3	2	±600	±180
54	Stretch_r_cornerlip_o	Horizontal displacement of right outer-lip corner	MW	B	Right	8	4	2	±600	±180
55	Lower_t_lip_lm_o	Vertical displacement of midpoint between left corner and middle of top outer lip	MNS	B	Down	8	5	2	±600	±180

(*continued overleaf*)

Table 2.7 *(continued)*

#	FAP name	FAP description	Units	Uni-orBidir	Pos motion	G r p	FDP subgrp num	Quant step size QP	Min/Max I-Frame quantized values	Min/Max P-Frame quantized values
56	Lower_t_lip_rm _o	Vertical displacement of midpoint between right corner and middle of top outer lip	MNS	B	Down	8	6	2	±600	±180
57	Raise_b_lip_lm_o	Vertical displacement of midpoint between left corner and middle of bottom outer lip	MNS	B	Up	8	7	2	±1860	±600
58	Raise_b_lip_rm_o	Vertical displacement of midpoint between right corner and middle of bottom outer lip	MNS	B	Up	8	8	2	±1860	±600
59	Raise_l_cornerlip_o	Vertical displacement of left outer-lip corner	MNS	B	Up	8	3	2	±600	±180
60	Raise_r_cornerlip_o	Vertical displacement of right outer-lip corner	MNS	B	Up	8	4	2	±600	±180

61	Stretch_l_nose	Horizontal displacement of left side of nose	ENS	B	Left	9	1	1	±540	±120
62	Stretch_r_nose	Horizontal displacement of right side of nose	ENS	B	Right	9	2	1	±540	±120
63	Raise_nose	Vertical displacement of nose tip	ENS	B	Up	9	3	1	±680	±180
64	Bend_nose	Horizontal displacement of nose tip	ENS	B	Right	9	3	1	±900	±180
65	Raise_l_ear	Vertical displacement of left ear	ENS	B	Up	10	1	1	±900	±240
66	Raise_r_ear	Vertical displacement of right ear	ENS	B	Up	10	2	1	±900	±240
67	Pull_l_ear	Horizontal displacement of left ear	ENS	B	Left	10	3	1	±900	±300
68	Pull_r_ear	Horizontal displacement of right ear	ENS	B	Right	10	4	1	±900	±300

3

MPEG-4 Face Animation Conformance

Eric Petajan

face2face Animation, Inc., Summit, New Jersey

3.1 INTRODUCTION

All MPEG standards are collections of media coding tools that can be combined in profiles to address specific applications. Each profile also specifies distinct levels that limit the decoder complexity needed to fully decode bit streams into objects and streams of audio/visual data. A given decoder is MPEG-4-compliant when it conforms to a particular profile and level as specified in the Conformance part of the standard. MPEG-4 Face Animation (FA) profiles and levels are described in this chapter. The philosophy behind FA conformance is explained and detailed examples are provided.

3.2 MPEG CONFORMANCE PRINCIPLES

A standard is a set of information that is disseminated to potential users before deployment. For example, a lightbulb manufacturer (let us call it Bulbco) must know the standard specifications for existing light fixtures before successfully designing and selling lightbulbs. The lightbulb standard allows other companies to compete with Bulbco but it also increases the size of the bulb market by ensuring that all receptacles can accommodate bulbs from Bulbco. The same principal applies to media and communication standards. In the case of FA, the Face Animation Parameters (FAPs) are the bulbs and the face models are the receptacles. The MPEG-4 FA Standard allows any MPEG-4-compliant face model to be animated by any stream of FAPs.

If Bulbco invented the lightbulb and was the first to sell bulbs and sockets then it could establish a *de facto* standard. Once a large number of sockets is installed, the cost of socket replacement inhibits competing bulb companies from defining nonstandard sockets for the sole purpose of capturing market share. Sometimes two companies invent similar new products simultaneously and dual standards divide the market for some period of time. The consumers who pick the losing standard may be left without a supply of product (see VHS vs Betamax). Again, the same principles apply to multimedia communication. The production of multimedia content is expensive and the owners of the content depend on the player, storage and manipulation technology to persist for a very long time. While the downloading of new player technology is possible, the player update process has proven to be problematic and time consuming. Consumers want their media on demand and with a minimum of delay.

3.3 MPEG-4 PROFILE ARCHITECTURE

The MPEG-4 standard contains a variety of component codecs, object definitions and interface specifications, which can be deployed individually or in sets to form an application. The standard is divided into visual objects, audio objects, graphics objects [Binary Format for Scenes (BIFS)], systems and Delivery Multimedia Integration Framework (DMIF). The FA standard contains a FAP codec, Face object, text-to-speech (TTS) interface and other objects that provide enhancements to the animated face. The FAP definition and codec is described in the visual part of the standard because it describes the frame-based motion of a visual object (the face). Since FAPs and Body Animation Parameters (BAPs) are encoded together and share many coding tools, the coded FAPs are called *face and body animation* (FBA) streams regardless of the presence of BAPs in the stream.

3.4 THE MINIMUM FACE

The author or owner of animated content wants to have some assurance that the content will appear with the quality that was intended. The attributes of an animated face subject to possible quality degradation include the rendered frame rate, the spatial resolution of the face region, the resolution of rendered facial textures and the motion of the face. The Simple Face Visual Profile specifies a minimum face as it appears on the output display of the MPEG-4 decoder, which conforms to the Simple Face Profile. FAPs are defined to be normalized and completely independent of the Face Model. The FBA codec compresses and delivers FAPs independently from the Face model. The Face Model may be resident in the terminal and have an appearance that is unknown to the FBA encoder. The model independence of the FBA stream limits the conformance specification and has resulted in a definition of the minimum face that is unconventional yet practical. The minimum face that conforms to the Simple Face Profile must display the following attributes:

- the face definition parameter (FDP) points on the face must move according to the FAP definitions;
- the neutral face FDP points must be separated by a minimum number of screen pixels as specified in Table 3.1;
- a minimum rendered frame rate as specified in Table 3.1.

Wireface is a minimal MPEG-4 FA program that is included in Reference Software (Part 5) of version 1 of the standard (ISO/IEC 14496). Wireface (shown in Figure 3.1) is intended to demonstrate a face model that satisfies the minimum face requirements with

Table 3.1 FBA conformance tests

Test	Description	Bit stream files (extension mp4)	Decoded FAP files (extension dec.fap)
1	Frame-based coding of min/max FAP values.	MinMaxFAPs	MinMaxFAPs
2	Exercise the use of start codes in I and P frames.	Marco15	FAPStartCode
3	Decoding of various MPEG-4 FAP bit streams.	Wow25, Marco30, opvis3, Expressions, Emotions, Opossum, Allfaps	Wow25, Marco30, opvis3, Expressions, Emotion, Opossum, Allfap
4	Test the ability of the FAP decoder to process bit streams with bit rates that are just under the maximum bit rate for each level (16 kbps for level 1 and 32 kbps for level 2).	FAPMaxBitrate16, FAPMaxBitrate32	FAPMaxBitrate16, FAPMaxBitrate32
5	Test left-right and right-left interpolation of FAPs after decoding only one side (left or right) of FAPs.	AllFaps_rlint	AllFaps.rlint
6	Test the decoding of high-level FAPs 1 and 2 (visemes and expressions). If the Viseme or Expression BIFS nodes are required by other MPEG-4 profiles, the complete implementation and interpretation of FAPs 1 and 2 is required in the terminal. If the Viseme and Expression BIFS nodes are not required in the terminal, the decoded values associated with FAPs 1 and 2 may be ignored. In all cases, the decoder must output the correct values.	VisExp	VisExp
7	Test the default min/max range values of the arithmetic decoder when no min/max values are included in the bit stream.	NoFAPMinMax	NoFAPMinMax
8	Test the use of the DCT coding mode with I and P frames.	FAPDCT	FAPDCT
9	Test the use of frame-based and DCT-based coding modes in the same sequence.	FrameDCT	FrameDCT
10	Test multiple faces.	FourFaces	
11	Test FIT node interpolation if BIFS is included in the terminal.	FIT	FIT

Note: DCT discrete cosine transform; FIT FAP Interpolation Table

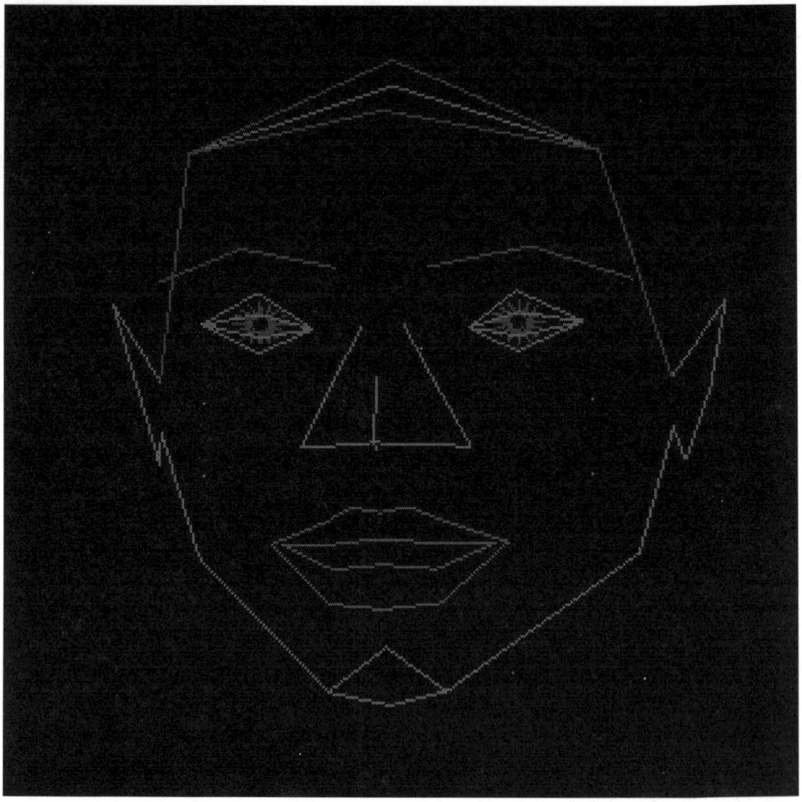

Figure 3.1 Wireface reference face model

a minimum of rendering complexity. Since Wireface is only composed of 3-D lines, all FDP feature points (FPs) are visible without occlusion from polygonal surfaces.

An MPEG-4 Profile must specify Levels in order to guarantee Quality of Service (QoS). While the Profiles indicate which decoders must be present in the player, the levels specify the maximum decoding operations per unit of time that a conforming player must perform. In general, the FBA stream must be contiguously decoded because of the use of temporal prediction and variable length coding. After the frames of FAP data are decoded, the rendering and display of the face model using the FAP data can occur at a lower frame rate than the FAP decoding rate.

Level 1 of the Simple Face Profile requires a conforming player to animate one face with a rendered frame rate of at least 15 Hz. The maximum FAP decoding frame rate is limited to 72 Hz. The maximum bit rate of the FBA stream is 16 kbits s^{-1}. Level 2 of the Simple Face Profile requires a conforming player to animate up to four faces with a combined rendered frame rate of at least 60 Hz and each face must be rendered at a rate that is not required to exceed 30 Hz. The combined face rendering rate limit is designed for conversations between faces where all the faces are not usually talking simultaneously. The maximum total bit rate is 32 kbits s^{-1}. The maximum FAP decoding frame rate is limited to 72 Hz (total shared between faces).

3.5 GRAPHICS PROFILES

The Systems (Part 1) of the MPEG-4 standard specifies descriptors of 3-D graphics objects and the transformation and animation of these objects. The Face Object is the parent node of a hierarchy of face-related objects that affect the appearance or the movement of the face. MPEG-4 players that conform to the Complete Graphics Profile must instantiate one or more Face nodes in response to BIFS commands in the MPEG-4 bit stream.

3.6 CONFORMANCE TESTING

The Conformance part of the MPEG-4 standard (ISO/IEC 14496-44) describes a set of tests that decoder vendors can use to guarantee QoS. For each test, a set of bit streams are provided, which can be decoded back into source material. The Simple Face decoder outputs FAP data that is written to the FAP node if the terminal contains a Face node as specified by the Complete Graphics Profile. If an alternate FA module is used, the rendered face must behave in correct response to the FAP data. Table 3.1 contains the conformance tests for the Simple Face Profile.

3.7 SUMMARY

A standard must include enough information to ensure interoperability between different vendors. The conformance part of the MPEG standards provides detailed decoder performance requirements and test bit streams to ensure interoperability between encoders and decoders from different vendors. The FA standard in MPEG-4 does not specify the rendered face on a pixel-by-pixel basis due to the nature of the FAPs. Instead, the minimum rendered face is defined and source code for the wireface model is provided. The Simple Face profile (MPEG-4 version 1) and the Simple FBA profile (version 2 includes the body) have the same requirements for the face. Level 1 is designed for wireless devices and PDAs (personal digital assistants), while level 2 is designed for PCs, set-top boxes and game consoles. Implementers of the FBA specification should buy the standard from *www.iso.org* referring to ISO/IEC 14496. In addition, the MPEG-4 Industry Forum (*www.m4if.org*) is dedicated to the promotion of MPEG-4 commercialization and provides the latest news and technical information for users and implementers of the standard.

PART 3
IMPLEMENTATIONS

From the leading practitioners in the field, we have collected an array of technologies relevant for implementing applications based on the MPEG-4 FA specification.

Different implementations of FA decoders (animation engines) are presented by Pandzic (interpolation approach, Chapter 4); Lavagetto and Pockaj (animation rules approach, Chapter 5); Hong, Wen and Huang (motion units approach, Chapter 7) and Pelachaud (pseudomuscular approach, Chapter 8).

In Chapter 9, Tsapatsoulis *et al.* provide insight into the rapidly expanding research area of recognition and synthesis of emotions based on MPEG-4 FAPs.

Major sources of facial animation data are treated in separate chapters. Ahlberg treats video analysis in Chapter 6 with model-based coding in mind. Hong, Wen and Huang present speech-driven FA in Chapter 7. Pelachaud concentrates on visual text-to-speech in Chapter 8.

We cover several high-level architecture approaches. A framework for delivering facial animation to a wide range of platforms and applications, in particular, Web and mobile, is presented by Pandzic in Chapter 4. Lavagetto and Pockaj present a comprehensive tool set and architecture for facial animation applications in Chapter 5. Chapter 10 describes the platform for interactive virtual characters, particularly suited for human–computer interfaces. The platform was developed by a number of partners in the Interface project supported by the Information Society Technologies (IST) programme of the European Commission.

This part provides a wealth of practical information on most important aspects of facial animation systems implementation.

4

MPEG-4 Facial Animation Framework for the Web and Mobile Applications

Igor S. Pandzic

Linkoping University, Linkoping, Sweden

ABSTRACT

We present a flexible framework for production and delivery of facial animation (FA). The goal of this framework is to be able to support a wide area of applications involving FA on a broad range of platforms, in particular (but not limited to) the Web and mobile platforms. This goal is achieved through the connectivity of all tools within the framework using standards and through the portability of the end user content delivery component. The framework consists of (1) a lightweight, portable, MPEG-4-compatible Facial Animation Player, (2) a system for fast automatic production of ready-to-animate, MPEG-4-compatible face models, including the facial motion cloning (FMC) method for automatic copying of morph targets from one face to another and (3) a plethora of MPEG-4-compatible tools for Facial Animation content production. We believe that this kind of approach offers enough flexibility to rapidly adapt to a broad range of applications involving FA on various platforms. We present the framework, the components that are specific to it and the integration of other standard components as well as their comparative overview. We believe that this kind of framework can satisfy a broad range of application classes and delivery platforms, ranging from interactive services on the Web and mobile platforms all the way to games, film and TV productions.

4.1 INTRODUCTION

Facial animation can be used in a broad range of application areas [1] such as film and TV, games, communications and human–computer interfaces (HCI). These applications may run on a wide variety of different hardware and software platforms, as illustrated in Figure 4.1. The same figure illustrates how FA may come from a variety of sources and how the MPEG-4 standard [2] can be used to connect the sources to the target platforms. This allows for great flexibility in reusing tools, produced contents and skill

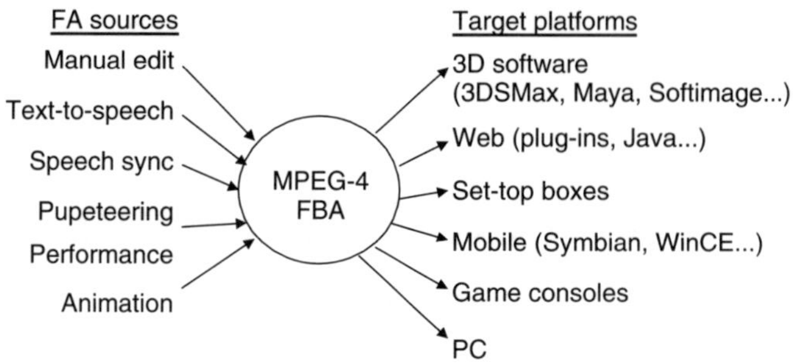

Figure 4.1 Sources and target platforms for facial animation

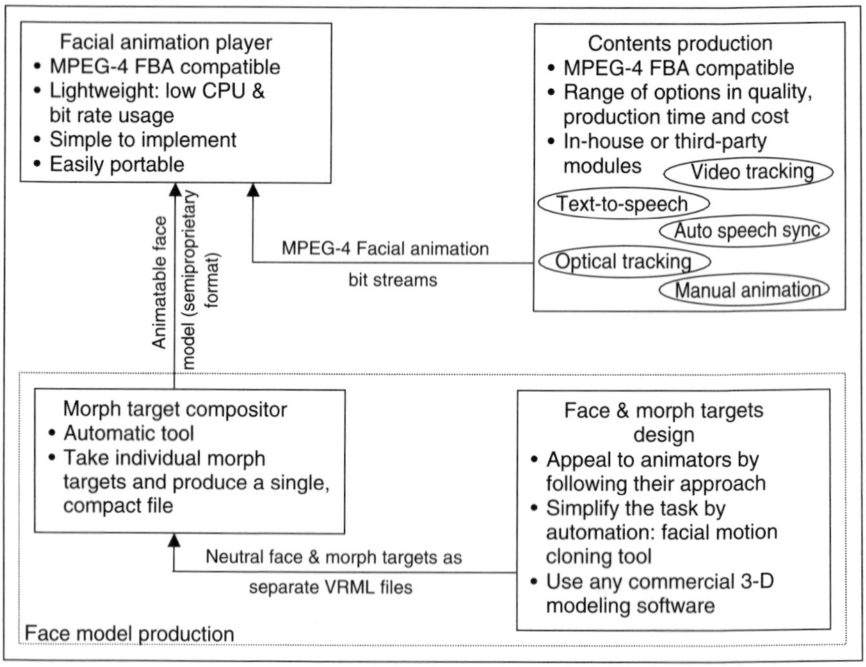

Figure 4.2 The facial animation framework (Reprinted from Igor S. Pandzic, "Facial Animation Framework for the Web and Mobile Platforms", *Proceedings of Web3D'02*, © 2002, with permission from Association for Computing Machinery) [3]

across different platforms and applications. For such a scenario, we propose the Facial Animation Framework, schematically presented in Figure 4.2.

The Facial Animation Player is the central part of the framework. It plays facial animation input in the MPEG-4 face and body animation (FBA) format from a wide variety of sources. Because of its simple and computationally modest algorithm, it is easy to implement as well as port and adapt to any platform. Current implementations include a Java applet and a prototype on the Symbian Quartz tablet communicator reference design [4]. The Facial Animation Player is presented in Section 4.2.

Section 4.3 presents our approach to producing animatable face models, that is, face models with the additional information necessary for them to be animated correctly in the Facial Animation Player. The face model production is based on an approach adopted by the MPEG-4 facial animation tables (FATs), which follows the widely used and well-known principle of morph targets, but extends it to include low-level facial actions rather than just high-level expressions and visemes. We also significantly simplify the task of the animator by introducing the FMC method to automatically copy a whole set of morph targets (or FATs) from one face model to another. We describe this method in some detail in Section 4.4.

In Section 4.5 we present a comparative overview of alternatives for Facial Animation content production ranging from text-to-speech (TTS) to facial tracking, all feeding content to the Facial Animation Player. Finally, in the concluding section we summarize the advantages of our approach.

4.2 THE FACIAL ANIMATION PLAYER

As Figure 4.2 shows, the Facial Animation Player is the cornerstone of the Facial Animation Framework. It includes an MPEG-4 FBA decoder and a facial animation system that animates the face based on the decoded FAPs. The MPEG-4 decoding process, based on integer arithmetic, is very low in CPU time consumption and easy to port to various platforms. In order to achieve similar simplicity in the facial animation system, we use the morph target approach widely used in computer animation, essentially the same as the MPEG-4 FAT approach (see Chapter 2 for explanation of FAT). A morph target is a variation of an object that has the same mesh topology, but different vertex positions. Essentially, it is a deformed object. Smooth morphing between the neutral object and the morph target is achieved by simple interpolation of vertex positions. Interpolation has been used extensively for facial animation as mentioned in References 5 to 7. We prefer it to procedural approaches like those mentioned in References 8 to 11, and certainly to the more complex muscle-based models like those mentioned in References 12 to 14 for a number of reasons:

- Its implementation simplicity ensures fairly easy portability to various platforms.
- Low CPU requirements ensure good performance even on modest platforms.
- Morph targets are intuitive and already in widespread use by animators.

In computer facial animation, morph targets are typically used for visemes and expressions. This represents a fairly easy way of obtaining speech animation. We extend the use of morph targets from visemes and expressions to all MPEG-4 FAPs. Each FAP (both low- and high-level) is defined as a morph target. Another extension is that in our implementation the morph target defines not only the new position of vertices but also changes in transforms in the scene, that is, translations and rotations. The morph target is defined for a particular value of the FAP. The movement of vertices and transforms for other values of the FAP are then interpolated from the neutral face and the morph target. This is essentially the same as the FAT approach, except that the FAT defines a piecewise linear interpolation for the change of each vertex position and transform parameter. In practice, simple linear interpolation seems to be sufficient for reasonable

FA in all situations encountered so far. Each FAP produces movement of vertices and transforms by a separate interpolation. They are then added together for all FAPs to produce the final shape of the face.

Because of its simplicity and low requirements, we expect that the Facial Animation Player will be easy to implement on a variety of platforms using different programming languages. We describe an implementation written as a Java applet and based on the Shout3D rendering engine [15]. An implementation on the Symbian Quartz [4] tablet communicator reference design (Figure 4.3) is in an early prototype phase and is not described in detail here.

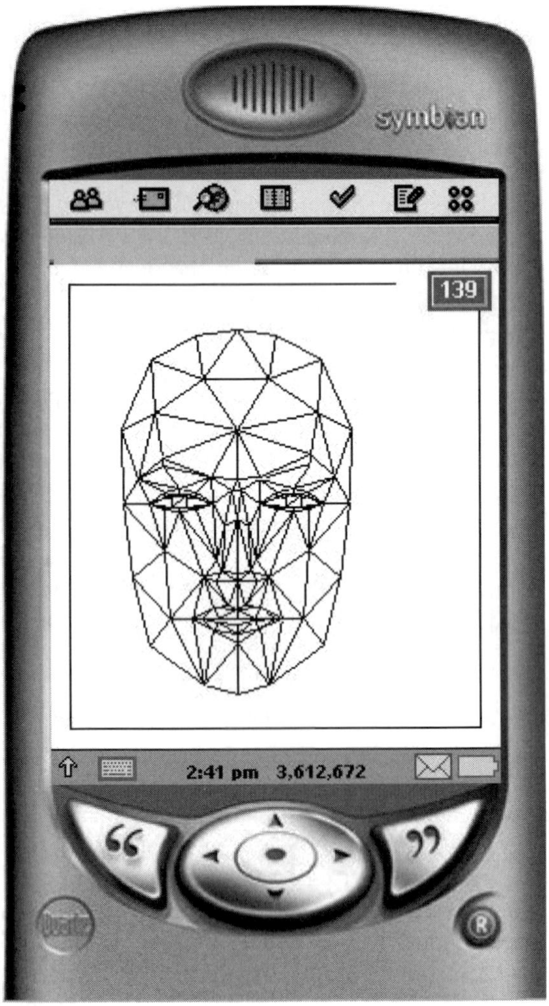

Figure 4.3 The facial animation player early prototype implementation on Symbian Quartz using the Candide face model from Linköping University (Reprinted from Igor S. Pandzic, "Facial Animation Framework for the Web and Mobile Platforms", *Proceedings of Web3D'02*, © 2002, with permission from ACM) [3]

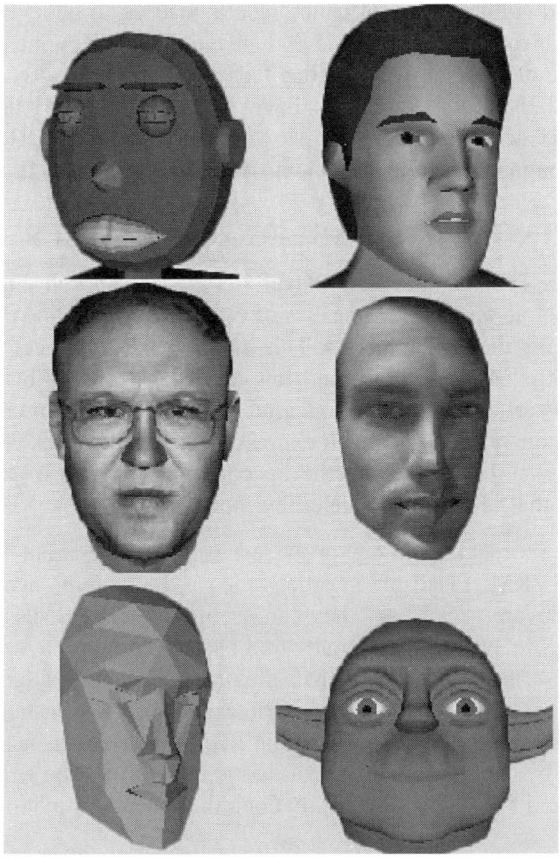

Figure 4.4 Examples of face models animated using the facial animation player

We have successfully tested the MPEG-4 Facial Animation player with several face models as illustrated in Figure 4.4. We can achieve interactive frame rates with models containing several thousands of polygons. The player has correctly interpreted the test FBA bit streams (Marco, Wow, Emotions, Opossum etc. available on the Web, see Appendix 2) as well as the bit streams produced by the TTS system. As the demonstration Web page [3] shows, the applet is fully controllable from the Web page by Javascript, making all interactions possible.

The bandwidth requirements of this implementation are fairly modest [3]. The size of the applet, currently 156 K, might be problematic. The reason is mostly in the 3-D renderer implementation, while the actual MPEG-4 FBA decoder and facial animation engine (FAE) take only 26 K. The face models vary in size but good ones can be implemented within 50 K.

We performed performance measurements for different face models [3] and have shown that the implementation can animate facial models of medium complexity (2000–3000 polygons) at interactive frame rates on fairly modest hardware (like PIII/500).

Bandwidth and performance results lead us to believe that our implementation is accessible to a widest possible audience of Web users, including the ones connected through simple modems and using fairly low performance PCs. Informal tests with small numbers of modem users have initially confirmed this, as the users did not complain of access problems or slowness of the service. However, more systematic testing through practical deployment of the system is still in planning phase.

4.3 PRODUCING ANIMATABLE FACE MODELS

Because the Facial Animation Player is based on morph targets, the production of a new animatable face model consists of two steps: producing the face in its neutral state and producing the morph targets. This means producing a deformed face corresponding to each of the MPEG-4 high- and low-level parameters. This concept is intuitive, and already very often used in commercial 3-D software for producing speech animation. The advantage of modeling each expression is that the artist can precisely control every detail, for example, stretching the lip corner may result in particular skin creasing in the cheeks that is exactly modeled by the artist.

The morph targets are produced each in a separate Virtual Reality Modeling Language (VRML) [16] file in order to leave the system open for the users of all commercial software packages – they can usually export VRML. The Morph Target Compositor (Figure 4.2) is an automatic tool used to compose a facial model with animation based on a series of separate VRML files containing morph targets. The final model that is used by the player is in a semiproprietary format. It is a standard VRML file containing the animation information stored in Interpolator nodes in a compact way that can be interpreted by the player. This means that the model will be visible in any VRML player, albeit static; in order to view animations the Facial Animation Player has to be used.

The obvious disadvantage of this approach is the time needed to create all necessary morph targets for each new face model. With low- and high-level FAPs, the total number of necessary morph targets is 86! We therefore propose a method called Facial Motion Cloning (FMC) that copies the complete range of morph targets, both low- and high-level, from one face to another. With FMC an artist can produce one very detailed face with all morph targets, then use it to automatically produce the full set of morph targets for a new face. The time saved in the production of the new face is dramatic, and the freedom to fine-tune the automatically produced morph targets is still there. Libraries of specific morph targets may become available or be offered commercially. For example, a particular morph target library may produce many creases and wrinkles to be particularly adapted to simulations of older persons. Another library may be suited for children and so on.

The FMC method is similar in goal to the Expression Cloning [17]. However, our method additionally preserves the MPEG-4 compatibility of cloned facial motion and it treats transforms for eyes, teeth and tongue. It is also substantially different in implementation.

4.4 THE FACIAL MOTION CLONING METHOD

FMC can be schematically represented by Figure 4.5. The inputs to the method are the source and target face. The source face is available in neutral position (*source face*) as

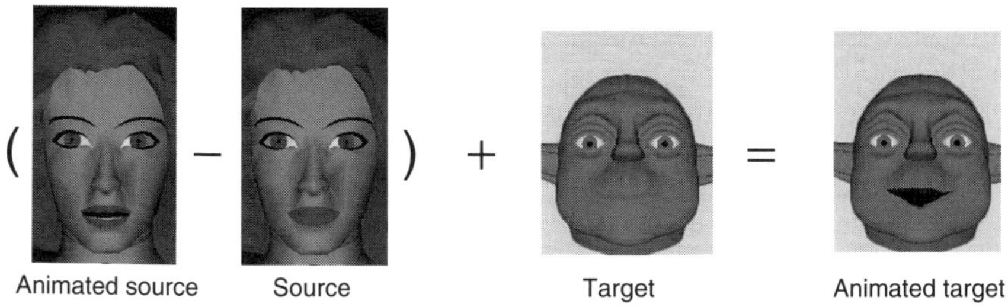

Figure 4.5 Overview of facial motion cloning

well as in a position containing some motion that we want to copy (*animated source face*). The target face exists only as neutral (*target face*). The goal is to obtain the target face with the motion copied from the source face – the *animated target face*.

To reach this goal we first obtain *facial motion* as the difference of 3-D vertex positions between the animated source face and the neutral source face. The facial motion is then added to the vertex positions of the target face, resulting in the animated target face.

In order for this to work, the facial motion must be normalized, which ensures that the scale of the motion is correct. In the *normalized facial space*, we compute facial motion by subtracting vertex positions of the animated and the neutral face. To map the facial motion correctly from one face to another, the faces need to be aligned with respect to the facial features. This is done in the *alignment space*. Once the faces have been aligned, we use interpolation to obtain facial motion vectors for vertices of the target face. The obtained facial motion vectors are applied by adding them to vertex positions, which is possible because we are working in the normalized facial space. Finally, the target face is denormalized. The whole process is described in the next five subsections.

When applying mathematical algorithms to real-world problems, we often have to treat exceptions. With the human face this seems to be the case even more than usual. In the last three subsections of this section we deal with the aliasing problem when mapping between meshes of different densities and with the parts of the face that cannot be fully treated by the basic algorithm: the lip region, eyes, teeth and tongue.

4.4.1 Interpolation from 2-D Triangle Mesh

Two steps of our method, alignment and motion vector mapping, can be reduced to the problem of interpolating function values from a number of points in a 2-D triangular mesh with known function values. We therefore begin by explaining how this interpolation method works. The inputs to the method are

- N input points $\mathrm{P}_i(x_i, y_i)$ in 2-D space.
- N function values at input points $\mathrm{F}_i = f(\mathrm{P}_i)$; note that F_i can be vectors.
- Triangulation of input points defined as M triangles $\mathrm{T}_j(a_j, b_j, c_j)$, $1 < a_j, b_j, c_j < N$, a_j, b_j, c_j defining the vertices of triangles as pointers into the array of input points.

From these inputs, we want to interpolate the value of the function $f(\mathrm{P})$ for any point $\mathrm{P}(x, y)$ in the 2-D space.

We first determine if P is within any of the triangles T_j. To do this, we test each triangle first with a simple bounding box test, then by computing barycentric coordinates b_1, b_2, b_3 [18] of the point P with respect to the triangle T_j. If $0 < b_k < 1$, P is inside T_j. If this is the case, we can compute the interpolated value:

$$f(\mathrm{P}) = \sum_{k=1}^{3} b_k \cdot f(\mathrm{P}_k) \tag{4.1}$$

If point P is lying within more than one triangle T_j, this means that the triangles are overlapping. In this case the interpolation method needs some application-specific way of resolving the conflict and choosing one of the triangles. The function value is then interpolated from the chosen triangle using Equation (4.1).

If P is lying outside all triangles, we use the interpolation value from the closest point on the closest triangle to P. As the measure of distance from P to a triangle, we use the sum of the excess barycentric coordinates, that is, any amount of b_k above 1 or below 0.

After identifying the triangle T with the minimum distance, we find the barycentric coordinates b'_k of the point P′, which is the closest point to P on the triangle T. This is done by truncating each b_k to values between 0 and 1, then dividing them by their sum to make sure that the sum of the new coordinates equals 1.

Finally, we apply Equation (4.1) to b'_k and obtain the interpolated value.

4.4.2 Normalizing the Face

The goal is to transform a face into a normalized facial space, in which all faces have the same key proportions, for example, same distance between the eyes and so on. The benefit is that the magnitude of motion in the normalized space is the same for different faces and we can therefore apply motion from one face to another by simple addition. Figure 4.6(b) shows a normalized face.

To define the proportions of the face we use the Facial Animation Parameter Units (FAPU, refer to Part 2 for details), as specified in the MPEG-4 standard, and the MPEG-4 Feature Points (FP, refer to Part 2 for details). Our method requires that the coordinates of key FPs be known for both source and target face. The required FPs are indicated in Figure 4.7. The regions of the face are determined from the proximity to the FPs. The appropriate FAPUs are then used as normalization factors for each vertex. This amounts to scaling the face model with a regionally changing scaling factor. The normalization factors are FAPUs (see Chapter 2). By normalizing with FAPUs we make sure that the motion of the FPs in the normalized space corresponds to the MPEG-4 FAPs.

4.4.3 Computing Facial Motion

The facial motion is defined as the difference in vertex positions between the animated and neutral face. It is expressed by an array of *facial motion vectors*, each vector corresponding to one vertex of the face. The facial motion is always computed in the normalized facial space, thus the facial motion vectors are correctly normalized and can be transferred from one face to another. To compute the facial motion vectors we simply subtract the 3-D position of each vertex in the neutral face from the position of the corresponding vertex in the animated face.

4.4.4 Aligning Source and Target Ace

When transferring facial motion from source to target face, we need to transfer the motion to the corresponding facial region. In order to achieve this, we compute the motion mappings using the *alignment space*. This is a 2-D space in which the source and target face are aligned with respect to the FPs.

Figure 4.6 shows the process of projecting source and target face into the 2-D space. The initial face is shown in Figure 4.6(a). After normalization, as explained in the previous subsection, we obtain the normalized face in Figure 4.6(b).

The normalized faces are then mapped into 2-D space using cylindrical projection around y-axis. The center of projection is computed in such a way that the angular width of the mouth remains constant. The zero angle is always aligned with the tip of the nose. These two rules, together with previous normalization, ensure that the projected faces, shown in Figure 4.6(c), are already roughly aligned.

In the final alignment, we move the FPs of the target face onto the corresponding FPs of the source face. The vertices that are not FPs (the nonFPs) are *pulled* by the FPs so that facial regions remain intact.

The pulling of nonFPs into their regional position is achieved by the previously described triangle mesh interpolation method. The input points for interpolation are the FPs. The interpolated function is the movement vector of the FP obtained by subtracting its original position from its new position (i.e. the position of the FP in the source face). The triangulation, necessary for the interpolation algorithm, is produced manually, as illustrated in Figure 4.7.

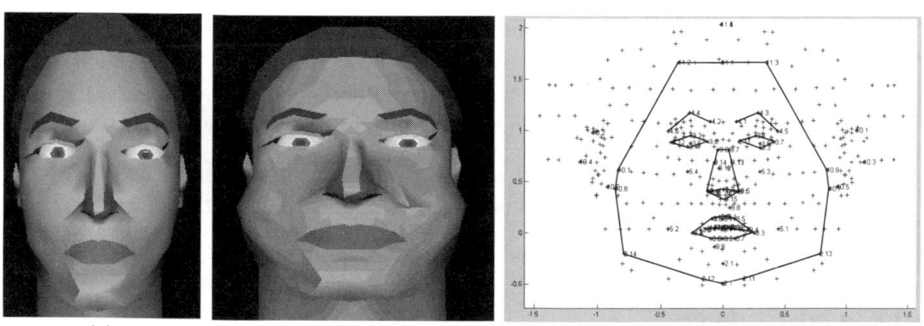

(a) (b) (c)

Figure 4.6 Normalizing and projecting the face. (a) The face in initial state; (b) normalized face; (c) the face projected into 2-D. Facial features are outlined for better visibility (Reprinted from Igor S. Pandzic, “Facial Animation Framework for the Web and Mobile Platforms” © 2002, with permission from ACM) [3]

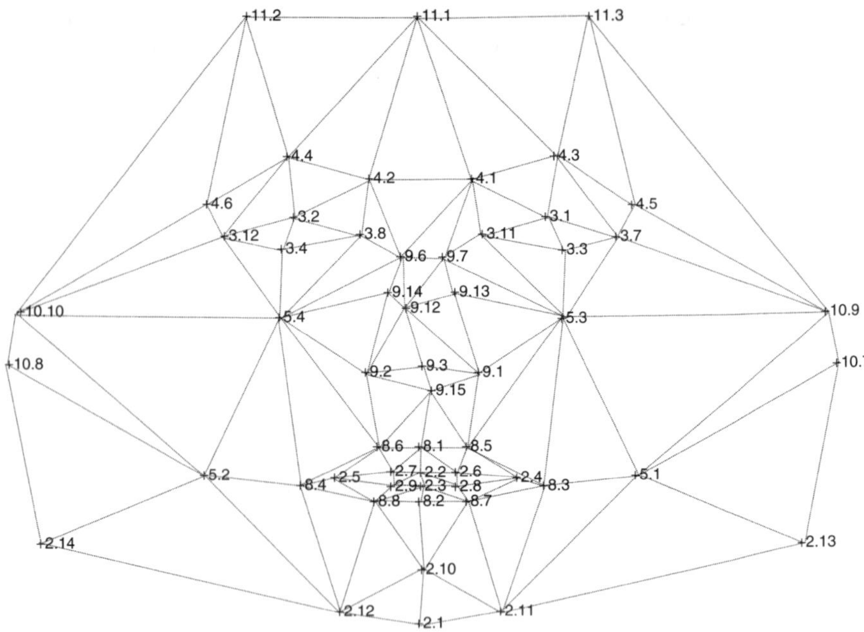

Figure 4.7 Triangulation of the feature points used for interpolation of non-FPs movements for alignment. This figure also illustrates which feature points are required for the facial motion cloning. In addition to the ones illustrated in the figure, feature points 7.2 (neck base), 3.5 and 3.6 (left and right iris centers) are also required

4.4.5 Mapping Facial Motion

We know the facial motion vector for each vertex of the source face, and we have mapped these vertices into the 2-D alignment space. To obtain the motion vectors for vertices of the target face, we again use the previously described triangle mesh interpolation method. The input points are all points of the source face. The input function values are the motion vectors at input points. The triangulation is directly available from the source face model. As this triangulation is defined in 3-D, there is a high possibility of overlapping triangles in 2-D. To resolve this conflict we choose the triangle with the smallest depth at the interpolation point. The depth at the interpolation point is obtained by interpolating depth of the three vertices of the triangle.

Once we have the facial motion vectors for each vertex of the target face, we simply add them to the vertex positions and denormalize the target face. This is the animated target face.

4.4.6 Antialiasing

In our interpolation method, the values of the motion vector are known at the points of the source face mesh. This is equivalent to sampling a (presumably smooth) function at a number of discrete, nonevenly distributed points. The density of the source face mesh is the equivalent of the sampling frequency, which usually changes for different regions of the face. From these values, we interpolate – or, in effect, resample – the value of the function at a different set of nonevenly distributed points, those of the target face. Again, the density of the target face mesh is equivalent to the frequency after resampling. A difference in these two sampling frequencies, that is, in mesh densities, leads to an aliasing problem manifesting itself in irregularities in the target face motion when mapping from a face with a higher density to a face with a lower density.

We solve this problem by interpolating the value at additional points around the original point, then averaging the result, in effect applying a low-pass filter to the interpolation function.

4.4.7 Treating the Lip Region

The lip region poses a particular problem for facial motion mapping because the points in the upper and lower lip are very close in the alignment space, but their motion is typically opposite. If we apply the above-described mapping algorithm, the motion of the lower lip would pull the upper lip as well, and vice versa, producing severe artifacts.

We therefore classify all vertices in the face into one of the three categories: upper-lip region, lower-lip region and out-of-lip region. When we calculate the motion vector for target face vertices belonging to the lower-lip region, we ignore the vertices of the upper-lip region of the source face, and vice versa.

We determine the lip regions from the upper- and lower-lip FPs, as well as the FPs of the nose and the chin. The regions are defined in the projected 2-D space (Figure 4.6c). The upper-lip region is determined as the polygon defined by the FPs of the inner upper-lip contour and the sides of the nose. The lower-lip region is determined as the polygon defined by the FPs of the inner lower-lip contour and the sides of the chin.

4.4.8 Treating Eyes, Teeth, Tongue and Global Motion

Eyes, teeth and tongue are typically modeled as separate polygon meshes in a face model and inserted into the scene hierarchy tree under separate transform nodes. In particular, the MPEG-4 FAT allow for specification of the motion of transform nodes. We therefore attempt to automatically identify the transform nodes responsible for the motion of eyes, teeth, tongue and the global motion, then map these motions onto appropriate transform nodes in the target face.

The strategy for finding the appropriate transform nodes is to look for the hierarchically lowest transform node that contains some motion, and encompasses the polygon meshes containing the FPs defining the wanted regions of the face, for example, FP 3.5 for left eye, 6.1 for the tongue and so on (see Chapter 2 for detailed illustration of FPs).

In the target face, we find the appropriate transform as the hierarchically lowest transform node that encompasses the polygon meshes we look for. In case of eyes and global transform, we also make sure that their centers of rotation are appropriate (centers of eyes and base of the neck, respectively). If they are not, we install new centers of rotation.

When these correspondences are established, we map rotations of eyes and global motion and translations of teeth and tongue, from source to target.

A proper way to treat the tongue would obviously require it to be flexible. The main FMC algorithm could be applied separately to the tongue. However, at the current stage we had no access to source facial models that actually animate the tongue as a flexible object and we limited the method to rigid object motion for the tongue.

4.4.9 Facial Motion Cloning Results

We will present the results of cloning on three face models shown in Figure 4.8: Candide, Data and Yoda. All motions were cloned from each model to each other model, producing a full grid of cloned animated models for each motion. One such grid is shown for the surprise expression in Figure 4.9, and another for the viseme I in

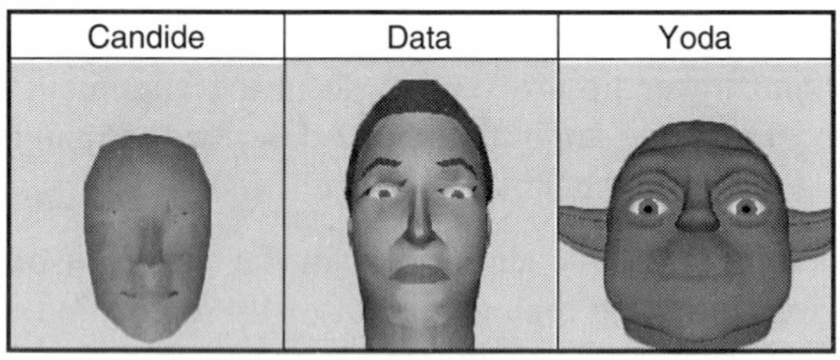

Figure 4.8 The face models used for testing, shown in neutral position. *Sources* for the models are Linköping University for Candide, Avatara VR (*www.avatara.com*) for Data and Yoda

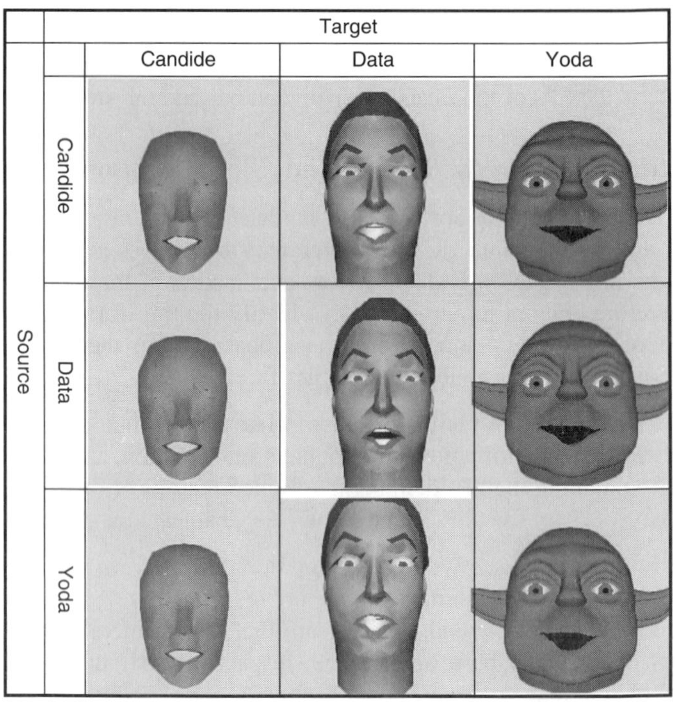

Figure 4.9 Cloning grid for surprise expression (Reprinted from Igor S. Pandzic, "Facial Animation Framework for the Web and Mobile Platforms" © 2002, with permission from ACM) [3]

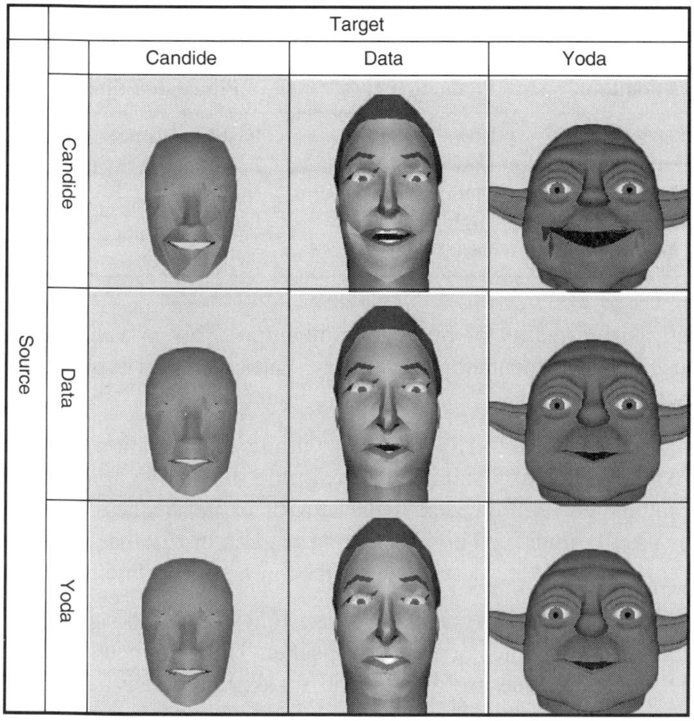

Figure 4.10 Cloning grid for the viseme I

Figure 4.10. In these grids, the main diagonal contains the source faces, and the rest of each row shows the results of cloning from that source face to all other face models. Therefore, looking at each row shows how an expression is cloned from one face to all other faces; looking at each column shows how the expression is cloned onto the same face from different sources.

In terms of visual results shown, even though only a small subset of them could be presented here, we believe that most facial movements for expressions and low-level FAPs are copied correctly to the target faces in most cases. Certain problems still appear, in particular when using low-level FAPs. This is due to the process of combining low-level FAPs to form expressions. If the low-level FAPs are not very carefully designed, their combination leads to irregularities in the combined expressions. Such irregularities, if existing in the source face, are carried over to the cloned face.

We believe that FMC offers a dramatic timesaving potential for producing facial animation based on morph targets or MPEG-4 FAT.

4.5 PRODUCING FACIAL ANIMATION CONTENT

A particular advantage of a facial animation framework based on MPEG-4 is the ability to switch between a variety of tools for facial animation content production. Table 4.1 shows the foreseen classes of solutions with a comparison of production cost and time

Table 4.1 Comparison of facial animation content production tools (Reprinted from Igor S. Pandzic, "Facial Animation Framework for the Web and Mobile Platforms" © 2002, with permission from ACM)

Content production tool	Production cost	Production time	Animation quality
Visual text-to-speech	Low	Fast (real time)	Fair
Automatic speech sync	Low	Fast (near real time)	Fair
Video tracking	Medium	Medium	Medium
Optical tracking	High	Long	High
Manual animation	Very high	Very long	Very high

as well as the quality of produced animation. This is a very general estimation and the specific implementations may vary, particularly as tools become more and more advanced.

Text-to-speech (TTS) combined with facial animation, usually termed visual TTS (VTTS), is particularly suitable for real-time interactive applications because it can be fully automated. A drawback is the lack of expressiveness as only lips are animated. This is usually improved by introducing random or rule-based facial motion, eye blinks and expressions. More details on VTTS can be found in Chapter 8.

Automatic speech synchronization relies on audio signal processing to synchronize the lips to the voice of a human speaker. This eliminates the sound quality problem of TTS. However, the use of a human speaker obviously eliminates the possibility of automatic content production. The expressiveness problem is similar to the VTTS case. These techniques are covered more closely in Chapter 7.

Video tracking of facial features has been a challenging problem for a long time, and still remains without a full solution, though recent implementations [19] approach real-time tracking while keeping a good tracking quality. Ahlberg discusses video tracking and presents his contribution in Chapter 6.

Optical tracking, using reflective markers glued on the face and a battery of calibrated cameras, can give high tracking quality. It involves expensive equipment and usually off-line processing increasing production time.

Manual animation still results in the highest quality when produced by a skilled artist. The high cost and production time mostly limit its application scope to high-budget film and TV productions.

4.6 CONCLUSION

We have presented a framework for implementing applications involving talking virtual characters particularly suited to the Web and mobile platforms. We believe that this framework

- should appeal to the computer animators, which is important for quality content;
- should easily adapt to a very broad range of applications with varying requirements;
- is simple, robust and has low resource requirements on the target platform;
- is easy to port to many different platforms.

ACKNOWLEDGMENTS

This research is partly supported by the VISIT programme of the Swedish Foundation for Strategic Research (SSF).

REFERENCES

[1] I. S. Pandzic, "Life on the web", *Software Focus Journal*, **2**(2), 52–59 (2001).
[2] Moving Picture Experts Group, ISO/IEC 14496 – MPEG-4 International Standard, *www.cselt.it/mpeg*.
[3] I. S. Pandzic, "Facial animation framework for the web and mobile platforms", *Proceedings of Web3D Symposium 2002*, Tempe, Arizona, February, 2002; demonstration at *www.icg.isy.liu.se/~igor/MpegWeb*.
[4] Quartz Version 6.0, Symbian Technical Paper, Symbian Developer Network, *www.symbiandevnet.com/techlib/techcomms/techpapers/papers/v6/over/quartz/index.html*.
[5] F. I. Parke, A Parametric Model for Human Faces, Ph.D. Thesis, University of Utah, Salt Lake City, 1974, UTEC-CSc-75-047.
[6] F. I. Parke, "Parametrized models for facial animation", *IEEE Computer Graphics and Applications*, **2**(9), 61–68 (1982).
[7] K. Arai, T. Kurihara, K. Anjyo, "Bilinear interpolation for facial expressions and metamorphosis in real-time animation", *The Visual Computer*, **12**, 105–116 (1996).
[8] N. Magnenat-Thalmann, N. E. Primeau, D. Thalmann, "Abstract muscle actions procedures for human face animation", *Visual Computer*, **3**(5), 290–297 (1988).
[9] J. E. Chadwick, D. R. Haumann, R. E. Parent, "Layered construction for deformable animated characters", *Computer Graphics (SIGGRAPH'89 Proceedings)*, **23**(3), 243–252 (1989).
[10] P. Kalra, A. Mangili, N. Magnenat-Thalmann, D. Thalmann, "Simulation of facial muscle actions based on rational free form deformation", *Proceedings of the Annual Conference of the European Association for Computer Graphics-Eurographics '92*, Cambridge, UK, September, 1992, pp. 65–69.
[11] M. Escher, I. S. Pandzic, N. Magnenat-Thalmann, "Facial deformations for MPEG-4", *Computer Animation 98*, IEEE Computer Society Press, Philadelphia, Pa., 1998, pp. 138–145.
[12] S. M. Platt, N. I. Badler, "Animating facial expressions", *Computer Graphics*, **15**(3), 245–252 (1981).
[13] K. Waters, "A muscle model for animating three-dimensional facial expressions", *Computer Graphics (SIGGRAPH'87)*, **21**(4), 17–24 (1987).
[14] D. Terzopoulos, K. Waters, "Physically-based facial modeling, analysis and animation", *Journal of Visualization and Computer Animation*, **1**(4), 73–80 (1990).
[15] Shout 3D, Eyematic Interfaces Incorporated, *http://www.shout3d.com/*.
[16] VRML, ISO/IEC 14772-1:1999, *www.web3d.org/fs_specifications.htm*, 1999.
[17] J. Noh, U. Neumann, Expression cloning, *Proceedings of SIGGRAPH 2001*, Los Angeles, Calif., August, 2001.
[18] D. F. Watson, "Contouring – a guide to the analysis and display of spatial data", *Computer Methods in Geosciences*, Vol. 10, Pergamon Press, New York.
[19] J. Ahlberg, "Using the active appearance algorithm for face and facial feature tracking", *2nd International Workshop on Recognition, Analysis and Tracking of Faces and Gestures in Realtime Systems (RATFFG-RTS)*, Vancouver, Canada, July, 2001, pp. 68–72.

5

The Facial Animation Engine

Fabio Lavagetto[1] and Roberto Pockaj[2]

[1]*Università degli Studi di Genova, Genova, Italia,* [2]*EPTAMEDIA srl, Genova, Italia*

5.1 INTRODUCTION

The Facial Animation Engine (FAE) is a software package for real-time animation of virtual faces. It has been designed and implemented in order to provide a simple and fast way to develop and animate 3-D virtual facial models according to the specifications of the MPEG-4 standard.

The main functionality of the FAE is its capability to deform virtual faces in response to MPEG-4 Facial Animation Parameters (FAPs) according to 'animation rules' automatically computed by the software itself. To create an animated face, in fact, the user has only the task of designing a face model and attaching it to very simple 'semantic' information used by the FAE to compute the animation rules. For these reasons, the FAE can be considered a high-level interface for the design of MPEG-4-compliant animated faces.

The FAE is a modular program, in which each module can be easily plugged in or unplugged from the whole software. Some of these modules are written in compliance with the MPEG-4 standard, some add new functionalities, some others are necessary to make the FAE a stand-alone application built on top of the MPEG-4 modules.

The FAE has been developed by founders of Eptamedia S.r.l. (*http://www.eptamedia.com*), a small Italian company, founded at the beginning of 2001 as a spin-off from the Department of Communication, Computer and System Science of the University of Genoa.

5.2 THE FAE BLOCK DIAGRAM

A brief description of the FAE is presented in this section, with the help of the functional block diagram reported in Figure 5.1. A more detailed discussion of the FAE can be found in the rest of this chapter.

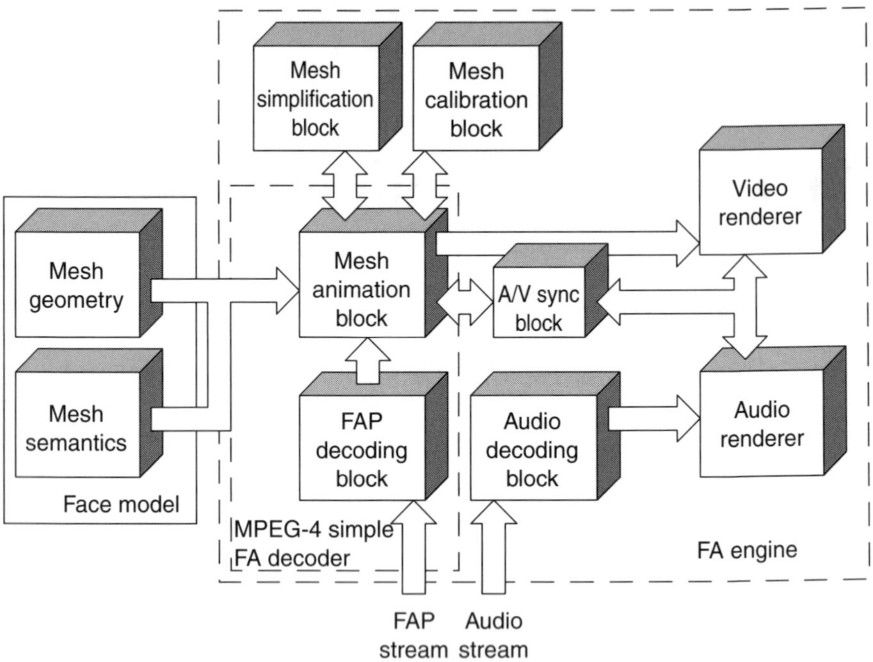

Figure 5.1 FAE functional block diagram

The *Face Model* is clearly not contained within the FAE software. The FAE can in fact load different face models with their associated semantic. The *Mesh Geometry* simply represents the description of the 3-D mesh, while the *Mesh Semantic* is used by the Mesh Animation Block to convert FAP values into mesh deformations.

The MPEG-4-compliant part of the code is composed of the *FAP Decoding Block* and by the *Mesh Animation Block*. The two blocks, together with the face model, represent a Simple Facial Animation Decoder, that gets an FAP stream as input and provides a deformed mesh as output. The *Mesh Calibration Block* is responsible for model calibration based only on feature points (FPs), while the *Mesh Simplification Block* is responsible for model complexity reduction when the current hardware is not powerful enough to guarantee a reasonable frame rate. Finally, the *A/V Sync Block* is in charge of maintaining a good synchronization between the *Video* and the *Audio Renderer*; it can request the Animation Block to skip mesh deformation and the Video Renderer to skip model rendering, in order to guarantee an appropriate A/V sync.

5.3 THE FACE MODEL

5.3.1 Mesh Geometry Description

The Mesh Geometry consists of a conventional VRML-like representation of a net of polygons stored in a proprietary format. It contains the coordinates of the vertices and the topology of the wireframe representing the face, together with the color and texture information associated with each polygon.

The face model can be designed by means of almost any 3-D authoring tool available nowadays. At the end of the design process, the model geometry can be exported in such a way that it can be handled by the FAE. There are very few constraints imposed by the FAE to the model geometry: the main one is that the mesh has to be composed of triangular patches only. Other constraints concern the mesh triangulation criteria because it has to allow mesh deformation in a reasonable way (for instance, upper and lower lips must not be directly connected by triangles because this would not allow for mouth opening), but these considerations are rather obvious.

The Mesh Geometry also contains a cross-reference table to specify the correspondence of the 84 FPs defined in MPEG-4 to specific mesh vertices.

5.3.2 Mesh Semantics Description

Differing from the Mesh Geometry description and based on existing VRML-like languages, the semantics description has been specified according to a proprietary format containing high-level information such as:

- the *look-up* table for mapping high-level FAPs (visemes and facial expressions) in terms of a vector of low-level FAPs; each face model will then reproduce visemes and expressions in its unique way;
- the displacement region, defining the mesh region affected by each FAP;
- for rotational FAPs, the corresponding rotation axis;
- the indexes of the vertices affected by the displacement of the feature point associated with each specific FAP;
- the criterion for weighting the amplitude of vertices displacement with respect to the displacement of the *feature points;*
- the predefined movement that describes how the vertices affected by each FAP has to be moved in space.

It should be noted how all this information is expressed in a simple way, as a function of the indexes of the *wireframe* vertices, instead of their spatial coordinates or of their topology (Figure 5.4).

5.3.3 The Model Authoring Tool

In order to make the creation of animated models even easier, a tool for the authoring of the Mesh Semantics has been developed (Figure 5.2). This tool provides a graphical interface for vertex labeling and reduces the time to obtain a new animated model, starting from the Mesh Geometry description, to the order of few hours of work. Some more hours of work are eventually necessary to further refine the quality of the animation.

It has to be stressed that by simply changing the Mesh Semantics, it is possible to use the FAE to animate a face model driven by a set of parameters different from the one defined by MPEG-4 or to remap the meaning of certain animation parameters:

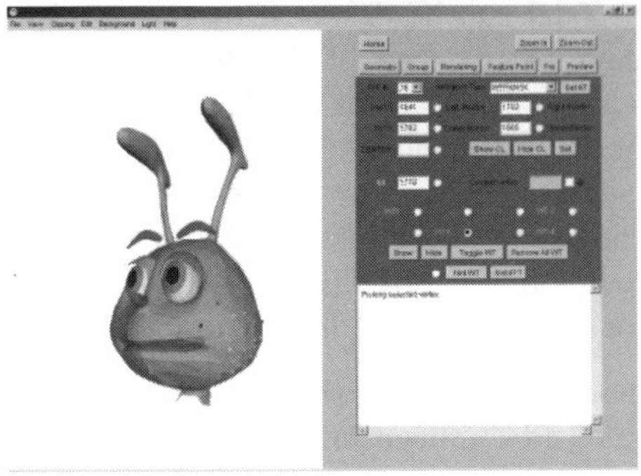

Figure 5.2 A screenshot of the graphical authoring tool for Mesh Semantics information design

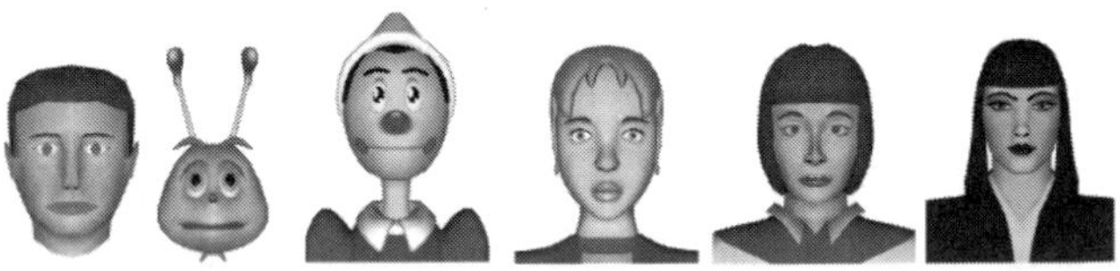

Figure 5.3 Some face models; from left to right: Mike (750 poly, no textures), Anty (3988 poly, no textures), Pinocchio (6988 poly, 4 textures), Asia (8392 poly, no textures), Sarah (3270 poly, no textures) and Francy (6138 poly, 5 textures) – Pinocchio is a copyright of Fondazione Nazionale Carlo Collodi

for instance, it is possible to use FAP 45 (thrust_tongue_tip) to stretch the nose tip and so on (although, strictly speaking, this is not allowed by MPEG-4...).

5.3.4 Sample Face Models

Several face models, either cartoonlike or humanlike, have been developed using the Authoring Tool. All the models can be animated with the FAE. Figure 5.3 shows all the models available at the time of writing this chapter. The complexity of the model (i.e. the number of polygons it is composed of and the number of texture images) obviously affects the perceived quality.

5.4 THE MESH ANIMATION BLOCK

With the term 'animation rule' we refer to the full set of rules that control the movement of a set of vertices, as a function of the value of each FAP.

The Animation block is responsible for mesh deformation in response to FAP and is therefore the block that computes the animation rules. It loads the Mesh Semantics once and subsequently, at each frame, uses this information to move the model vertices in the space [1].

During the loading phase, all the weights associated with the animated vertices are computed according to the specifications in the Mesh Semantics and stored.

It is clear the criterion used to assign weights to a set of vertices together with predefined movements, represents a critical key issue for generating the animation rules. In the FAE implementation, the displacement of each vertex associated with a feature point can be equal or proportional (according to a weight of value equal or less than one) to the displacement of the corresponding feature point. The computation of the weights is done automatically on the basis of the information contained in the semantic file through a two-step procedure.

In the first step, the cylindrical projection of the model vertices is computed with respect to an axis parallel to the coordinated axis y and vertically aligned in the middle of the head; the displacement domains associated with each FAP are then determined.

In the second step, the corresponding weighting function, specified in the Mesh Semantics, is applied. The weighting function, defined on the $[XY]$ projection plane, provides indications on the weight associated with each vertex.

The weighting functions have null value at least along two boundary sides of the displacement domain. In this way, a unitary weight will always be assigned to the *feature point*, while decreasing weights will be assigned to the remaining vertices approaching the boundary of the displacement domain. Currently, eight different weighting functions have been defined and implemented.

After having calculated all the vertex weights, for every frame the Animation Block computes the displacement of each animated vertex, by applying one out of fifteen predefined movements (e.g. a rotation, a translation or more complex movements). The number of predefined movements is lower than the number of FAPs, since the same action can be associated with different FAPs.

To better understand this procedure, let us discuss an example. Let us consider, among the many movements that can affect the left eyebrow, which is controlled by FAP31 (raise_l_i_eyebrow), which controls the vertical displacement of the eyebrow's internal region.

From the Mesh Semantics the Animation Block recognizes the vertices whose displacement is controlled by FAP31 and identifies the region onto which the displacement is mapped. In addition, the Animation Block knows what is the predefined movement associated with FAP31. Let us now assume that the predefined movement is a simple vertical translation along the y-axis (this is not the truth, indeed, because the eyebrow movement requires a more complex deformation, i.e. a more complex predefined movement where the z-coordinate of the feature point and of the other affected vertices of the domain must vary in an anatomically consistent way, depending on the particular skull shape).

In Figure 5.4 the feature point associated with FAP31 is identified (circle) on the wireframe region around the left eyebrow, together with the vertices (small squares) that are affected by the predefined movement associated with FAP31, and the bounding points (big squares), which limit the domain of the displacement region.

For each vertex that must be moved, the Animation Block computes a specific weight that defines the magnitude of the vertex displacement with respect to the feature

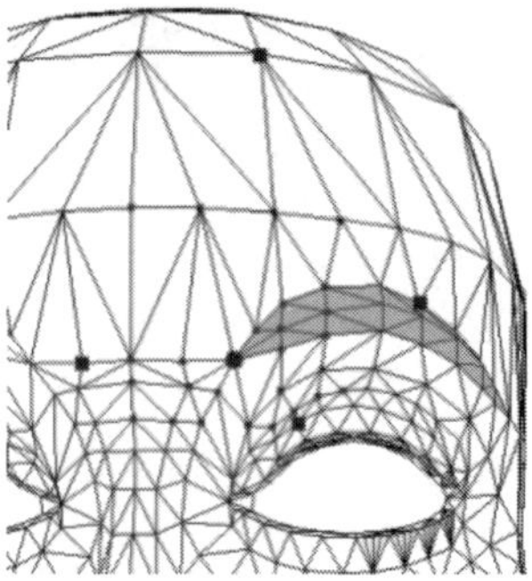

Figure 5.4 Information necessary to compute the animation rule controlled by FAP31. This is a graphical representation of part of the information contained in the Mesh Semantics for FAP3. The big squares indicate the limits of the displacement region, small squares the animated vertices and the circle represents the feature point (the type of the weighting function and the type of predefined movement cannot obviously be represented in the figure)

point. The weight is defined as a function of the vertex position within the displacement region, and the type of weighting function is specified in the Mesh Semantics.

Using the very simple weighting function reported in Equation (5.1) as an example, we can compute the weight associated with each vertex moved by FAP31:

$$W_i = \left(\frac{X_i - X_{\min}}{X_F - X_{\min}}\right) \cdot \left(\frac{Y_i - Y_{\min}}{Y_F - Y_{\min}}\right) \text{ for } X_i < X_F \text{ and } Y_i < Y_F \tag{5.1}$$

where, X_F and Y_F represent the coordinates of the associated feature point F, X_i and Y_i represent the coordinate of the ith vertex for which the weight W_i is computed and $X_{\min}$, $X_{\max}$, $Y_{\min}$, $Y_{\max}$ are the coordinates of the bounding points, which limit the displacement region. These parameters are better understood with reference to Figure 5.5.

Once a weight has been associated with each vertex, the specific predefined movement is applied accordingly. In the case of FAP31, as stated before, if we assume a simple vertical translation along the y-axis, we will obtain as a result of the animation process, a variation of the y-coordinate of each animated vertex. The amplitude of the

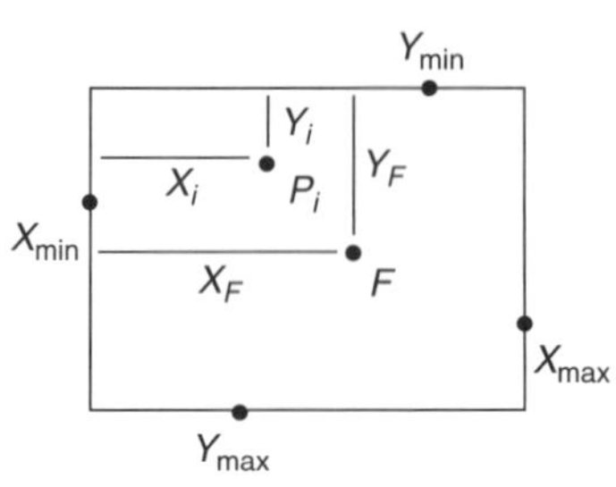

Figure 5.5 Parameters for the algorithm computing the weights associated with vertices

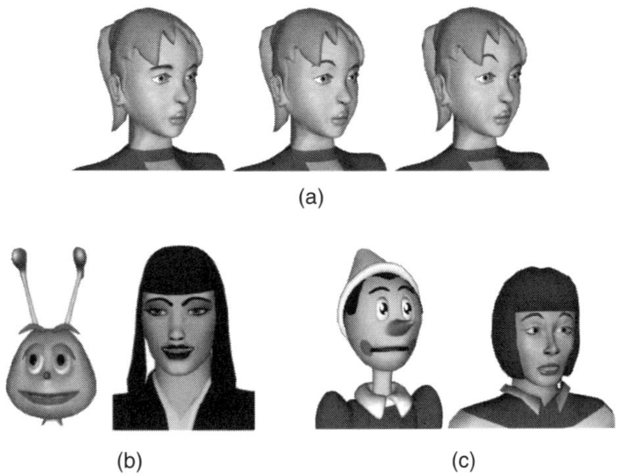

Figure 5.6 Sample movements and expressions: (a) Right eyebrow raising on model 'Asia'; (b) joy expression on models 'Anty' and 'Francy' and (c) surprise expression on models 'Pinocchio' and 'Sarah'

variation for each vertex will be proportional (according to the associated weight) to the displacement of the feature point.

It must be noticed how the displacement regions of each feature point can overlap each other (this means that a vertex of the wireframe not coinciding with a feature point can be affected by the movements associated with different FPs).

5.4.1 Animation Results

In the following pictures, some examples of elementary movements and some facial expressions are presented. The orientation of the head has been chosen appropriately each time to give more evidence to the specific movement applied. Figure 5.6a shows the effect of predefined movements (right eyebrow raising) on the model 'Asia'. Figures 5.6b,c show some facial expressions on different models. It can be noticed that, though the same frame of FAP has been applied to a couple of rather different models, the expression is reproduced in a consistent way.

5.5 THE MESH CALIBRATION BLOCK

The task of the Calibration Block is to reshape and modify the appearance of a generic face model used by the FAE by using the information of a specific 3-D face model to reshape (a process referred to as 'calibration process'). In particular, the work described in the present section makes use of the 3-D coordinates of the FPs and of the texture of the specific model (the target) to deform the generic model. Future work in this area will help to improve the result of the calibration process by using the entire mesh of the target model to reshape the generic one.

The quality of the generated faces depends on how the unconstrained vertices are moved (the goal is to avoid distortions of the calibrated face as much as possible). The employed algorithm is based on the theory of Radial Basis Functions (RBF), as discussed in the following.

5.5.1 Multilevel Calibration with RBF

As we pointed out in the preceding text, there are relatively few (MPEG-4 defines 84 FP) points that can be used to calibrate the generic model with respect to its total number of vertices (which, depending on its complexity, varies from a few hundred to a few thousand points). Model calibration can therefore be considered as a problem of 'scattered data interpolation' [2].

Let us assume the proprietary model to be composed of a finite set of points $X = \{x_1, \ldots, x_N\} \in \Re^3$, including a subset $S = \{s_1, \ldots, s_n\} \subset X$ representing the ensemble of FPs. Let us associate to S a corresponding set $Y = \{y_1, \ldots, y_n\} \in \Re^3$ indicating the subset of calibration FPs transmitted by the encoder.

The problem to solve is to define an interpolation function $F : \Re^3 \rightarrow \Re^3$ such that

$$F(x_i) = y_i, \quad 1 \leq i \leq n \tag{5.2}$$

The approach that has been followed to construct the interpolation function models F is

$$F(x) := \sum_{j=1}^{n} w_j \Phi(x - x_j) \tag{5.3}$$

where $w_j \in \Re^3$ represents the weight associated with the $\Phi(x - x_j)$, and the function Φ: $\Re^3 \rightarrow \Re$ is definite positive on $\Re^3$, in the sense that for any finite set of points $S = \{s_1, \ldots, s_n\} \in \Re^3$ the matrix

$$A = [\Phi(x_k - x_j)]_{1 \leq j,k \leq n} \tag{5.4}$$

is definite positive.

In this way the solution to the system is guaranteed:

$$F(x_k) = \sum_{j=1}^{n} w_j \Phi(x_k - x_j) = y_k, \quad 1 \leq k \leq n \tag{5.5}$$

The family of functions of the kind $\Phi : \Re^3 \rightarrow \Re$ employed in this approach are radial and can be expressed as

$$\Phi(x) = \phi(\|x\|_2), \quad x \in \Re^3 \tag{5.6}$$

with ϕ: $\Re_{\geq 0} \rightarrow \Re$ and the Euclidean norm computed in $\Re^3$.

The interpolation function obtained in this way has the limitation of not including the affine transformation. This limitation, however, can be overcome by adding a polynomial term [3] defined in space P_m^3, of maximum order m, that generalizes the expression (5.3) to

$$F(x) = \sum_{j=1}^{n} w_j \Phi(x - x_j) + \sum_{l=1}^{M} v_l p_l(x) \tag{5.7}$$

with $M = \dim P_m^3$, $v_1, \ldots, v_M \in \Re^3$ and $p_1, \ldots, p_M$ polynomials of P_m^3 introducing the term

$$\sum_{i=1}^{n} w_i p_l(x_i) = 0 \quad 1 \leq l \leq M \tag{5.8}$$

The system of equations can be expressed as

$$\begin{bmatrix} A & P^T \\ P & 0 \end{bmatrix} \cdot \begin{bmatrix} w \\ v \end{bmatrix} = \begin{bmatrix} y \\ 0 \end{bmatrix} \tag{5.9}$$

and its solution exists and is unique in the hypotheses formulated in Reference 4.

For the specificity of the application, it has been considered appropriate to choose, among the multitude of possible RBF's, those with the property of being monotonically decreasing to zero for increasing values of the radius, such as Gaussian functions and inverse multiquadrics. The disadvantages of such functions are mainly due to the global effect produced by the interpolation function that is obtained and by the smoothness of the surface that is originated. Unfortunately, despite the peculiarity already mentioned, experimental results show an excessive interaction between the various RBF's that contribute to generate the interpolation function. This is not in accordance with the usual hypothesis made in the calibration that each feature point (on which the RBF's are centered) influences a limited region of the wireframe. To make an example, it is quite reasonable to assume that the FPs of the mouth do not interact with the characteristic points of the ears and of the upper part of the head.

A particular kind of RBF's, called Compactly Supported (RBFCS), whose properties are described and discussed in References 4 and 5, have been considered. This type of RBF is widely used in a variety of applications because of the significant computational advantages they offer and the good properties they exhibit: among them, the locality of the RBFCS domain is a particularly important characteristic in our case, in order to bound the reshaping action applied on the model.

From a careful analysis of the position of the FPs, it can be noticed that some subsets of points (eyes and mouth for instance) are very clustered in a small area of the wireframe, while other subsets are very sparsely distributed. This nonuniformity of feature point distribution makes it complicated to build an interpolation function capable of satisfying both the requirements of precision and smoothness.

The basic idea [6] of the method we propose to use to solve the problem (at least partly) originated by these requirements is to subdivide the ensemble of FPs into a number of subsets S_i:

$$S_0 \subset S_1 \ldots \subset S_t = S \tag{5.10}$$

such that, in each of them, data are distributed as uniformly as possible, at least for the smallest subsets.

The calibration process consists of t levels of interpolation. At the jth level, the residual of the previous level is interpolated by means of the RBFCS ϕ_j of support σ_j.

After t steps, the resulting interpolation function is:

$$F: = \sum_{k=1}^{t} F_k \tag{5.11}$$

In order to preserve the global surface smoothness, low-order RBFs are employed at the first iterations, while high-order RBFs are applied in the last iterations to add local somatic details.

5.5.2 Calibration with Texture

Besides the model calibration, the Calibration Block includes the possibility to manage also the texture information and the texture coordinates associated with each feature point to allow the texture adaptation on the model surface.

The first problem to be solved for adding the texture information to the calibrated model is that of mapping the complete set of 3-D model vertices onto the 2-D-texture domain. To implement this operation a two-step algorithm has been adopted:

- 2-D projection of the calibrated model onto the texture domain;
- mapping of the 2-D projection of the FPs on the texture coordinates.

For what concerns the mapping of the FPs on the texture coordinates, the same algorithm based on RBFs, already used for the calibration of the 3-D model is used. The 2-D projections of the FPs are forced to coincide with the texture coordinates of the same FPs. In this case, instead of using a multilevel reshaping algorithm, the data have been processed at one shot. It is apparent that RBF in $\Re^2$ are suitable in this case.

5.5.3 Calibration Results

In Figures 5.7 and 5.8 some examples of mesh calibration with texture are shown, including the animation of the calibrated model.

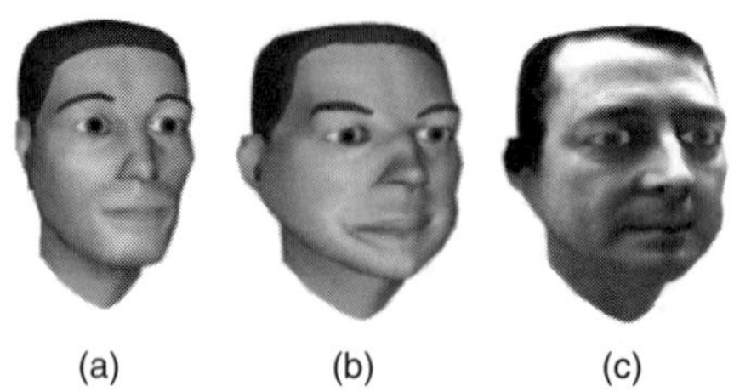

Figure 5.7 Examples of model calibration: (a) original model; (b) calibrated model and (c) with texture

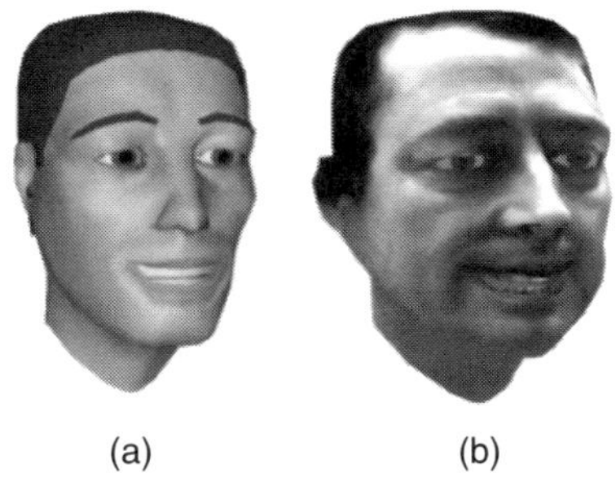

Figure 5.8 Example of animation of a calibrated model: (a) original model and (b) calibrated and animated model

5.6 THE MESH SIMPLIFICATION BLOCK

Recently, the interest on mesh simplification has increased significantly. The idea of building a virtual 3-D model at different resolutions is not new: Clark [7] introduced the concept more than 25 years ago.

However, the most significant achievements in this area have been obtained in recent years when various techniques have been proposed by many researchers. In this chapter, a simplification technique is applied to MPEG-4 animated faces to prove how it is possible to reduce the model complexity even during an animation, still preserving the correct interpretation of the FAPs defined by MPEG-4.

5.6.1 Iterative Edge Contraction and Quadric Error Metric

The model simplification technique uses the algorithm proposed by Garland and Heckbert in Reference 8, which is based on two main concepts: the Iterative Contraction of Edges and the Quadric Error Metric. The algorithm processes the original surface and iteratively removes vertices and faces from the model. At each step, a single operation, consisting of the contraction of a pair of vertices, is performed (Figure 5.9).

This operation, denoted in Reference 8 as $(\mathbf{v}_i, \mathbf{v}_j) \rightarrow \mathbf{v}$, where $\mathbf{v}_i$ and $\mathbf{v}_j$ are the original vertices and $\mathbf{v}$ is the contraction result, modifies the model surface in three steps:

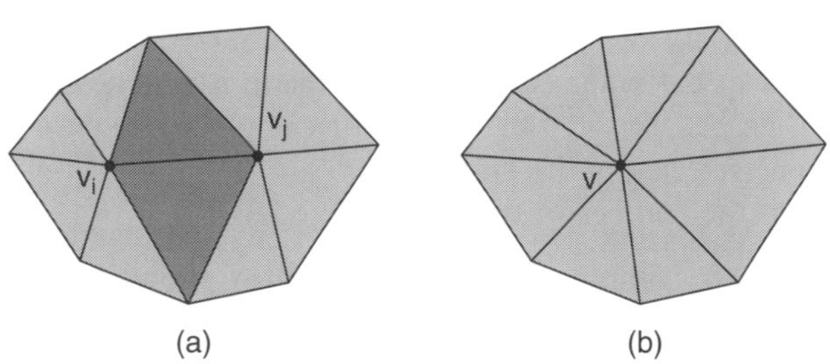

Figure 5.9 Contraction of two vertices with a common edge: (a) before and (b) after contraction

1. $\mathbf{v}_i$ and $\mathbf{v}_j$ are moved to the new position $\mathbf{v}$
2. All the occurrences of $\mathbf{v}_j$ are replaced with $\mathbf{v}_i$
3. $\mathbf{v}_j$ and all the degenerated faces are removed.

The first step actually modifies the surface geometry, the second modifies the surface topology and the last one removes elements that are not necessary anymore. The algorithm is initialized by first selecting all the possible pairs of vertices candidate for the contraction (for instance, all the vertices that share a common edge), and then assigning a cost for the contraction to every pair. Hence, the contractions are performed starting from the one with the minimum associated cost, obviously updating all the remaining contraction costs at every step.

Many algorithms based on the iterative edge contraction have the structure described above. A cost value is associated with each pair of vertices according to various different criteria. Each contraction introduces an approximation error on the mesh, which is typically proportional to the cost associated with the contraction.

Evidently, the definition of the contraction cost is a key issue. In the literature, many methods for estimating the error introduced by a surface approximation have been proposed.

However, the metric that best fits the purpose of simplifying a facial model in real time is once again suggested in Reference 8. The proposed technique, the quadric error metric, satisfies both the requirements of real-time performances and of the quality of the simplified surface.

The term quadric is introduced by Garland *et al.* because, if we define $Q(\mathbf{v})$ the function representing the error introduced after a vertex pair is collapsed, the surfaces defined by $Q(\mathbf{v}) = \varepsilon$ are quadric surfaces. With this metric, when a pair of vertices ($\mathbf{v}_i$, $\mathbf{v}_j$) is collapsed, the resulting quadric for the new vertex $\mathbf{v}$ is given by $Q = Q_i + Q_j$, Q_i and Q_j being the quadrics associated with $\mathbf{v}_i$ and $\mathbf{v}_j$ respectively.

The cost of the contraction $(\mathbf{v}_i, \mathbf{v}_j) \rightarrow \mathbf{v}$ is then computed as $Q(\mathbf{v}) = Q_i(\mathbf{v}) + Q_j(\mathbf{v})$.

It is clear why the described metric is very efficient in terms of computation burden and in terms of memory requirements: in fact, each vertex is associated with a quadric and this is the only information needed to evaluate the approximation error.

The position $\mathbf{v}$ of the new vertex introduced as a consequence of a collapse is optimal if, for a given quadric Q, $Q(\mathbf{v})$ is minimized. The position of the new vertex is suboptimal in the other cases. Among them, it is relevant to the case when, after contracting the pair ($\mathbf{v}_i$, $\mathbf{v}_j$), the new vertex is placed in $\mathbf{v}_i$ or $\mathbf{v}_j$. With this approach, that is the one selected in the proposed work, the simplified model is in fact composed of a subset of the vertices of the original model.

Finally, the simplification algorithm also allows preserving discontinuities in the original model, like open edges or borders between different face materials. These elements have a significant visual importance, because they often represent the most significant details of the model. In order to preserve them, the algorithm adds a large penalty factor to the quadric of each vertex of the edge/border to be preserved.

5.6.2 Simplification of MPEG-4 Animated Faces

In order to comply with the MPEG-4 specifications on both the original and simplified models and also to guarantee a successful integration of the animation and simplification algorithms, the following requirements have been imposed:

1. the properties of the simplified model must be preserved in order to allow its animation, that is, the animation rule that the FAE automatically computes for the original model must still be valid without additional computations for the simplified model;
2. the feature points must be preserved, as well as borders between different parts of the face (like lip contours);
3. when possible, the symmetry of the face has to be exploited to reduce the simplification time (once a vertex pair is selected for contraction, it is likely that the next pair selected is the symmetric pair).

The simplification criteria are controlled by only two parameters: the desired number of faces in the simplified model and the maximum error allowed during a contraction. While the first parameter has a clear meaning, it is worthwhile to spend a few words on the second one. The error threshold proved to be a rather good objective criterion for defining the subjective quality of a face model and it is therefore useful to avoid the removal of a very high number of faces during the simplification process.

Since the FAE computes the animation rules as a function of the vertex spatial positions and not as a function of the mesh topology, it was decided to adopt a suboptimal strategy for the placement of a new vertex after a pair has collapsed. Specifically, given the contraction $(\mathbf{v}_i, \mathbf{v}_j) \rightarrow \mathbf{v}$, $\mathbf{v}$ is located in $\mathbf{v}_i$ or in $\mathbf{v}_j$; in other words, during a contraction, one of the two vertices is removed, all the others remaining unchanged.

After the FAE has loaded the original model and has computed the animation rules, it is possible to start the simplification process without the need to compute the animation rules again, since the position of the model vertices is not modified (Figure 5.12).

Preserving edges between regions of different colors (or materials) is particularly important when dealing with the model of a human face. In Figure 5.10 a simplification of the Oscar model, in use at the DSP Lab of DIST is shown, without preservation of borders. From the reported images it is evident how much the borders between regions affect the subjective quality even at the very first steps of the simplification process.

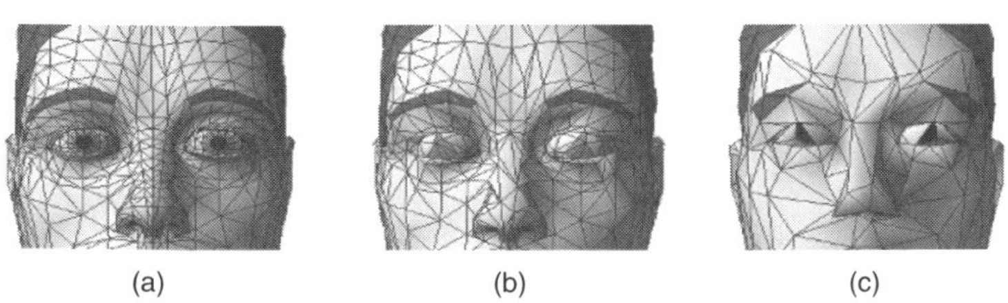

Figure 5.10 Simplification of the Oscar model: (a) original with 2444 polygons; (b) simplified to 1944 polygons and (c) simplified to 840 polygons, without border preservation

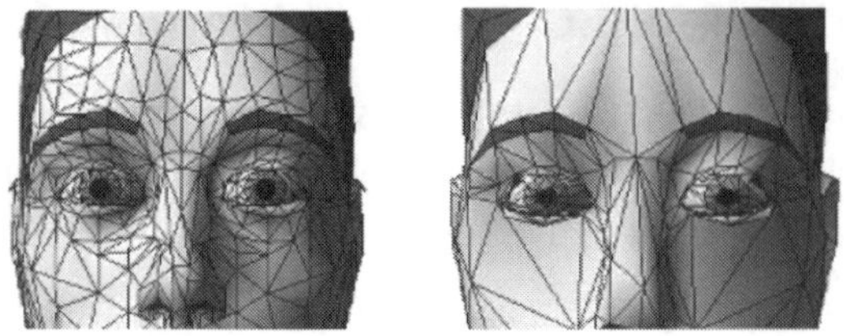

Figure 5.11 Simplification of the Oscar model: (a) simplified to 1944 polygons and (b) simplified to 840 polygons, with border preservation

For instance, in the central image of Figure 5.10, the loss of the pupil shape represents an unacceptable degradation of the subjective quality of the simplified model.

For these reasons, it has been decided to preserve, the borders of regions such as irises, pupils, lips, nostrils and eyebrows, during the simplification process. In addition to that, vertices associated with FPs and open edges like the neck, have been preserved as well. In Figure 5.11 the Oscar model is simplified again, with the same number of polygons as in Figure 5.11, but this time preserving FPs and borders.

Preserving the FPs is necessary not only to provide a satisfactory subjective quality of the simplified model, but also to guarantee a consistent animation. In fact, to the FPs are applied the model deformations described by means of the FAPs [9] defined in MPEG-4.

5.6.3 Simplification with Textures

Let us now consider texture mapping. The texture is an image applied to the model surface in order to increase the realism of the virtual object; it does not modify or increase the complexity of the mesh itself but simply adds to it photographic details. The process of applying a texture to a surface is known as texture mapping. To map a texture onto a polygon, it is necessary to assign to each vertex of the polygon a pair of texture coordinates.

In the literature, the problem of simplifying a textured model has been faced by Hoppe [10, 11] and Gardland [12]. In their works, they extend the concept of quadric metric to take into account material properties like texture. Although they often use an optimal criterion to position vertices generated by a collapse, the algorithm described in this paper still uses a suboptimal criterion, to guarantee real-time performances.

Therefore, as we have seen for the geometry simplification, if we consider the pair $(\mathbf{v}_i, \mathbf{v}_j)$ and if the new vertex is located in $\mathbf{v}_i$, its texture coordinate will be that of $\mathbf{v}_i$; the same applies if the new vertex is located in $\mathbf{v}_j$.

5.6.4 Simplification Results

In the following picture, some examples of simplified models, textured and not textured, are presented, as well as an example of the animation of a simplified model

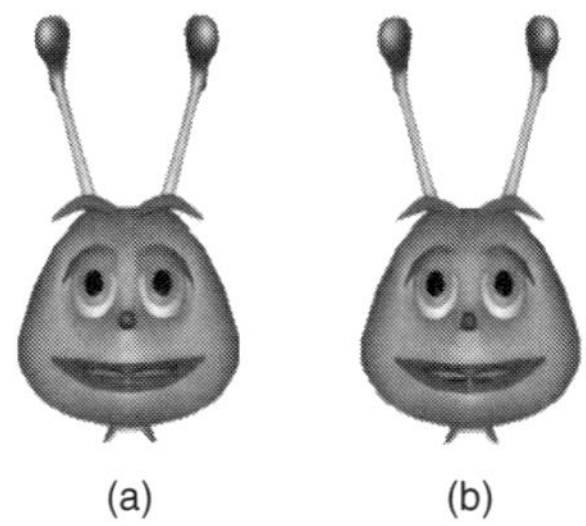

Figure 5.12 Example of mesh simplification and animation: (a) model Anty simplified to 40% of the original number of polygons compared to the full resolution model and (b) while performing a joy expression

Figure 5.13 Example of mesh simplification: (a) model Anty simplified to 40% of its complexity and (b) model Francy with texture simplified to 40%

(Figure 5.12). Figure 5.13a shows an example of simplified textured model, in which this latter is almost indistinguishable from the original.

5.7 THE FAP DECODING BLOCK

The FAP Decoding Block contains the standard MPEG-4 Facial Animation bit stream decoder, whose syntax has already been described in this book (see Chapter 2). It also contains a proprietary animation stream decoder, to be used in those applications that cannot be compliant with the standard. In both cases, the bitrate required for animations at 25 frames per second is on the order of 2 kbps.

It is not worthwhile to further describe the standard decoder here, it being more interesting to discuss some particular aspects of the MPEG-4 specification that, in the opinion of the authors, can be exploited in order to reach the above-mentioned compression of 2 kbps.

5.7.1 FAP Interpolation

As in any conventional approach to lossy video compression, also in facial animation the most straightforward solution to reduce the bit rate associated with the transmission of FAPs is increasing the quantization scaling factor used for parameter encoding. The evident disadvantage of this approach is that of progressively degrading the animation quality with the introduction of jerky facial movements that are usually very annoying. In synthetic facial animation, the difference from natural video coding is the coarse quantization of parameters does not affect the rendering quality of each individual frame but the smoothness of the synthesized facial movements. However, another possibility to reduce the bit rate due to FAP encoding is based on the exploitation of the *a priori* knowledge about the object that is animated, namely a human face. Through the analysis of the time correlation characterizing the facial movements of a person while talking it is possible to reduce much of the FAP redundancy. In fact, instead of coding and transmitting the complete set of 66 low-level FAPs at each frame, it is possible to encode only a significant subset of them and to let the decoder generate the missing parameters, thus achieving very low bit rate animation.

This procedure, called FAP interpolation represents the core issue discussed in the following text in this section.

MPEG-4 specifications concerning facial animation uses the term 'FAP interpolation' to indicate the procedure through which a generic Face Animation decoder autonomously defines the values of those FAPs that have not been encoded and transmitted. On the basis of the knowledge of only a limited number of FAPs, the objective of the interpolation procedure is to estimate a variable number of the missing ones. In this case, as both the known and the estimated FAPs belong to the same frame, the estimation procedure is performed intraframe and no interframe FAP prediction is performed. In this respect, therefore, we should talk of an actual FAP extrapolation rather than of interpolation. Having pointed out this terminology ambiguity, let us now examine the related MPEG-4 specifications, focusing the attention on some points that are not of clear and immediate interpretation. In MPEG-4 specifications, FAPs are clustered in a few groups depending on the region of the face they are applied to, as reported in (Chapter 2) with the objective of optimizing the compression efficiency. Together with I-Frames, encoded without any prediction, two different hierarchical masks are transmitted, i.e. fap_mask_type and fap_group_mask, to specify the FAP subset that will be transmitted in the current I-Frame and in the following P-Frames.

The fap_mask_type is a 2-bit mask that indicates whether a group is transmitted, not transmitted or partially transmitted. The fap_group_mask, which is encoded only in case the value associated with fap_mask_type is 01 or 10, specifies which FAPs of those in the group are actually transmitted and which of them can be interpolated by the decoder (i.e. for which of them the decoder is free to autonomously decide the value).

For the analysis of FAP correlation, only a few of the ten FAP groups defined in MPEG-4 have been considered, and the analysis carried out has been oriented to model the FAP trajectories related to the human facial movements associated with normal speech production. The modeling objective is twofold: reducing the number of FAPs to be encoded and transmitted and providing a suitable FAP interpolation mechanism at the decoder. Should a particular nonhuman facial movement be rendered for a specific application, such as for the animation of a cartoonlike character, it will be enough to modify the FAP mask associated with the transmission of an I-frame and to transmit all the FAPs encoding that particular animation for the time needed.

Table 5.1 List of FAP to transmit in normal speech production (tx means transmitted FAP, empty rows indicate FAP not necessary for typical animations)

		Transmitted FAP						
		FAP16	FAP17	FAP33	FAP51	FAP52	FAP53	FAP59
FAP3	Open_jaw					−0.650		
FAP4	Lower_t_midlip				1			
FAP5	Raise_b_midlip					1		
FAP6	Stretch_l_cornerlip						1	
FAP7	Stretch_r_cornerlip						1	
FAP8	Lower_t_lip_lm				0.854			
FAP9	Lower_t_lip_rm				0.854			
FAP10	Raise_b_lip_lm					0.728		
FAP11	Raise_b_lip_rm					0.728		
FAP12	Raise_l_cornerlip							1
FAP13	Raise_rLcornerlip							1
FAP14	Thrust_jaw					0.288		
FAP15	Shift_Jaw							
FAP16	Push_b_lip	tx						
FAP17	Push_t_lip		tx					
FAP18	Depress_chin							
FAP31	Raise_l_i_eyebrow			0.659				
FAP32	Raise_r_i_eyebrow			0.659				
FAP33	Raise_l_m_eyebrow			tx				
FAP34	Raise_r_m_eyebrow			1				
FAP35	Raise_l_o_eyebrow			0.564				
FAP36	Raise_r_o_eyebrow			0.564				
FAP37	Squeeze_l_eyebrow			−0.101				
FAP38	Squeeze_r_eyebrow			−0.101				
FAP39	Puff_l_check						0.477	
FAP40	Puff_r_check						0.477	
FAP41	Lift_l_check							0.310
FAP42	Lift_r_check							0.310
FAP51	Lower_t_midlip_o				tx			
FAP52	Raise_b_midlip_o					tx		
FAP53	Stretch_l_cornerlip_o						tx	
FAP54	Stretch_r_cornerlip_o						1	
FAP55	Lower_t_lip_lm_o				0.854			
FAP56	Lower_t_lip_rm_o				0.854			
FAP57	Raise_b_lip_lm_o					0.728		
FAP58	Raise_b_lip_rm_o					0.728		
FAP59	Raise_l_cornerlip_o							tx
FAP60	Raise_r_cornerlip_o							1

At the end of a statistical analysis of more than 22 000 FAP frames, whose details are described in Reference 13, a strong correlation between several FAPs has been found. As a final result of this study, a table has been designed to identify which FAPs are worth being transmitted and which of the others can be interpolated frame by frame from the transmitted ones. The very simple, but effective interpolation criterion sets the interpolated FAP proportional to one of the transmitted ones, according to Equation (5.12):

$$\mathrm{FAP}i_n = \alpha \cdot \mathrm{FAP}j_n \tag{5.12}$$

where n represents the nth frame of a sequence, FAPi is the parameter to be interpolated and FAPj the parameter actually transmitted.

Table 5.1 reports the list of FAP to transmit (indicated by tx in the table) and the value of α for the interpolated FAP.

5.8 THE AUDIO DECODING BLOCK

The audio decoding block is typically machine dependent (while all the codes described up to here is mainly machine-independent). In order to test the real-time performances of the FAE, several audio decoders have been plugged into the Audio Decoder Block (from G.726 to ACELP codecs). All of them have proved to be able to work in real time with the FAE.

5.9 THE IMPLEMENTATION

The core code implementing the FAE is organized in six different libraries, as sketched in Figure 5.14:

- The *Input/Output* lib (I/O) is responsible for file I/O and visual stream decoding when necessary
- The *Mesh Animation* lib (MA) interprets the Mesh Semantics and is responsible for the actual mesh deformation
- The *Mesh Calibration* lib (MC) takes a Mesh Geometry as input and provides a calibrated mesh as output
- The *Mesh Simplification* lib (MS) transforms an input Mesh Geometry into an output simplified mesh

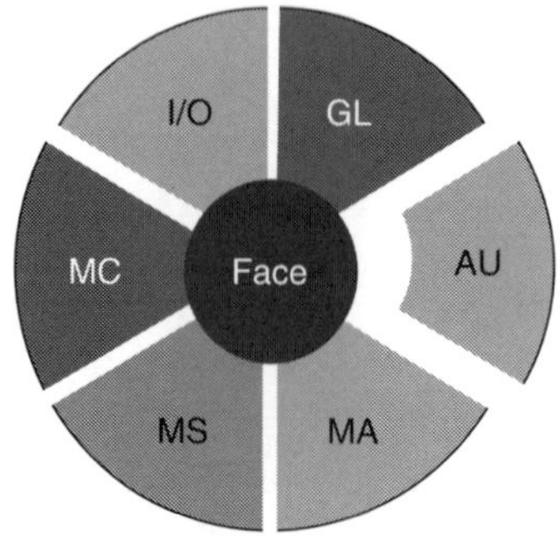

Figure 5.14 The core implementation of the FAE, with the six libraries

- The *Graphic* lib (GL) takes the scene (composed of a face with an optional static background) and renders it
- The *Audio* lib (AU) is in charge of audio decoding and rendering.

All the libraries are written in ANSI C, with the exception of the Graphic and Audio Libraries. It is worth noticing that the GL lib makes use of the OpenGL Graphic Libraries, thus allowing the porting of the code on several different platforms (so far, the FAE has been compiled on MS Windows™ platforms, Linux and Irix™ platforms.

A roughly one-to-one mapping exists between the FAE libraries and the Blocks described in Figure 5.1. All the libraries access the Face Data (i.e. Mesh Geometry and Mesh Semantics) with the exception of the Audio library. This means that all the libraries that process the face data can perform their computation in any order with respect to the other ones or can perform no computation at all. Typically, mesh animation is done on a frame-based time scale, mesh calibration is performed at the beginning of a session and mesh simplification, only if a reduction of the rendering time is necessary. It may be noticed that the animation library can still perform its task after the calibration and/or simplification libraries have modified the original Mesh Geometry.

The core of the FAE has been encapsulated in two outer layers: the Graphical User Interface (GUI)/Sync Layer and the ActiveX Control Layer, as shown in Figure 5.15. The GUI/Sync Layer is responsible for the A/V synchronization and for the User Interface. At this level, in fact, the decoding and rendering time is checked and a frame-skipping mechanism is activated in order to maintain a proper synchronization with the rendered audio. Also the GUI (menus, buttons, visualization etc.), in case of stand-alone application, is handled at this level.

Notice that the number of application programming interfaces (APIs) available from the core code is rather high, because of the number and the nature of the operations that are required.

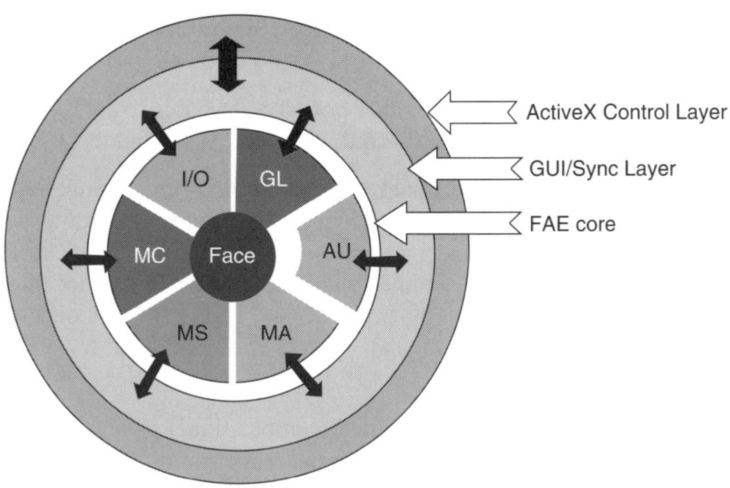

Figure 5.15 GUI/Sync Layer and ActiveX Control Layer

The ActiveX Control Layer is the outermost layer. It is used to integrate the FAE into other applications and provides the calling application a limited set of APIs. The loading of a Model or of a background, or the beginning of the reproduction of a sequence are typical commands that can be handled by this layer.

With the ActiveX Control Layer, two major applications have been implemented so far. The first application has been the integration of the FAE into the EptaPublisher software, a program developed by Eptamedia Srl that can generate facial animation sequences starting from plain text. This software relies on any commercial speech synthesizer to create the synthetic audio and uses a proprietary module to create the lip movements corresponding to the pronounced text.

The second application is called *EptaPlayer* and basically represents a plug-in through which it is possible to render facial animation sequences within a web browser. A virtual guide or assistant can therefore enrich the content of the web pages, entertaining or helping the user in the navigation.

A more detailed description of the above applications, and of several others, will be given in Chapter 16.

As a side note, the I/O library (as far as the proprietary model loading is concerned) and the MA library (as far as the mesh deformation is concerned) have been recently integrated into an MPEG-4-compliant 3-D player developed by bSoft (*www.bsoft2000.com*). The two libraries together represent an MPEG-4-compliant Facial Animation Decoder (Version 1, Visual Simple F.A. profile).

5.9.1 Performances

The FAE has been developed with the objective to obtain real-time performances, wide portability and low computational requirements.

On an MS Windows 2000™ PC the FAE requires some 6 MB of memory for a model of medium complexity like Sarah (see Figure 5.3) and the size of the executable is approximately 500 kbytes.

The code has been successfully ported on MS Windows, Linux and Irix platforms and the porting on palm-size PCs equipped with MS Windows CE™ is underway.

The rendering performances ensure real-time animation on even low-performance hardware. The Table 5.2 provides figures (including decoding and mesh animation time) of the frame rate achieved on different hardware and with different models. It is clear that even rather complex models like Pinocchio can be rendered without accelerated graphic board on entry level PCs (consider that, for most applications, 18 frame per second on a 352×288 window are enough for good perceived quality).

On hardware systems with lower computational power, the Mesh Simplification Block can be enabled in order to reduce the rendering time. It has been verified that the simplification algorithm can halve the number of polygons of a 5000 polygon model in 600ms on a Pentium II 233, decreasing the animation and rendering time by 40%.

Table 5.2 Rendering time on different PC equipped with MS Windows 2000™

Processor	Model Sarah – frame rate [fps] (3300 poly, no textures, smooth shading rendering)		Model Pinocchio – frame rate [fps] (7000 poly, 4 textures, texture mapped)	
	Window size: 352 × 288	Window size: 1024 × 768	Window size: 352 × 288	Window size: 1024 × 768
PII @ 600 MHz	28.6	8.1	16.6	6.8
Amd @ 1 GHz	45.1	12.7	24.2	9.2
Amd @ 1 GHz with graphic accelerator	60.1	60.1	40.4	32.6

REFERENCES

[1] P. Alfeld, "Scattered data interpolation in three or more variables", *Mathematical Methods in CAGD III*, in M. Daelhen, T. Lyche, L. Shumacker, eds., Academic Press, New York, 1989.

[2] P. Alfeld, "Scattered data interpolation in three or more variables", *Mathematical Methods in CAGD III*, in M. Daelhen, T. Lyche, L. Shumacker, eds., Academic Press, New York, 1989.

[3] R. Schaback, "Creating surfaces from scattered data using radial basis functions", *Mathematical Methods in CAGD III*, in M. Daelhen, T. Lyche, L. Shumacker, eds., Academic Press, New York, 1995, pp. 1–21.

[4] C. A. Micchelli, "Interpolation of scatterd data: distance matrices and conditionally positive definite functions", *Constrained Approximation*, **2**, 11–22 (1986).

[5] FBA Test Data Set, *"http://www-dsp.dist. unige.it/snhc*.

[6] F. J. Narcowich, R. Schaback, J. D. Ward, "Multilevel Interpolation and Approximation", *International Journal on Applied and Computational Harmoci Analysis*, Academic Press, New York; in press.

[7] J. Clark, E. Catmull, "Recursively generated B-spline surfaces on arbitrary topological meshes", *Computer-Aided Design*, **10**(6), 350–355 (1978).

[8] M. Garland, P. Heckbert, "Surface simplification using quadric error metrics", *Proceedings of SIGGRAPH 97*, August, 1997, pp. 209–216.

[9] F. Lavagetto, R. Pockaj, "The facial animation engine: towards a high-level interface for the design of MPEG-4 compliant animated faces", *IEEE Transactions on Circuits and Systems for Video Technology*, **9**(2), 277–289 (1999).

[10] H. Hoppe, "New quadric metric for simplifying meshes with appearance attributes", *IEEE Visualization 1999*, October, 1999, pp. 59–66.

[11] H. Hoppe, "Progressive mesh", *Proceedings of SIGGRAPH '96*, August, 1996, pp. 99–108.

[12] M. Garland, Quadric-based Polynomial Surface Simplification, Ph.D. thesis, Carnegie Mellon University, Pittsburgh, Pa., 1999.

[13] F. Lavagetto, R. Pockaj, "An efficient use of MPEG-4 FAP interpolation for facial animation at 70 bit/frame", *IEEE Transactions on Circuits and Systems for Video Technology"*, **11**(10), 1085–1097 (2001).

6

Extracting MPEG-4 FAPs from Video

Jörgen Ahlberg

Linköping University, Linköping, Sweden

6.1 INTRODUCTION

The encoding of Facial Animation Parameters (FAPs) from a video sequence consists of two processes: the extraction of the parameters from the video (i.e. to understand what to encode) and then the encoding itself (i.e. the compression). The encoding itself in MPEG-4 is mostly a matter of choice of compression parameters; the compression techniques used are described in Chapter 2. The extraction of facial parameters from video, on the other hand, is not a completely solved problem. The related problems of face detection, identification, recognition and tracking are subject to research in a large number of universities and companies and have been an active research topic for decades.

In this chapter, we present a system for finding and tracking a face and extract animation parameters from video. The system uses an initial color-processing step for finding a rough estimate of the position, size and in-plane rotation of the face, followed by a refinement step driven by the active appearance model (AAM) search algorithm. The latter step refines the previous estimate, and also extracts local animation parameters. The system is able to track the face and some facial features in near real time, and can compress the result to a bit stream compliant to MPEG-4 face and body animation (FBA).

A more general overview of different kinds of face tracking systems is given in Chapter 11.

6.2 METHODS FOR DETECTION AND TRACKING OF FACES

May be the most intuitive and simple way of finding a face in an image is to look for some characteristic features of the human face. Such features could be color, shape, motion or combinations thereof. Feature-based methods have the advantage that fast implementations are possible and the disadvantage that they are not very robust.

Appearance-based methods regard the face as a pattern in the image, and try to model its appearance by a statistical model, for example, eigenfaces. Eigenfaces have been used for face recognition and face detection for some years, and the main problems have turned out to be nonlinearity of the face space and (for certain applications) the computational complexity.

Several methods for handling the nonlinearity have been proposed, including geometrical normalization [1], isometric feature mappings (Isomap) [2], local linear embedding (LLE) [3], and neural networks and support vector machines (SVM) [4, 5]. In our work, we use geometrical normalization to create a shape-free image, as described later.

To handle the computational complexity problem, fast search algorithms are needed. Two examples are simulated annealing (SA) [6] and the AAM search algorithm. AAM search will be briefly described in this chapter, since it is fundamental to our current system.

Face tracking is a special case of 3D object tracking, which of course has been subject to heavy research because of its many applications. Examples of current research can be found in [5, 7, 8], as well as in Chapters 11 and 15 in this book.

6.3 ACTIVE AND STATISTICAL MODELS OF FACES

Deformable models describing the shape of a class of objects (for example, faces) have been in use for some time, evolving from the snakes or active contour models introduced in 1988 [9] to statistical models such as *point distribution models* (PDMs) and *active shape models* (ASMs) [10, 11]. The shape of a statistical shape model is typically controlled by adding a linear combination of shape/deformation modes to an average shape $\bar{s}$ according to

$$s = \bar{s} + S\sigma$$

where σ is a vector of shape/deformation parameters, and the columns of S contain the deformation modes. The resulting vector s contains the 2D or 3D coordinates of the model's control points, edge points or vertices.

In parallel, statistical models describing the texture of objects have been developed, starting with Turk and Pentland introducing eigenfaces for face recognition in 1991 [12]. The main problem with the eigenfaces soon turned out to be that the face distribution in the image space is highly nonconvex and complex and thus it is not well modeled by a linear basis. Two popular ways of handling this problem are to create either a nonlinear model or geometrically normalized eigenfaces, introduced by Ström *et al.* who used it for image compression [1]. Geometrically normalized eigenfaces are also called *shape-free eigenfaces* or *eigentextures* or *texture modes*. The parameterized (and geometrically normalized) texture of a model is described by

$$x = \bar{x} + X\xi$$

where the texture parameters ξ put weights on the texture modes (columns of X) and $\bar{x}$ is the average texture.

Appearance models [13] combine shape and texture into one statistical model, in which a number of appearance modes are controlled by the model parameters. Such a model would consequently be expressed as

$$s = \bar{s} + A_s p$$

$$x = \bar{x} + A_x p$$

that is, the same parameters control shape and texture.

In 1998 the AAMs were introduced by Edwards *et al.* [13], by combining an appearance model with a fast search algorithm for matching the model to an image. The AAMs have been the target for quite some research since and extensions include view-based face spaces [14], modeling the aging of faces [15], face detection [16] and constrained AAM search [17] to pick a few. A comparison between ASMs and AAMs can be found in Reference 18 and a combination in Reference 19. For an introduction to AAMs, see Reference 20. A more in-depth treatment can be found in Reference 21. The AAM search algorithm is briefly described below.

6.3.1 The Active Appearance Model Search Algorithm

Given an input image i and a model parameter vector p, we can map the image onto the model, reshape the model to the standard shape, and thus create a normalized input image

$$j = j(i, s(p))$$

From now on we fix the input image and regard j as a function of p only. We compute the residual image as

$$r(p) = j(p) - x(p)$$

and our goal is to find the optimal adaptation of the model to the input image, that is, to find the p that minimizes the error measure

$$\mathrm{e}(p) = \|r(p)\|^2$$

Assuming that we have a rough estimate of the model parameters, we would like to find the update vector Δp that minimizes $e(p + \Delta p)$. Following Reference 20, we Taylor-expand r around $p + \Delta p$ getting

$$r(p + \Delta p) = r(p) + D\Delta p + O(\Delta p)$$

where

$$D = \frac{\partial}{\partial p} r(p)$$

We approximate $r(p)$ with the first terms and regard it as a linear function. Thus, we want to minimize

$$e(p + \Delta p) = \|r(p) + D\Delta p\|^2$$

of which the least square solution is

$$\Delta p = -(D^T D)^{-1} D r(p)$$

The vector Δp gives us a probable update in the search space, and we compute a new parameter vector and a new error measure:

$$p' = p + \Delta p$$
$$e' = e(p')$$

If $e' < e$, we update p accordingly ($p' \to p$) and iterate until convergence. If $e' > e$, we try smaller update steps (0.5 and 0.25). If neither of these improves the error measure, we declare convergence.

6.3.2 Training for Active Appearance Model Search

From a set of training data (models adapted to images) we can estimate the gradient matrix D, and from the estimated D compute the update matrix U as the negative pseudoinverse of D:

$$U = -D^* = -(D^T D)^{-1} D^T$$

To be able to use the AAM search, we should consequently estimate the gradient matrix D. We do that by perturbating p from the set of (manually) adapted models, parameter by parameter, step by step. The jth row in D can thus be estimated as

$$D_j = \sum_k [r(p + \Delta p_{jk}) - r(p)]$$

where Δp_{jk} is a vector that perturbs p in the jth component to the amount of $k \cdot c$ for some suitable constant c.

6.4 AN ACTIVE MODEL FOR FACE TRACKING

The active model used here is a simplification of the AAM, and we will here describe the model and how it is parameterized. As a starting point, we use the wireframe face model Candide, which has been popular in video coding research for many years. The third variant of this model, Candide-3 [22], is also compliant to MPEG-4 Face Animation.

The Candide model has manually been adapted to a set of images by varying a set of 12 parameters:

- The first six parameters are the global motion (or pose) parameters; 3D rotation, 2D translation and scale. We call the pose parameter vector $\pi = [r_x, r_y, r_z, t_x, t_y, z]^T$.
- The remaining parameters are activation levels for the six Action Units (AUs) (from FACS, the Facial Action Coding System [23]) controlling the lips and eyebrows. They are contained in the vector sigma, and control the shape of the head according to $s = \bar{s} + S\sigma$.

We collect those parameters in a 12D vector $p^T = [\pi^T, \sigma^T]$, which parameterizes the geometry of the model. Thus, the geometry is described by

$$g(p) = R(z + 1)(\bar{s} + S\sigma) + \tau$$

where the columns of S are AU definitions from Candide-3, σ is the vector of AU activation levels, $\bar{s}$ is the standard shape of the Candide model, $\tau = \tau(t_x, t_y)$ is a function of the x and y translations, $R = R(r_x, r_y, r_z)$ is a rotation matrix created from the three rotation parameters and z is a scaling factor.

The image under the model has, for each image in the training set, been mapped onto the model, creating a texture-mapped wireframe model. The model has then been normalized to a standard shape, size and position (i.e. $p = 0 \Rightarrow g = \bar{s}$), in order to collect a geometrically normalized set of textures. On this set, a principal component analysis (PCA) has been performed and the texture modes (geometrically normalized eigenfaces) have been computed.

We can now describe the complete appearance of the model by the geometry parameters p and an N-dimensional texture parameter vector ξ, where N is the number of texture modes we want to use for synthesizing the model texture.

6.4.1 Analysis – Synthesis

As in Section 6.4 we can create the normalized input image $j(p)$. The texture parameters minimizing the model error $e(p)$ are then given by projecting the normalized input image $j(p)$ on the texture modes, that is,

$$\xi = X^T[j(p) - \bar{x}]$$

Thus, p is the only necessary parameter in our case:

$$g = g(p)$$
$$x = x(p)$$

where

$$x(p) = \bar{x} + XX^T[j(p) - \bar{x}]$$

The entire process from input image to normalized image is illustrated in Figure 6.1.

Note that our model is not a complete appearance model in the sense of Reference 13, since we parameterize the geometry only and let the texture depend on the input image as well. For appearance models, a PCA is performed to find the suitable subspace of appearance modes combining deformation modes and texture modes. In our application, we only parameterize the model in terms of deformation (including global motion) since we know in advance what kind of parameters we are interested in extracting. If we want to extract AUs (the parameters typically used for Candide) we simply parameterize and train our model on those parameters (or deformations spanning the same subspace). We can still use the AAM search exactly as described above.

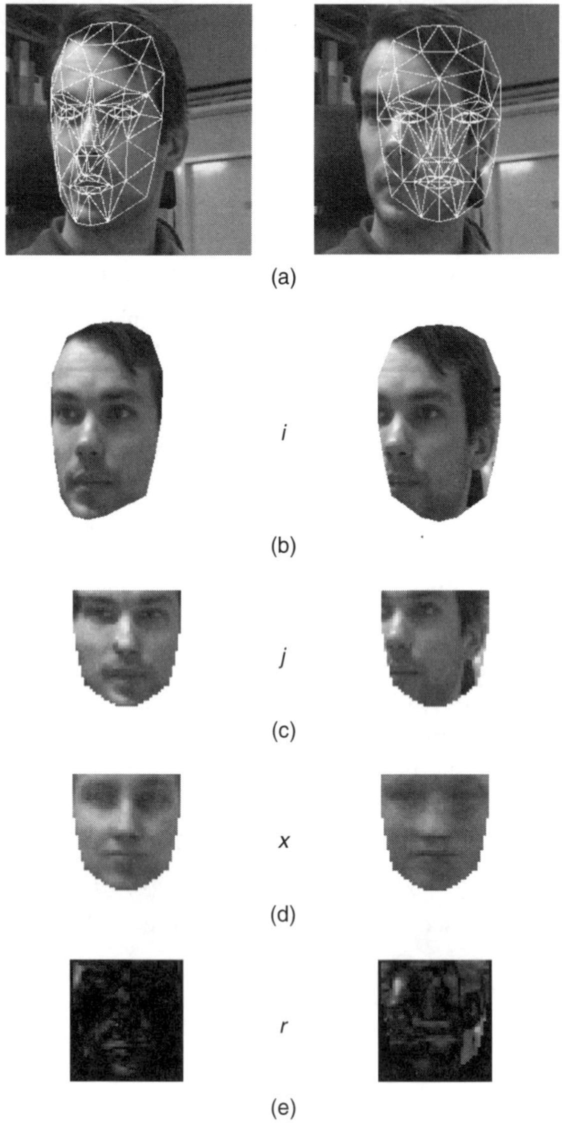

Figure 6.1 The model matching and texture approximation process. (a) A good and a bad (left and right column, respectively) model adaptation is shown; (b) the image is mapped onto the model; (c) the model is reshaped to the standard shape, producing the image j; (d) the normalized texture is approximated by the texture modes, producing the image x; (e) the residual image r is computed. The better the model adaptation, more similar are the images j and x

6.4.2 Collecting Training Data

To try out this scheme, the model has been adapted (manually in the beginning, then semiautomatically) to more than three hundred images of six different persons from different angles and with different facial expressions. The following AUs from the

FACS have been chosen as deformation parameters: *Jaw drop*, *Lip stretcher*, *Lip corner depressor*, *Upper lip raiser*, *Eyebrow lowerer* and *Outer eyebrow raiser*.

The adapted model has for each image been normalized to a standard shape with the size 40×42 (or 80×84) pixels, as shown in Figure 6.1c, and a PCA has been performed on the resulting training textures to compute the texture modes. The mean texture and the DC-level have been subtracted from the training textures prior to the PCA.

With the texture modes available, all the parameters have been perturbed, one by one and for each image, in several small steps, and the matrix D estimated. From D, the update matrix U has been computed.

6.4.3 Tracking a Face with the Active Model

The AAM search efficiently adapts the model to the image provided that the initial estimate of the model parameters is good enough. The better the initial estimate is, the smaller the risk of the AAM search getting stuck in a local minimum not corresponding to the correct model adaptation. In a video sequence, the changes between each frame are quite small (provided that the frame rate is high enough), so the adaptation from the previous frame can be used as the initial estimate. However, for the first frame, some other technique has to been used, as discussed in the next section.

6.5 THE COLOR-BASED FACE-FINDING ALGORITHM

To quickly find the approximate location of the face in the input image, the pixels are traversed and each given a likelihood value of being a skin colored pixel. This likelihood value is based on *à priori* collected statistics, to which a mixture of Gaussian distributions has been adapted using the Expectation-Maximization (EM) algorithm. The resulting image of likelihood values is blurred and then thresholded and of the remaining objects in the image, the largest one is selected as the most probable face candidate. The position, size and orientation of this 'blob' are used as the initial estimate handed over to the refinement step. Examples of resulting estimates are shown in Figure 6.2.

Figure 6.2 Examples of the color-based algorithm giving a rough initial estimate of the face location

This kind of algorithm has been chosen because it is fast and simple. The obvious drawback is that it needs recalibration for each camera and for differing lighting conditions.

6.6 IMPLEMENTATION

We have implemented a C++ library with routines for handling (active) face models and training them for face tracking. The implementation uses OpenGL for the texture mapping in the AAM search, utilizing the fact that modern graphics cards have specialized hardware for such tasks. The geometrical normalization of the input image (see Figure 6.1c) is thus performed in a very short time (less than 6 ms), and the speed of the algorithm is dependent more on the graphics bus than on the CPU.

The shape of the texture modes is determined by the standard shape of the Candide model (with the upper part of the head removed) scaled so that the size is 40×42 pixels (see Figure 6.1c,d). The texture modes and the update matrix U have been computed in color and in grayscale for comparison.

Since the vertices of the Candide-3 model correspond well to the facial feature points defined in MPEG-4, the extracted parameters can easily be converted to FAPs, described elsewhere in this book. The FAPs are measured in face-dependent scales, using different FAP Units (FAPUs). The FAPs are also measured relative to the neutral face, and thus a neutral face model is kept in memory. Using this neutral face model, the FAPUs and the FAPs are computed and then compressed using the MPEG-4 reference software. The entire process takes only about five milliseconds per frame and does not influence the real-time performance. The output is an MPEG-4-compliant bit stream that can be played in any MPEG-4 FBA player, for example, the web-based player described in Chapter 4 or the Facial Animation Engine described in Chapter 5.

6.7 RESULTS

The experiments presented here are performed on a PC with a 500-MHz Intel Pentium III processor and an ASUS V3800 graphics card with video input. The color-based algorithm runs in approximately 0.1 seconds, and the Active Appearance Search needs about 15 ms per iteration. Typically, less than 10 iterations are needed for each frame, and the closer the initial estimate is to the optimum the fewer the iterations needed. Thus, if a video sequence is recorded at a high frame rate (with small motion between each frame), the tracking will also run on a higher speed. Visual results are shown in Figure 6.3.

Using grayscale texture modes and update data, it turned out that the computation in the graphics card (which internally uses RGB) became almost 20% slower. However, the computations performed in the CPU became (as expected) about three times faster, and then only 20% of the total computing time is due to the CPU (the rest being computations in the graphics card).

Testing on a video sequence of a few hundred frames gave results according to Table 6.1 below. It is clear that the grayscale computations are preferable, since the visual results are equivalent.

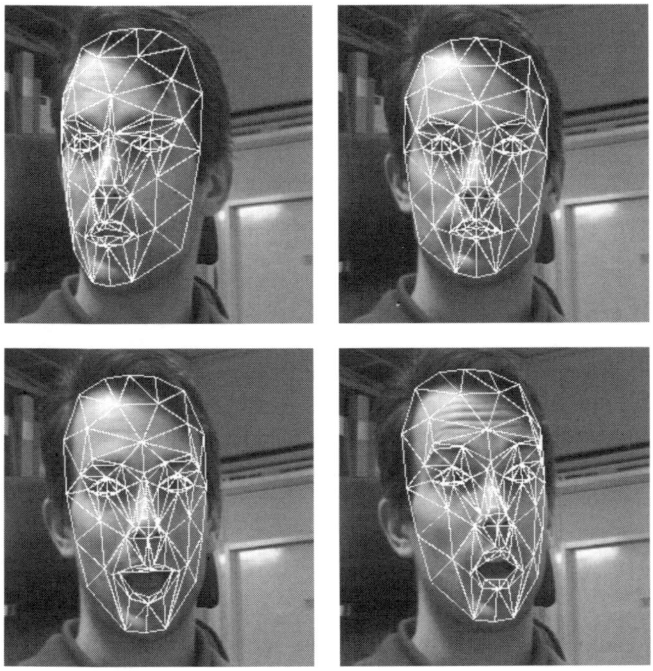

Figure 6.3 The model adapted to four frames of a video sequence

Table 6.1 Timing results (average over 341 frames)

Measurement	RGB	Grayscale
Iterations per frame	6.9	6.8
Total time per frame (ms)	94.1	69.1
Time per iteration (ms)	13.6	10.2
Time for computing Δp (ms)	7.2	5.05

6.8 IMPROVEMENTS

There are several ways that the system as it is described in this chapter can be improved. Four things to be considered are mentioned here:

First, the color-based algorithm is not robust enough, and should be complemented with some more simple and fast technique. For example, we could require that an area could be a face candidate only if there is some difference (due to motion) between the first and second frame.

Second, the computation time could be improved by precomputation of destination coordinates in the image warping process. However, this fact cannot be exploited by the graphics card, and the entire image warping would then need to be performed by the CPU. Using the vector instructions of modern CPUs, this might be faster than using the graphics card anyway.

Third, as all tracking systems, this system can lose track and therefore, some kind of reinitialization scheme is needed. One possible procedure is to invoke the color-based algorithm when the Active Model does not converge to a small error measure, giving a new initial estimate to the Active Model.

Fourth, our current model is a simplification of the AAM, in order to achieve real-time performance. The model should be extended to a full AAM and compared to the current one in terms of accuracy, speed and robustness.

6.9 CONCLUSION

We have presented a system that tracks a face and facial features in a video sequence. The resulting animation data is encoded using MPEG-4 Face Animation. The system works in near real time, and the experimental results are promising. With some further development and optimization, it should be possible to implement a real-time 3D face and facial feature tracker on consumer hardware.

ACKNOWLEDGMENT

This research was funded by the European Fifth Framework IST project InterFace and the Swedish SSF project VISIT (VISual Information Technology)

REFERENCES

[1] J. Ström, F. Davoine, J. Ahlberg, H. Li, R. Forchheimer, "Very low bit rate facial texture coding", *Proceedings of International Workshop on Synthetic/Natural Hybrid Coding and 3D Imaging*, Rhodes, Greece, 1997, pp. 237–240.

[2] J. B. Tenenbaum, V. de Silva, J. C. Langford, "A global geometric framework, for nonlinear dimensionality reduction", *Science*, **290**(5500), 2319–2323 (2000).

[3] S. Roweis, L. Saul, "Nonlinear dimensionality reduction by local linear embedding", *Science*, **290**(5500), 2323–2326 (2000).

[4] H. Wechsler, P. J. Phillips, V. Bruce, F. Fogelman Souilé, T. S. Huang, eds., *Face Recognition – From Theory to Applications*, *NATO ASI Series F: Computer and Systems Sciences*, Vol. 163, Springer-Verlag, Berlin, Heidelberg, 1998.

[5] S. Gong, S. J. McKenna, A. Psarrou, *Dynamic Vision – From Images to Face Recognition*, Imperial College Press, London, 2000.

[6] J. Ahlberg, Extraction and Coding of Face Model Parameters, Thesis No. 747, LIU-TEK-LIC-1999: 05, Department of Electrical Engineering, Linköping University, Sweden, 1999.

[7] S. Gokturk, J.-Y. Bouget, R. Grzeszczuk, A data-driven model for monocular face tracking, *Proceedings of 8th International Conference on Computer Vision*, Vancouver, Canada, July, 2001.

[8] M. La Cascia, S. Sclaroff, V. Athitsos, "Fast, reliable head tracking under varying illumination: an approach based on registration of texture-mapped 3D models", *Transactions on Pattern Analysis and Machine Intelligence*, **22**(4), 322–336, (2000).

[9] M. Kass, A. Witkin, D. Terzopoulos, "Snakes: active contour models", *International Journal of Computer Vision*, **1**, 321–331 (1988).

[10] T. F. Cootes, C. J. Taylor, "Active shape models – 'smart snakes' ", *Proceedings of British Machine Vision Conference*, Springer-Verlag, Leeds, UK, 1992, pp. 266–275.

[11] M. Sonka, V. Hlavac, R. Boyle, *Image Processing, Analysis, and Machine Vision*, 2nd ed., PWS Publishing, Boston, Mass., 1999.

[12] M. Turk, A. Pentland, "Eigenfaces for recognition", *Journal of Cognitive Neuroscience*, **3**(1), 71–86 (1991).

[13] G. J. Edwards, T. F. Cootes, C. J. Taylor, "Interpreting face images using active appearance models", *Proceedings of 3rd International Conference on Automatic Face and Gesture Recognition*, IEEE Computer Society Press, Nara, Japan, 1998, pp. 300–305.

[14] T. F. Cootes, K. N. Walker, C. J. Taylor, "View-based active appearance models", *Proceedings of International Conference on Face and Gesture Recognition*, Grenoble, France, 2000, pp. 227–232.

[15] A. Lanitis, C. J. Taylor, T. F. Cootes, "Modeling the process of ageing in face images", *Proceedings of International Conference on Computer Vision*, Vol. 1, IEEE Computer Society Press, Kerkyra, Greece, 1999, pp. 131–136.

[16] G. J. Edwards, T. F. Cootes, C. J. Taylor, "Advances in active appearance models", *Proceedings of International Conference on Computer Vision*, Vol. 1, IEEE Computer Society Press, Kerkyra, Greece, 1999, pp. 137–142.

[17] T. F. Cootes, C. J. Taylor, "Constrained active appearance models", *Proceedings of International Conference on Computer Vision*, Vol. I, IEEE Computer Society Press, Vancouver, Canada, 2001, pp. 748–754.

[18] T. F. Cootes, G. J. Edwards, C. J. Taylor, "Comparing active shape models with active appearance models", *Proceedings of British Machine Vision Conference*, Vol. 1, 1999, pp. 173–182.

[19] S. Mitchell, B. Lelieveldt, R. van der Geest, J. Shaap, J. Reiber, M. Sonka, "Segmentation of cardiac MR images: an active appearance model approach", *Proceedings SPIE Medical Imaging*, Vol. 3979, San Deigo, Calif., 2000, pp. 224–234.

[20] T. F. Cootes, G. J. Edwards, C. J. Taylor. "Active appearance models", *IEEE Transactions on Pattern Analysis and Machine Intelligence*, **23**(6), 681–685 (2001).

[21] T. F. Cootes, C. J. Taylor, Statistical Models of Appearance for Computer Vision, Draft report, Wolfson Image Analysis Unit, University of Manchester, October, 2001, *http://www.wiau.man.ac.uk*.

[22] J. Ahlberg, CANDIDE-3 - An Updated Parameterized Face, Report No. LiTH-ISY-R-2326, Department of Electrical Engineering, Linköping University, Sweden, 2001.

[23] P. Ekman, W. V. Friesen, *Facial Action Coding System*, Consulting Psychologist Press, Palo Alto, Calif., 1977.

7

Real-Time Speech-Driven Face Animation

Pengyu Hong, Zhen Wen and Thomas S. Huang

University of Illinois at Urbana-Champaign, Urbana, USA

ABSTRACT

This chapter presents our research on real-time speech-driven face animation. First, a visual representation, called *Motion Unit* (MU), for facial deformation is learned from a set of labeled face deformation data. A facial deformation can be approximated by a linear combination of MUs weighted by the corresponding MU parameters (MUPs), which are used as the visual features of facial deformations. MUs explore the correlation among those facial feature points (FP) used by the MPEG-4 Face Animation (FA) to describe facial deformations caused by speech producing. MU-based FA is compatible with MPEG-4 FA. We then collect a set of audiovisual (AV) training database and use the training database to train a real-time audio-to-visual mapping (AVM).

7.1 INTRODUCTION

Speech-driven face animation takes advantage of the correlation between speech and facial coarticulation. It takes speech stream as input and outputs corresponding FA sequences. Therefore, speech-driven face animation only requires very low bandwidth for 'face-to-face' communications. The AVM is one of the main research issues of speech-driven face animation. First, the audio features of the raw speech signals are calculated. Then, the AVM maps the audio features to the visual features that describe how the face model should be deformed.

Some speech-driven face animation approaches use phonemes or words as intermediate representations. Lewis [1] used linear prediction to recognize phonemes. The recognized phonemes are associated with mouth shapes that provide keyframes for FA. Video Rewrite [2] trains hidden Markov models (HMMs) [3] to automatically label phonemes in both the training audio tracks and the new audio tracks. It models short-term mouth coarticulation within the duration of triphones. The mouth image sequence of a new audio track is generated by reordering the mouth images selected from the training footage. Video Rewrite is an off-line approach. It requires a very large training database to cover all possible cases of triphones and needs large computational resources. Chen and Rao [4] train HMMs to parse the audio feature vector sequences of isolated words into state sequences. The state probability for each audio frame is evaluated by the trained HMMs. A visual feature is estimated for every possible state of each audio frame. The estimated visual features of all states are then weighted by the corresponding probabilities to obtain the final visual features, which are used for lip animation.

Voice Puppetry [5] trains HMMs for modeling the probability distribution over the manifold of possible facial motions from audio streams. This approach first estimates the probabilities of the visual state sequence for a new speech stream. A closed-form solution for the optimal result is derived to determine the most probable series of facial control parameters, given the boundary (the beginning and ending frames) values of the parameters and the visual probabilities. An advantage of this approach is that it does not require recognizing speech into high-level meaningful symbols (e.g. phonemes, words), with which it is very difficult to obtain a high recognition rate. However, the speech-driven face animation approaches mentioned in References 2, 4, 5 have relatively long time delays.

Some approaches attempt to generate the lip shapes using one audio frame via vector quantization [6], affine transformation [7], Gaussian mixture model [8] or artificial neural networks [9, 10]. Vector quantization [6] first classifies the audio feature into one of a number of classes. Each class is then mapped to a corresponding visual feature. Though it is computationally efficient, the vector quantization approach often leads to discontinuous mapping results. The affine transformation approach [7] maps an audio feature to a visual feature by a simple linear matrix operation. The Gaussian mixture approach [8] models the joint probability distribution of the audiovisual vectors as a Gaussian mixture. Each Gaussian mixture component generates an estimation of the visual feature for an audio feature. The estimations of all the mixture components are then weighted to produce the final estimation of the visual feature. The Gaussian mixture approach produces smoother results than the vector quantization approach does. In Reference 10, Morishima and Harashima trained a multilayer perceptron (MLP) to map the Linear Predictive Coding (LPC) Cepstrum coefficients of each speech frame to the mouth-shape parameters of five vowels. Kshirsagar and Magnenat-Thalmann [9] trained a MLP to classify each speech segment into the classes of vowels. Each vowel is associated with a mouth shape. The average energy of the speech segment is then used to modulate the lip shapes of the recognized vowels.

However, those approaches proposed in References 6 to 10 do not consider the audio contextual information, which is very important for modeling mouth coarticulation during speech producing. Many approaches have been proposed to train neural networks as AVMs while taking into account the audio contextual information. Massaro *et al.* [11] trained an MLP as the AVM. They modeled the mouth coarticulation by considering the speech context information of eleven consecutive speech frames (five backward, current, and five forward frames). Lavagetto [12] and Curinga *et al.* [13]

train time-delay neural networks (TDNNs) to map the LPC cepstral coefficients of speech signals to lip animation parameters. TDNN is a special case of MLP and it considers the contextual information by imposing ordinary time delay on the information units. Nevertheless, the neural networks used in References 11 to 13 have a large number of hidden units in order to handle large vocabulary. Therefore, their training phrases face very large searching space and have very high computational complexity.

7.2 MOTION UNITS – THE VISUAL REPRESENTATION

MPEG-4 FA standard defines 68 MPEG-4 FAPs (facial animation parameters). Among them, two are high-level parameters, which specify visemes and expressions. The others are low-level parameters that describe the movements of sparse FP defined on head, tongue, eyes, mouth and ears. MPEG-4 FAPs do not specify detail spatial information of facial deformation. The user needs to define the method to animate the rest of the face model. MPEG-4 FAPs do not encode the information about the correlation among facial FP. The user may assign some values to the MPEG-4 FAPs that do not correspond to natural facial deformations.

We are interested in investigating natural facial movements caused by speech producing and the relations among those facial FP in MPEG-4 standard. We first learn a set of MUs from real facial deformations to characterize natural facial deformations during speech producing. We assume that any facial deformation can be approximated by a linear combination of MUs. Principal Component Analysis (PCA) [14] is applied to learning the significant characteristics of the facial deformation samples. MUs are related to the works in References 15, 16.

We put 62 markers in the lower face of the subject (Figure 7.1). Those markers cover the facial FP that are defined by the MPEG-4 FA standard to describe the movements of the cheeks and the lips. The number of the markers decides the representation capacity of the MUs. More markers enable the MUs to encode more detailed information. Depending on the need of the system, the user can flexibly decide the number of the markers. Here, we only focus on the lower face because the movements of the upper face are not closely related to speech producing. Currently, we only deal with 2-D deformations of the lower face. However, the method described in this chapter can be applied to the whole face as well as the 3-D facial movements if the training data of 3-D facial deformations are available. To handle the global movement of the face, we add three additional markers. Two of them are on the glasses of the subject. The third one is on the nose. These three markers mainly have rigid movements and we can use them to align the data. A mesh is created on the basis of these markers for visualization purpose. The mesh is shown to overlap with the markers in Figure 7.1.

We capture the front view of the subject while he is pronouncing all English phonemes. The subject is asked to stabilize his head as much as possible. The video is digitized at 30 frames per second. Hence, we have more than 1000 image frames. The

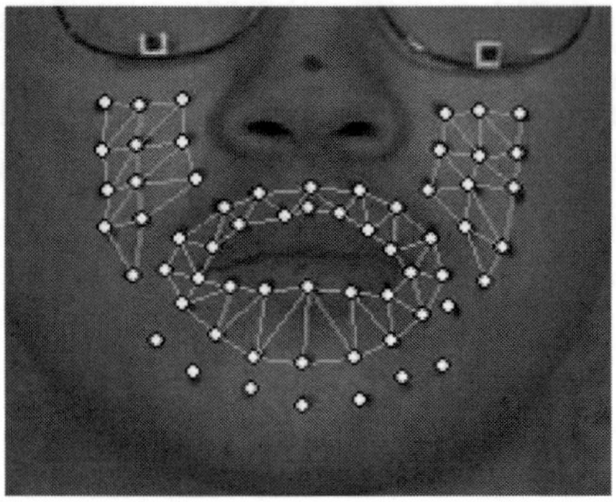

Figure 7.1 The markers and the mesh

markers are automatically tracked by template matching. A graphic interactive interface is developed for manually correcting the positions of trackers using the mouse when the template matching fails as a result of large facial motions. To achieve a balanced representation on facial deformations, we manually select facial shapes from these samples of image frames so that each viseme and the transitions between each pair of visemes are nearly evenly represented. To compensate the global face motion, the tracking results are aligned by affine transformations defined by those three additional markers.

After normalization, we calculate the deformations of the markers with respect to the positions of the markers in the neutral face. The deformations of the markers at each time frame are concatenated to form a vector. PCA is applied to the selected facial deformation data. The mean facial deformation and the first seven eigenvectors of the PCA results, which correspond to the largest seven eigenvalues, are selected as the MUs in our experiments. The MUs are represented as $\{\vec{m}_i\}_{i=0}^{M}$. Hence, we have

$$\vec{s} = \vec{m}_0 + \sum_{i=1}^{M} c_i \vec{m}_i + \vec{s}_0 \tag{7.1}$$

where $\vec{s}_0$ is the facial shape without deformations and $\{c_k\}_{k=1}^{M}$ is the MUP set. The first four MUs are shown in Figure 7.2. They represent the mean deformation and the local deformations around cheeks, lips and mouth corners, respectively.

MUs are also used to derive robust face and facial motion tracking algorithms [17]. In this chapter, we are only interested in real-time speech-driven face animation.

(a) $\vec{s}_0 + \vec{m}_0$ (b) $\vec{s}_0 + k\vec{m}_1$ (c) $\vec{s}_0 + k\vec{m}_2$ (d) $\vec{s}_0 + k\vec{m}_3$

Figure 7.2 Motion units. $k = \|\vec{m}_0\|$

7.3 MUPS AND MPEG-4 FAPS

It can be shown that the conversion between the MUPs and the low-level MPEG-4 FAPs is linear. If the values of the MUPs are known, the facial deformation can be calculated using Equation (7.1). Consequently, the movements of facial features in the lower face used by MPEG-4 FAPs can be calculated because MUs cover the FP in the lower face defined by the MPEG-4 standard. It is then straightforward to calculate the values of MPEG-4 FAPs.

If the values of MPEG-4 FAPs are known, we can calculate the MUPs in the following way. First, the movements of the facial features are calculated. The concatenation of the facial feature movements forms a vector $\vec{p}$. Then, we can form a set of vectors, for instance, $\{\vec{f}_0, \vec{f}_1, \ldots, \vec{f}_M\}$, by extracting the elements that correspond to those facial features from the MU set $\{\vec{m}_0, \vec{m}_1, \ldots, \vec{m}_M\}$. The vector elements of $\{\vec{f}_0, \vec{f}_1, \ldots, \vec{f}_M\}$ and those of $\vec{p}$ are arranged so that the information about the deformations of the facial FP is represented in the same order. The MUPs can be then calculated by

$$\begin{bmatrix} c_1 \\ \vdots \\ c_M \end{bmatrix} = (F^T F)^{-1} F^T (\vec{p} - \vec{f}_0) \tag{7.2}$$

where $F = [\vec{f}_1 \ \vec{f}_2 \ \cdots \ \vec{f}_M]$.

The low-level parameters of MPEG-4 FAPs only describe the movements of the facial features and lack detailed spatial information to animate the whole face model. MUs are learned from real facial deformations, which are collected so that they provide the dense information about facial deformations. MUs capture the second-order statistic information about the facial deformation and encode the correlation information of the movements of the facial FP.

7.4 REAL-TIME AUDIO-TO-MUP MAPPING

The nonlinear relation between audio features and the visual features is complicated, and there is no existing analytic expression that can be used to fully describe the relation. MLP, as a universal nonlinear function approximator, has been used to learn the nonlinear AVMs [9–11]. We also train MLPs as an AVM. Unlike other works using MLPs, we divide the AV training data into 44 subsets. An MLP is trained to estimate MUPs from audio features using each AV training subset.

The audio features in each group are modeled as a Gaussian model. Each AV data pair is classified into one of the 44 groups whose Gaussian model gives the highest score for the audio component of the AV data. We set the MLPs as three-layer perceptrons. The inputs of an MLP are the audio feature vectors of seven consecutive speech frames (3 backward, current and 3 forward time windows). The output of the MLP is the visual feature vector of the current frame. We use the error backpropagation algorithm to train the MLPs using each Action Unit's (AU) training subset separately. In the estimation phase, an audio feature vector is first classified into one of the 44 groups. The corresponding MLP is selected to estimate the MUPs for the audio feature vector. By dividing the data into 44 groups, lower computational complexity is achieved. In our experiments, the maximum number of the hidden units used in the three-layer perceptrons is only 25 and the minimum number of the hidden units is 15. Therefore, both training and estimation have very low computational complexity. A method using the triangular average window is used to smooth the jerky mapping results.

7.5 EXPERIMENTAL RESULTS

We videotape the front view of the same subject as the one in Section 7.2 while he is reading a text corpus. The text corpus consists of one hundred sentences that are selected from the text corpus of the DARPA TIMIT speech database. Both the audio and video are digitized at 30 frames per second. The sampling rate of the audio is 44.1 kz. The audio feature vector of each audio frame is its 10 Mel-Frequency Cepstrum Coefficients (MFCC) [18]. The facial deformations are converted into MUPs. Overall, we have 19 532 AV samples in the training database. 80% of the data is used for training. The rest of the data is used for test.

We reconstruct the displacements of the markers using MUs and the estimated MUPs. The evaluations are based on the ground truth of the displacements and the reconstructed displacements. To normalize the displacements of each marker, we divide the displacements by the maximum absolute ground truth displacement of the marker. We calculate the Pearson product–moment correlation coefficient and the related standard deviations using the normalized displacements. The Pearson product–moment correlation coefficient between the ground truth and the estimated data is

$$R = \frac{\mathrm{tr}\{E[(\vec{d}-\vec{\mu})(\vec{d}'-\vec{\mu}')^T]\}}{\sqrt{\mathrm{tr}\{E[(\vec{d}-\vec{\mu})(\vec{d}-\vec{\mu})^T]\}\mathrm{tr}\{E[(\vec{d}'-\vec{\mu}')(\vec{d}'-\vec{\mu}')^T]\}}} \tag{7.3}$$

where $\vec{d}$ is the ground truth, $\vec{\mu}=E(\vec{d})$, $\vec{d}'$ is the estimation result, and $\vec{\mu}'=E(\vec{d}')$. The average standard deviations are also calculated as

$$\begin{aligned} \nu_{\vec{d}} &= \frac{\sum_{r=1}^{\gamma}(C_{\vec{d}}[r][r])^{1/2}}{\gamma} \\ \nu_{\vec{d}'} &= \frac{\sum_{r=1}^{\gamma}(C_{\vec{d}'}[r][r])^{1/2}}{\gamma} \end{aligned} \tag{7.4}$$

where $C_{\vec{d}}=E[(\vec{d}-\vec{\mu})(\vec{d}-\vec{\mu})^T]$ and $C_{\vec{d}'}=E[(\vec{d}'-\vec{\mu}')(\vec{d}'-\vec{\mu}')^T]$. The Pearson product–moment correlation and the average standard deviations measure how good the global match between the shapes of two signal sequences is. The value range of the Pearson correlation coefficient is [0 1]. The larger the Pearson correlation coefficient, the better the estimated signal sequence matches with the original signal sequence. The mean-square errors (MSEs) are also calculated. The results are shown in Table 7.1.

Figure 7.3 illustrates the estimated MUPs of a randomly selected testing audio track. The content of the audio track is 'Stimulating discussions keep students' attention'. The figure shows the trajectories of the values of four MUPs (c_1, c_2, c_3 and c_4) versus the time. The horizontal axis represents the frame index. The vertical axis represents the magnitudes of the MUPs corresponding to the deformations of the markers before normalization. Figure 7.4 shows the corresponding y trajectories of the six lip FP (8.1, 8.2, 8.5, 8.6, 8.7, and 8.8) of the MPEG-4 FAPs.

Table 7.1 Numeric evaluation of the trained real-time AVM

	Training data	Testing data
R	0.981	0.974
$\nu_{\vec{d}}$	0.195	0.196
$\nu_{\vec{d}'}$	0.181	0.179
MSE	0.0025	0.0027

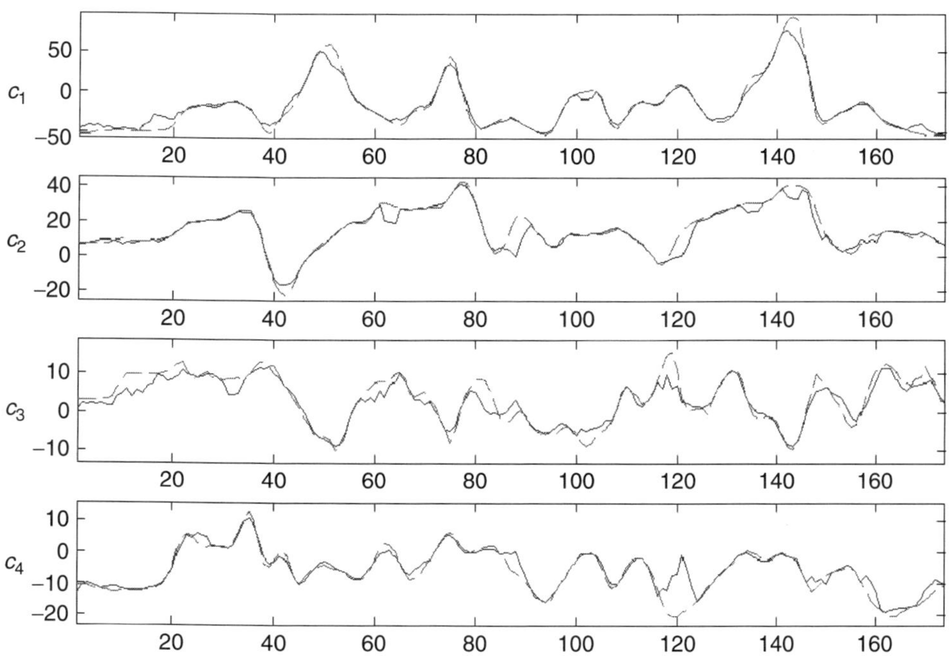

Figure 7.3 An example of audio-to-MUP mapping. The solid lines are the ground truth. The dash lines represent the estimated results. MUPs correspond to the deformations of the markers before normalization

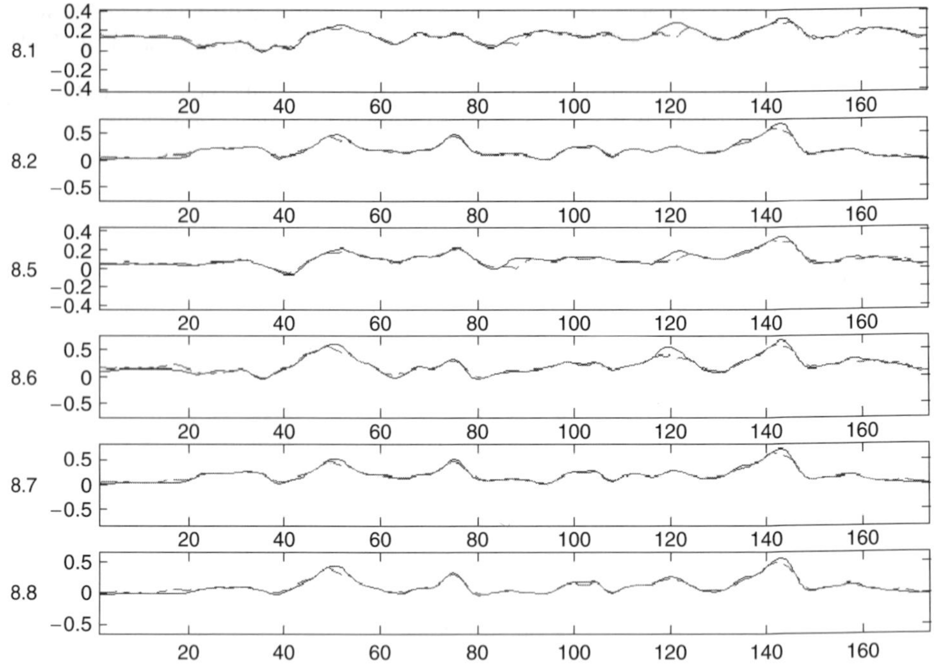

Figure 7.4 The trajectories of six MPEG-4 FAPs. The speech content is the same as that of Figure 7.3. The solid lines are the ground truth. The dash lines represent the estimated results. The deformations of the feature points have been normalized

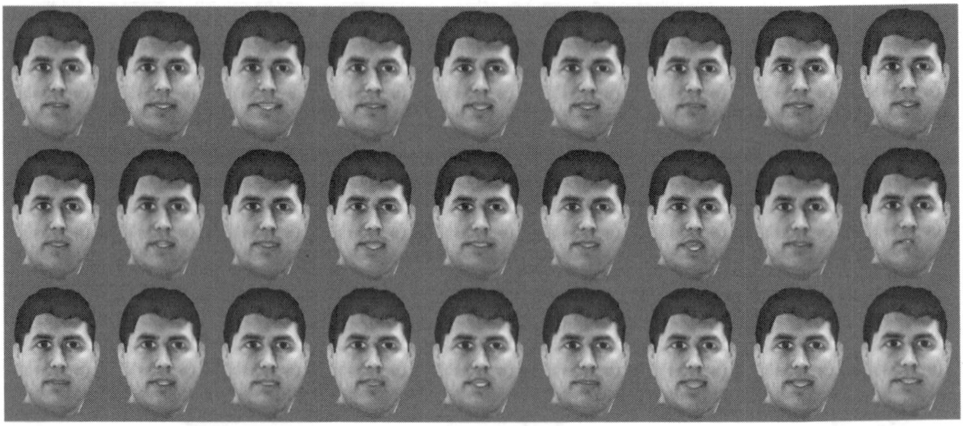

Figure 7.5 An example of the real-time speech-driven face animation of the iFACE system. The order is from left to right and from top to bottom

7.6 THE iFACE SYSTEM

We developed a face modeling and animation system – the iFACE system [19]. The system provides functionalities for customizing a generic face model for an individual, text-driven face animation and off-line speech-driven face animation. Using the method presented in this chapter, we developed the real-time speech-driven face animation function for the iFACE system. First, a set of basic facial deformations is carefully and manually designed for the face model of the iFACE system. The 2-D projections of the facial shapes of the basic facial deformation are visually very close to MUs. The real-time AVM described in this chapter is used by the iFACE system to estimate the MUPs from audio features. Figure 7.5 shows some typical frames in a real-time speech-driven face animation sequence generated by the iFACE system. The text of the sound track is 'Effective communication is essential to collaboration'.

7.7 CONCLUSION

This chapter presents an approach for building a real-time speech-driven face animation system. We first learn MUs to characterize real facial deformations from a set of labeled face deformation data. A facial deformation can be approximated by combining MUs weighted by the corresponding MUPs. MUs encode the information of the correlation among those MPEG-4 facial FPs that are related to speech producing. We show that MU-based FA is compatible with MPEG-4 FA. A set of MLPs is trained to perform real-time audio-to-MUP mapping. The experimental results show the effectiveness of trained audio-to-MUP mapping. We used the proposed method to develop the real-time speech-driven face animation function for the iFACE system, which provides an efficient solution for very low bit-rate 'face-to-face' communication.

REFERENCES

[1] J. P. Lewis, "Automated lip-sync: background and techniques", *Journal of Visualization and Computer Animation*, **2**, 118–122 (1991).

[2] C. Bregler, M. Covell, M. Slancy, Video rewrite: driving visual speech with audio, *SIGGRAPH'97*, August, 1997.

[3] L. R. Rabiner, "A tutorial on hidden Markov models and selected applications in speech recognition", *Proceedings of the IEEE*, **77**(2), 257–286 (1989).

[4] T. Chen, R. R. Rao, "Audio-visual integration in multimodal communications", *Proceedings of the IEEE*, **86**(5), 837–852 (1998).

[5] M. Brand, Voice puppetry, *SIGGRAPH'99*, August, 1999.

[6] S. Morishima, K. Aizawa, H. Harashima, "An intelligent facial image coding driven by speech and phoneme", *Proceedings of IEEE ICASSP*, Glasgow, UK, 1989, p. 1795.

[7] H. Yehia, P. Rubin, E. Vatikiotis-Bateson, "Quantitative association of vocal-tract and facial behavior", *Speech Communication*, **26**(1–2), 23–43 (1998).

[8] R. Rao, T. Chen, R. M. Mersereau, "Exploiting audio-visual correlation in coding of talking head sequences", *IEEE Transactions on Industrial Electronics*, **45**(1), 15–22 (1998).

[9] S. Kshirsagar, N. Magnenat-Thalmann, Lip synchronization using linear predictive analysis, *Proceedings of IEEE International Conference on Multimedia and Expo*, New York, August, 2000.

[10] S. Morishima, H. Harashima, "A media conversion from speech to facial image for intelligent man-machine interface", *IEEE Journal on Selected Areas in Communications*, **4**, 594–599 (1991).

[11] D. W. Massaro, J. Beskow, M. M. Cohen, C. L. Fry, T. Rodriguez, Picture my voice: audio to visual speech synthesis using artificial neural networks, *Proceedings of AVSP'99*, Santa Cruz, Calif., August, 1999.

[12] F. Lavagetto, "Converting speech into lip movements: a multimedia telephone for hard of hearing people", *IEEE Transactions on Rehabilitation Engineering*, **3**(1), 90–102 (1995).

[13] S. Curinga, F. Lavagetto, F. Vignoli, Lip movements synthesis using time-delay neural networks, *Proceedings of EUSIPCO-96*, Trieste, September, 1996.

[14] I. T. Jolliffe, *Principal Component Analysis*, Springer-Verlag, New York, 1986.

[15] T. F. Cootes, C. J. Taylor, D. M. Cooper, J. Graham, "Active shape models – their training and application", *Computer Vision and Image Understanding*, **61**(1) 38–59 (1995).

[16] P. Hong, Facial Expressions Analysis and Synthesis, MS thesis, Computer Science and Technology, Tsinghua University, Beijing, China, 1997.

[17] P. Hong, An Integrated Framework for Face Modeling, Facial Motion Analysis and Synthesis, Ph.D. Thesis, Computer Science, University of Illinois at Urbana-Champaign, Urbana-Champaign, 2001.

[18] L. R. Rabiner, B. H. Juang, *Fundamentals of Speech Recognition*, Prentice Hall, New York, 1993.

[19] P. Hong, Z. Wen, T. S. Huang, "iFACE: a 3D synthetic talking face", *International Journal of Image and Graphics*, **1**(1), 1–8 (2001).

8

Visual Text-to-Speech

Catherine Pelachaud

Università di Roma La Sapienza, Rome, Italy

ABSTRACT

Our goal is to create a natural talking face with, in particular, lip-readable movements. On the basis of real data extracted from a speaker with an optoelectronic system that applies passive markers on the speaker face, we have approximated the data using a neural network model. In this chapter we present our work on the construction of a VTTS in which we have included coarticulation rules. In particular, we describe our computational model of lip movements and our 3-D facial model compliant with MPEG-4 standard. Our experiment is based on some phonetic phonological considerations of the parameters defining labial orifice and of identification tests of visual articulatory movements.

8.1 INTRODUCTION

Speech intelligibility and speech perception by hearers improve when aided by visual signals as well as audio [1–3]. Communication between humans is done not only via verbal language but also through our use of intonation, gaze, hand gesture, body movement and facial expression. Some information may be transmitted by audio and visual signals and this redundancy of occurrence is exploited [4–6]. For example, a question can be marked by any one of the following signals: the voice pitch, a raised eyebrow or a head direction or by a combination of these signals. Also, the interpretation of a signal from one modality can be modulated by other co-occurring signals. A raised eyebrow coinciding with a high end of tone is interpreted as a question mark signal rather than as the emotional signal of surprise in American English [7].

Signals from visual and audio channels complement each other. The complementary relation between audio and visual cues helps in ambiguous situations. Indeed, some phonemes can be easily confused acoustically (e.g. /m/ and /n/) but can be easily differentiated visually (/m/ is done by lip closure, while /n/ is not). Looking at a face while talking improves human perception [8–10]. People, especially those hard of hearing, make use of gesture information to perceive speech. Similarly, speech recognition

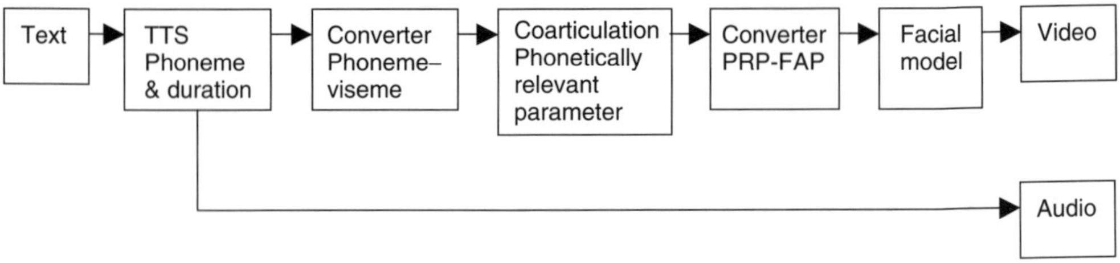

Figure 8.1 VTTS architecture

performance when combining the audio and visual channels is higher than when only one channel is used [1].

Our aim is to build a visual text-to-speech system (VTTS), that is, a system which combines text-to-speech technology with facial animation (FA). Figure 8.1 describes the general flow of our system. Our system follows the general framework of a VTTS but includes two modules: the coarticulation rules and the converter of the phonetically relevant parameters (PRPs) into facial action parameters (FAPs). The input to VTTS is an ASCII text that is to be acoustically and visually synthesized. This text is decomposed by the speech synthesizer in a sequence of phonemes with their duration. These temporal values is used to ensure a correct synchronization of the lip movements. From this sequence of phonemes the system computes the lip shape: it first looks at which visemes are involved and then applies coarticulation rules. Our system defines lip shapes using labial parameters that have been determined to be PRPs. These parameters need to be transformed into FAPs to drive the FA. Acoustically synchronized animation is obtained by composing the audio and animation streams.

This chapter presents an overview of the problems linked with the creation of a natural talking face whose facial model is based on MPEG-4 standard and has lip-readable movements. Our aim is to build a believable multimodal embodied agent that is able to communicate through synchronized verbal and nonverbal signals. We first give information on articulatory characteristics of lip movement during speech, looking at the definition of visemes and of phonologically and phonetically relevant labial parameters. We pay particular attention to the effect of coarticulation. We then provide some information on the tongue shape computation during speech. Finally, we present our 3-D facial model.

8.2 LIP SHAPES

The computation of lip shape during speech is not as straightforward as one might expect. Indeed, a simple mapping between single speech sounds and unique associated lip shapes does not exist. The same lip movements are involved in producing a variety of different speech sounds. As a consequence, many words look alike even though they sound different.

8.2.1 Visemes

Since many phonemes cannot be differentiated on the basis of audio signals (such as voiced, voiceless or nasal), Fisher [11] introduced the notion of visual phonemes (visemes). Phonemes are clustered on the basis of their visual similarity. A viseme may not be viewed as theoretically found but rather as an experimental concept – a distinction similar to the one existing between allophone and phoneme [12]. Research has been done to cluster vowels and consonants on the basis of their visual similarity. This relation of similarity does not necessarily correspond to the relation linking acoustic similar phonemes. Consonants with different voicing or nasal properties are easily confused visually, while consonants differing in place of articulation are quite recognizable [such as /p/ and /b/ (lip closure) as well as /f/ and /v/ (lower lip to upper teeth) (cited in Reference 12)]. But vowels are not so easily grouped by similarity. For some languages, rounding parameter is the most important parameter of distinction among vowels [12].

8.2.2 Labial Parameters

Research has shown that only few parameters are necessary to describe lip shapes [9, 13, 14]. These parameters are capable of describing each cluster of visually confused phonemes (visemes). Four parameters have been found to be phonetically and phonologically relevant: lip height (LH), lip width (LW), upper-lip protrusion (UP) and lower-lip protrusion (LP) [15].

8.3 AUDIOVISUAL SPEECH

Lip movements of a real speaker have been recorded at the Istituto di Fonetica e Dialettologia-C.N.R. of Padova by means of an optoelectronic motion analyzer, ELITE, (Figure 8.2). ELITE applies passive markers on the speaker's face [15–17]. It records the acoustic signal as well as the displacement and velocity of the markers, every 1/10th of a second, producing displacement curves (Figure 8.3). Markers have been placed on eight spots on the face: one marker on each lip corner, one on each ear, one on the chin, one on the mid upper lip, one on the mid lower lip and one on the tip of the nose. The first two markers allow one to get the LW; the next two to get information on the global head motion; the marker on the chin gives information on the jaw action; the last two markers provide information on the upper- and the lower-lip protrusion, respectively. The articulatory characteristics of Italian vowel and consonant targets in the /'VCV/ context were quantified from four subjects, repeating each item five times.

For the moment we have decided to concentrate on the four phonetically and phonologically relevant parameters: LH, LW, UP and LP. Our first step is to approximate the displacement curves of these four articulatory parameters over time.

Our approach is to approximate each curve by a mathematically described function. The original curves are read and stored in an array called $\text{Curve}_i(t)$. Each curve has three peak values (maxima or minima points) corresponding to the production of V, C and V. For each of these targets within each curve, we look for the time of these peaks (Figures 8.4–8.7). We gather these temporal values in an array called *time*. We notice that we may encounter asynchronies of the labial target over the acoustic signal, according to the parameter and/or the phoneme. Further, the different ranges of extension for different parameters have to be stressed, for example, the UP and LP

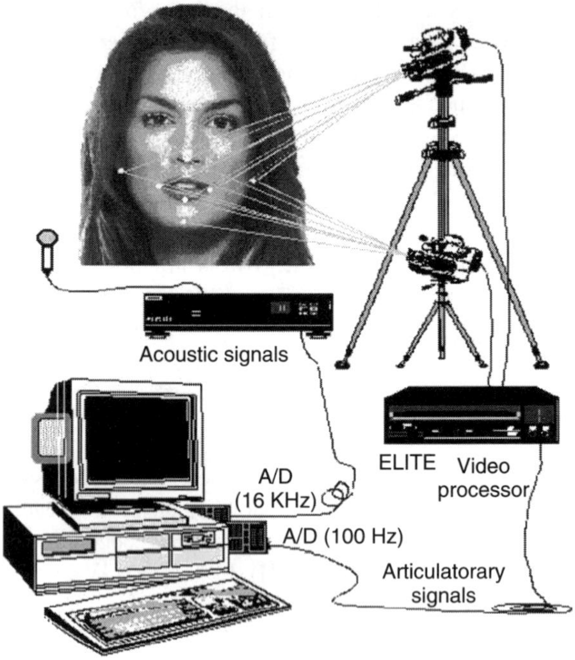

Figure 8.2 The ELITE system [From Magno-Caldognetto *et al.* (1995), (1997), (1998)] [15–17]

variations under 1 mm are evidently not so relevant (cf. Figures 8.6 and 8.7). We want the curve to be particularly well approximated around the three peak points. To ensure a better approximation, we consider two more points surrounding the peak (at time t): one point at time [time(t) − 1] and one point at time [time(t) + 1]. That is, we are considering nine points for each Curve_i in the approximation phase.

Using a neural network model, we have represented the curve as the weighted sum of radial basis functions (RBFs) f_i of the form

$$f_i(t) = \sum_{j=1}^{9} \lambda_j e^{-\frac{|t-\text{time}(t_j)|^2}{\sigma_j^2}}$$

where λ_j and σ_j are the parameters that define the RBF. The approximation method tries to minimize the equation:

$$\min[f_i(t) - \text{Curve}_i(t)]^2$$

that is, we have to find the λ_j and σ_j that best verify this equation. Each RBF is characterized by nine pairs (λ_j, σ_j). We use unconstrained nonlinear optimization method as the minimizing method using matlab. This approach uses a quasi-Newton algorithm

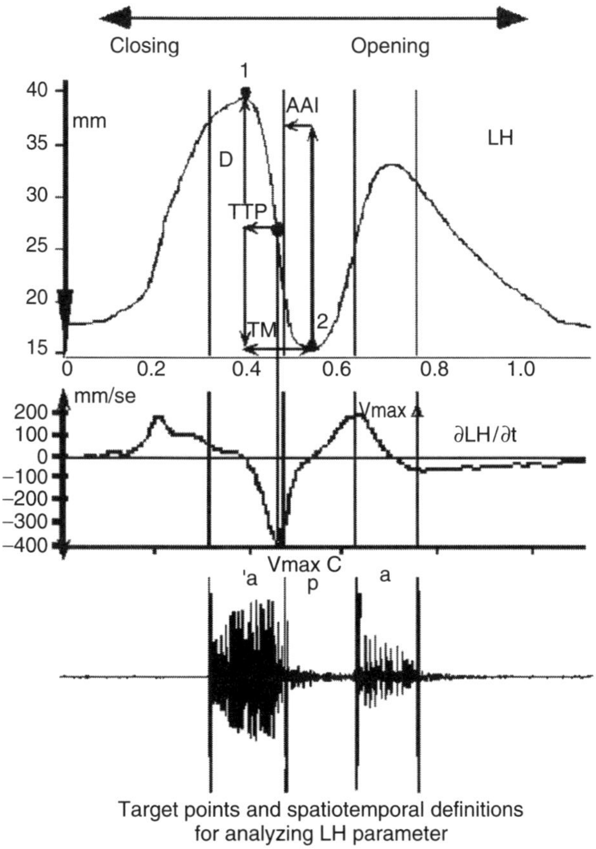

Figure 8.3 Spatiotemporal measurement of the markers on the speaker's face [From Magno-Caldognetto *et al.* (1995), (1997), (1998)] [15–17]

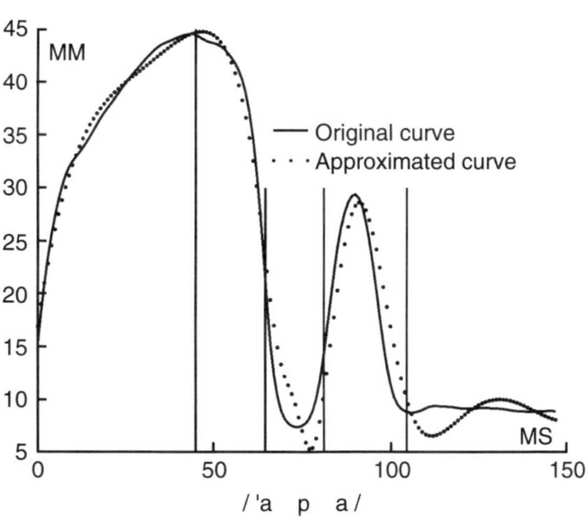

Figure 8.4 Lip-height approximation of the sequence /′apa/; vertical lines defined the acoustic segmentation. The values of LH parameter are nonnormalized

and requires the gradients in λ_j and σ_j:

$$\frac{\partial f_i(t)}{\partial \lambda_j} = \mathrm{e}^{-\frac{|t-\mathrm{time}(t_j)|^2}{\sigma_j^2}}$$

$$\frac{\partial f_i(t)}{\partial \sigma_j} = \lambda_j \mathrm{e}^{-\frac{|t-\mathrm{time}(t_j)|^2}{\sigma_j^2}} * 2 * \frac{|t-\mathrm{time}(t_j)|^2}{\sigma_j^3}$$

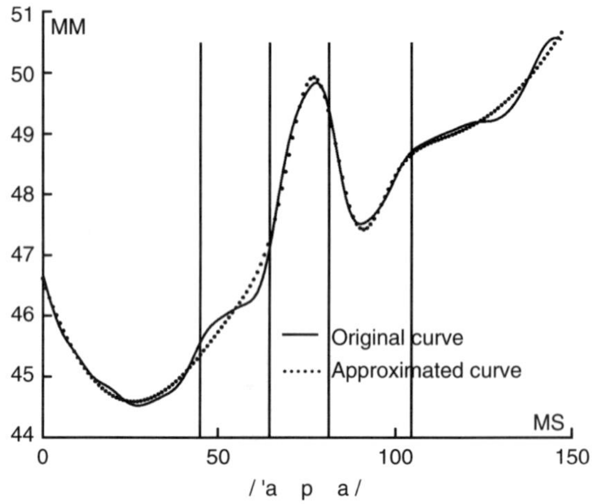

Figure 8.5 Lip-width approximation of the sequence /ˈapa/. The values of LW parameter are nonnormalized

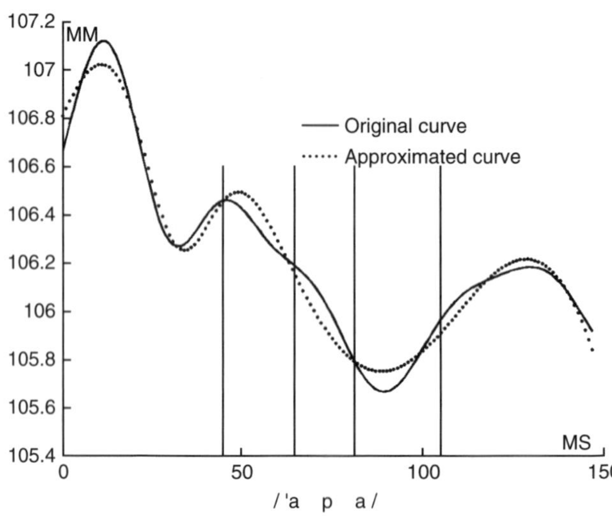

Figure 8.6 Upper-lip protrusion approximation of the sequence /ˈapa/. The values of UP parameter are nonnormalized

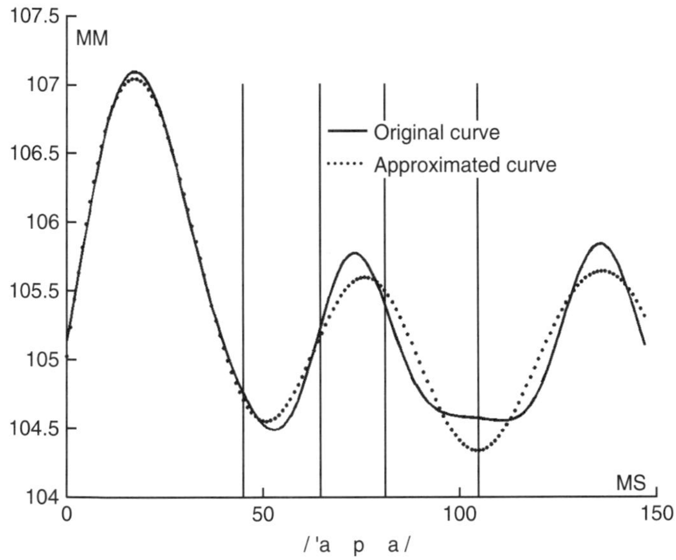

Figure 8.7 Lower-lip protrusion approximation of the sequence /′apa/. The values of LP parameter are nonnormalized

Results of the approximation of the original curves for several lip parameters are shown in the Figures 8.4–8.7.

Having found the parameters that best described the curves ′VCV for V, C and V, we are able to proceed to the first step toward animating the 3-D facial model. The original curves are sampled every 1/10th of a second. For animating a 3-D face we need a frame every 1/25th of a second at a minimum. Having a mathematical representation of ′VCV curve for each four articulatory parameters, it is easy to get a value every 1/25th of a second for these four parameters (LH, lip width, upper- and lower-lip protrusion). Finally, we need to convert these four parameters into parameters that drive the facial model, that is, in facial animation parameter set (FAPS) (Figures 8.8 and 8.9).

Figure 8.8 Lip shape of /′a/ in /′apa/

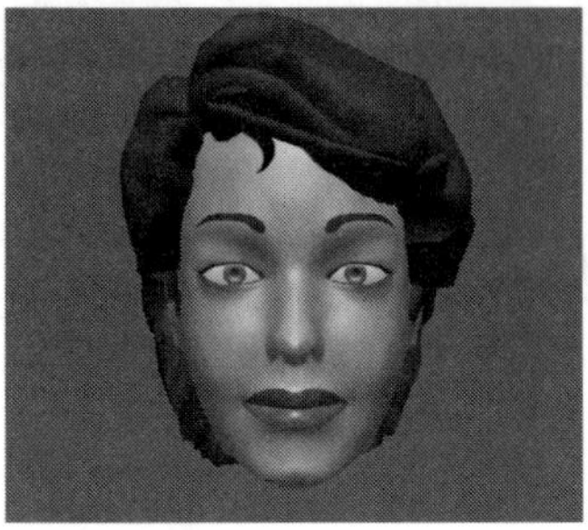

Figure 8.9 Lip shape of /p/ in /ˈapa/

8.4 COARTICULATION

Primary units of word are segments. A segment is the smallest indivisible entity. A succession of segments give the internal structure of a word, but there is no clear boundary between segments in a word. A word is not made of a succession of discrete units but rather as a continuous stream of segments. Finding a boundary in acoustics and in articulation is not possible. Only a distinguishable change in acoustics and/or in articulation distinguishes successive segments in a word. This effect is called *coarticulation*. Coarticulation arises from the temporal overlap of successive articulatory movements, which cause the boundaries between articulatory units to become less prominent.

As we have already mentioned, segments are not emitted sequentially but rather they affect one another in a nonlinear fashion [18]. Löfqvist [19] integrated this result in his model. The influence of a segment over an articulator can be represented by a dominance function [19]. A dominance function is established for each speech articulator. Such a function specifies the time-invariant influence (i.e. the dominance) that an articulator can have over the articulators involved in the production of preceding or succeeding segments. That is, it simulates the behavior of an articulator over time: before a certain time, the segment has no influence over the articulator; then it starts, the maximum influence arriving at the peak of the dominance curve; finally, the curve decreases, meaning that the segment loses effect over the articulator. The amplitude and field of influence of each dominance function depend on the segment and on the articulator. For a sequence of segments, the dominance functions for a given articulator may overlap in time.

8.4.1 Models of Coarticulation

The model of coarticulation proposed by Cohen and Massaro [20] implements Löfqvist's gestural theory of speech production [19]. The system uses overlapping dominance functions to specify how close the lips come to reaching their target value for each viseme. Dominance functions have the form of a negative exponential function scaled by a magnitude factor. For a given segment, different dominance functions can overlap. The final lip shape (target value) is obtained by taking the weighted average of all the co-occurrent dominance functions. Their model has been evaluated using different tests of intelligibility [8, 21]. Perceptual tests are performed for each combination of acoustic and visual modalities. Confusion matrices are calculated and statistical analyses are performed to study the contribution of intelligibility of the various modalities. The comparison of these matrices give the overall intelligibility of each phonemic item in each setting.

Several models [22, 23] modified the formula developed by Cohen and Massaro [20]. In particular, Le Goff and Benoit [22] extended this model to get an n-continuous function and proposed a method to automatically extract the parameters defining the dominance function from data measured on a real speaker. Their analysis method requires that a speaker has lips painted in blue. Then, they use chroma-key technique to extract lip shape. The principle of resolution consists in finding a solution and testing if it is satisfactory. The testing is done by measuring the Euclidean distance from the synthesized curve and the original one. If the distance is too large, the process is restarted again. A relaxation method is used to iteratively optimize the different variables. It uses the dichotomy principles applied to each parameter and looks for the minimum distance function between the synthesized and real curves. The computation of the minimum is done by looking at the partial derivative of the target parameter function for each parameter. They have evaluated their extraction method by following two approaches [24]: quantitative evaluation test, based on measurements to test if the movements produced are correct or not, and qualitative evaluation test, based on perceptual test to check how the visual information is perceived.

Some difficulties may be found with the above method [25], specially for the bilabial stops for which labial closure is necessary but is not always maintained if one uses the dominance function. To overcome them, Reveret *et al.* [25] adapt Öhman's coarticulation model [26]. This model suggests that there is a dominant stream of vowel gestures on which consonant gestures are superimposed. The authors first determined the appropriate gestures for each vowel and each consonant. Then, any articulatory parameter is expressed by an algebraic sum of consonant and vowel gestures [25]. Their model is based on the analysis of real data of a speaker with several dots on his face that are tracked through time. They use the principal component analysis (PCA) technique to define few phonetically relevant parameters. These parameters are then transformed into FAP to drive a MPEG-4-compliant facial model [27].

The approach by Pelachaud *et al.* [28] implements the look-ahead model. Lip movements are viewed according to a notion of speech as a sequence of key positions (corresponding to phonemes belonging to nondeformable clusters) and transition positions (corresponding to phonemes belonging to deformable clusters). The problem is to determine the shape computation of the transition position for which the look-ahead model is well suited. Indeed, this model predicts that any articulatory adjustment starts immediately after a key position and lasts until the next one. The transition position receives the same shape as the 'strongest' key position ('strongest' meaning lip shapes belonging to the least deformable clusters). Both forward and backward look-ahead rules are considered by the algorithm. To solve particular problems (certain visual transitions between segments) that cannot be solved by these two rules, a three-step algorithm is employed. In the first step, the forward and backward coarticulation rules are applied to all clusters defined as context-dependent. The next step considers the relaxation and contraction times of the involved muscles. Finally, the program examines the manner in which two consecutive actions are performed geometrically.

8.5 TONGUE MOVEMENT

The most highly perceivable movements in speech production involve the lips and the jaw. The tongue and teeth, however, are also occasionally visible and should be considered. The human tongue plays a significant part in speech production. Sounds are differentiated, amongst other factors, by the position of the tongue related to the palate and by the curvature and contraction of the tongue. The tongue is a highly flexible organ composed of muscles, fat and connective tissue that can be bent, twisted and tensed [29]. As one might expect, consonants and vowels exhibit quite different characteristics. For vowels, both the entire tongue surface and its curvature are important, while for consonants it is mainly the points of contact between the tongue and the palate or teeth that matter. For consonants, the tongue touches the palate with more tension than for vowels. In most graphics systems, the tongue is represented by a parallelepiped that can move inward, outward, upward and downward [20, 30, 31].

Pelachaud and his coworkers' approach [32] uses soft object techniques to model tongue movement. In modeling tongue movement, Pelachaud *et al.* [32] apply the look-ahead model similar to the one used to compute lip shapes [28], which ensures that for some highly deformable phonemic segments, the tongue shape will be influenced by the surrounding segments. Collision between the tongue and the palate is also handled.

Other models have been proposed [33, 34]. They modeled very accurately the tongue shape and movement on the basis of magnetic resonance imaging (MRI) data and Electropalatography (EPG) data. The 3-D model of the tongue is trained-fitted to the data of a real speaker using error minimization algorithms [33]. Moreover, the EPG data is used to detect any collision between the palate and the tongue by specifying the contact location between the two elements. Highly realistic tongue movement is thus obtained. Such a model is being used for language training in particular for hearing impaired people.

However, such a sophisticated model uses a large set of parameters, much larger than the one specified by the MPEG-4 standard.

8.6 FACIAL MODEL

Our facial model is compliant with MPEG-4 standard [35, 36]. The facial model is the core of an MPEG-4 decoder and is based on the MPEG-4 face and body animation (FBA) specification. Two sets of parameters describe and animate the 3-D facial model: FAPS and facial definition parameter (FDP). The FDPs define the shape of the model, while FAPS define the facial actions. When the model has been characterized with FDP, the animation is obtained by specifying for each frame the values of FAPS. As our goal is to compute lip movements from data, we do not consider the first FAP that defines visemes, rather we are proposing a method to define them as exposed in this chapter. The FAP corresponding to expressions is not considered either; we also use here our own set of expressions [37]. All other FAPS (the remaining 66) have been implemented.

The model uses a pseudomuscular approach [38]. The muscle contractions are obtained through the deformation of the polygonal network around feature points (FPs).

Each feature point follows MPEG-4 specifications. Furthermore, the model has been divided into regions (Figure 8.10) that correspond to muscle contraction action [39]. Points within a single region may be modified by several FAPS, but they can react differently depending on the considered FAP. For example, given a region R and two FAPS, FAP_i and FAP_j, both of which act on R (or part of it), FAP_i may have a greater influence on each point of the region R (or its subpart) than FAP_j.

The deformation due to a FAP is performed in a zone of influence that has an ellipsoid shape whose centroid is the feature point (Figure 8.11). The displacement of points within this area of influence obeys a deformation function that is a function of the distance between the points and the feature point (Figures 8.12 and 8.13). The

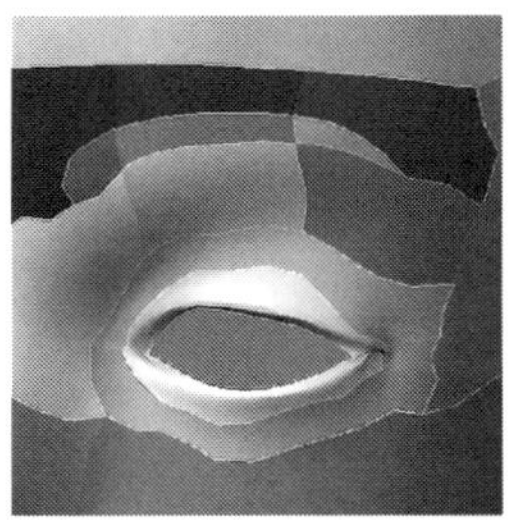

Figure 8.10 Region subdivisions of the eyebrow

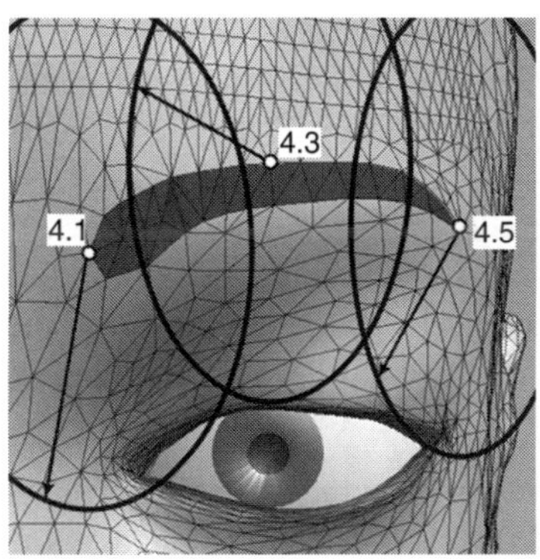

Figure 8.11 Feature points and their area of influence around the eyebrow

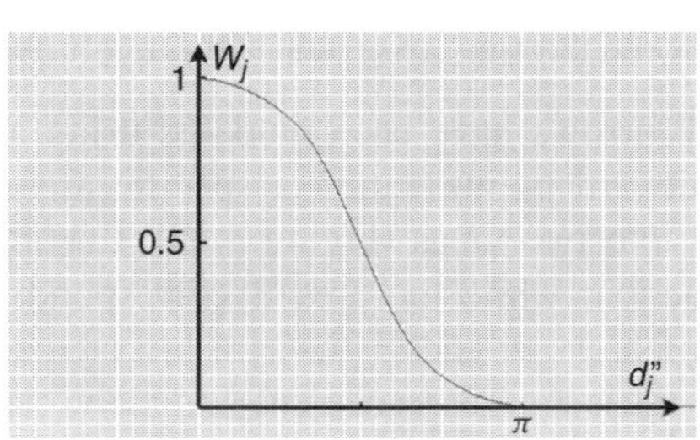

Figure 8.12 Deformation function

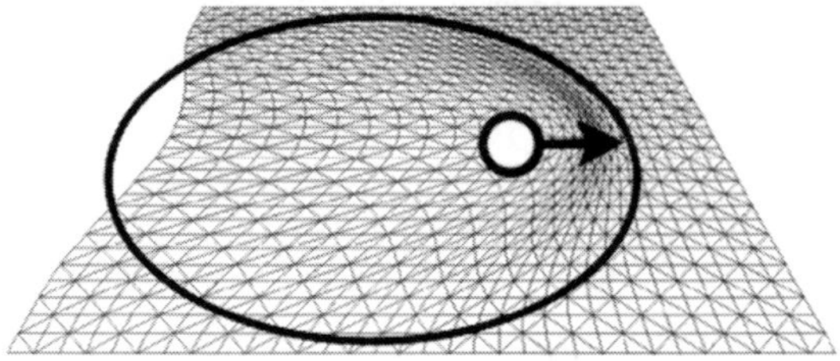

Figure 8.13 Skin deformation in the area of influence

displacement of a point depends on the intensity of the FAP it is subject to, on its position in the area of influence of the FAP and also on which region it belongs to and how this region is affected by a given FAP. The displacement ΔP_j of a point P_j that belongs to the area of influence of the FAP_i and a region R_k is given by

$$\Delta P_j = F_i * W_j * W_{ki} \tag{8.1}$$

where F_i is the intensity of FAP_i, W_j is the value of the deformation function at the point P_j and W_{ki} represents the weight of deformation of FAP_i over the region R_k. W_j depends on the distance between P_j and the feature point of the area of influence. Of course this value is equal to zero for all points outside this area of influence. This allows us to modify only the points belonging to a given area of influence of a FAP without modifying the other points of the facial model. On the other hand, W_{ki} specifies how the region R_k is affected by FAP_i. This factor can be set to zero if a region is not to be affected by a given FAP. In Figure 8.11 we see the zone of influence of the three FAPS (all have ellipsoid shape) and the three FPs where the FAPS are applied. In Figure 8.10 the regions over the same part of the face are shown. To be sure that under the FAPS acting on the eyebrow the points within the eyelid region will not be affected, all W_{ki} factors between the eyelid region and the FAPS for the eyebrow are set to zero. Therefore the eyelid will have a null displacement under these particular FAPS.

The facial model also includes particular features such as wrinkles and furrows to enhance its realism. Wrinkles are modeled using bump mapping technique based on per-vertex technique of the OpenGL library. Wrinkles correspond to lines of discontinuity and are perpendicular to muscle contraction. The intensity of the deepening of the wrinkles depends on the value of the FAP. We have implemented only the wrinkles appearing on the forehead that arise when the eyebrows are raised (Figure 8.14).

Bulges and furrows have been modeled using a specialized displacement function that move outward points within a specific area. The region of each bulge has been decomposed into two areas (Figure 8.15): the area A corresponds to the area of the muscle contraction, while the area B corresponds to the area of skin accumulation. The points of areas A and B that are affected by muscular contraction will be deformed by the muscular displacement function, while the points of area B (area of the bulge/furrow) will be moved outward to simulate the skin accumulation and bulging (Figures 8.15, 8.17 and 8.18), obeying a displacement function as defined in

Figure 8.14 Wrinkles appearing under eyebrow raising

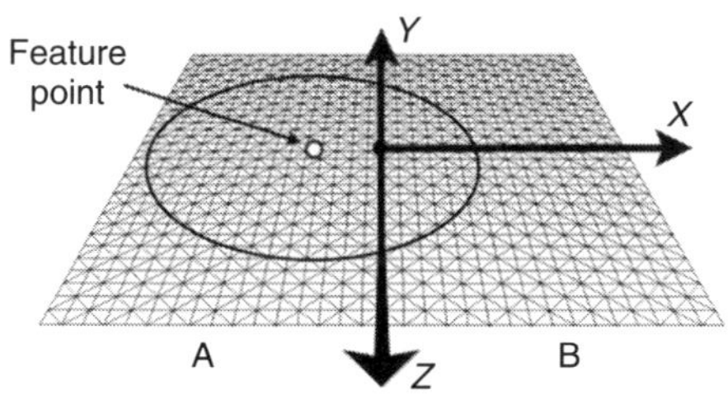

Figure 8.15 Within area of influence, the two zones A (muscular traction) and B (accumulation)

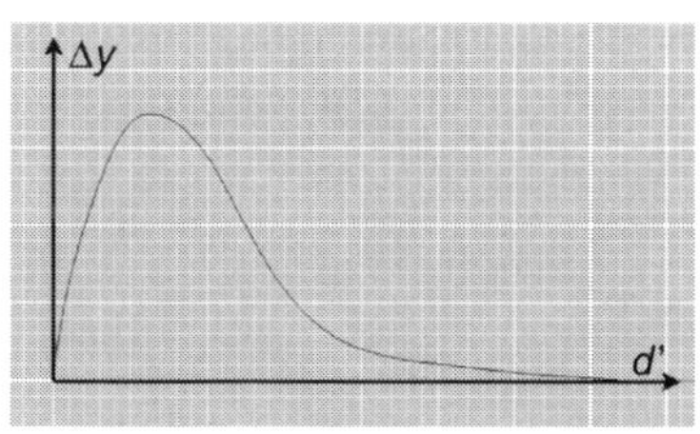

Figure 8.16 Displacement function for bulge computation

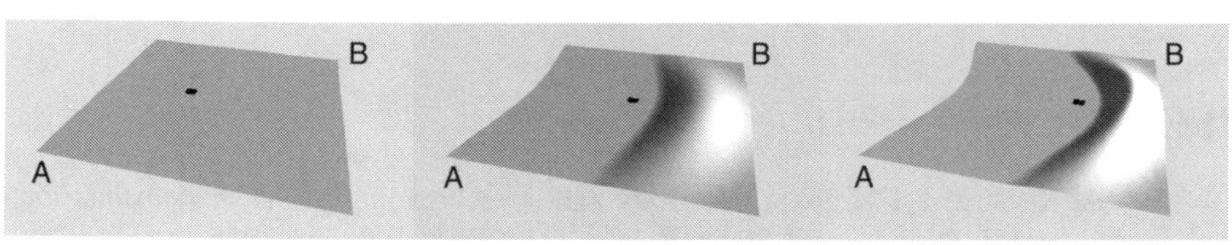

Figure 8.17 Bulge creation

Figure 8.16. The outward displacement ΔP_j^y of a point P_j in the area B of the bulge under the action of FAP_i is computed as

$$\Delta P_j^y = \Delta P_j^x * K_i * W2_j \tag{8.2}$$

Figure 8.18 Simulation of the nasolabial furrow

ΔP_j^x is the displacement of the point P_j due to FAP$_i$ as defined in Equation (8.1). ΔP_j^x depends on the distance between the point P_j and the feature point of the area of influence; $W2_j$ is the value of the deformation function and depends on the distance between the point P_j and the boundary between the area A and the area B (as defined in Figure 8.15). K_i is a constant that characterizes the bulge height. Examples of bulge creation are given in Figures 8.17 and 8.18.

8.7 CONCLUSION

In this chapter we have presented a 3-D facial model that includes particular facial features, wrinkles and bulges, to enhance its realism. We have also introduced a method to approximate lip movements on the basis of real data extracted from human speakers. The approximation is done using RBFs. The choice of the articulatory parameters is obtained through phonetic and phonological considerations. The problem of coarticulation as well as several approaches that were implemented in the construction of talking face have been introduced.

ACKNOWLEDGMENT

I am very grateful to Emanuella Magno-Caldognetto, Piero Cosi and Claudio Zmarich of the Istituto di Fonetica e Dialettologia-CNR of Padova for their helpful collaboration in this project and for providing the data collected with ELITE. I would like to thank Stefano Pasquariello for having implemented the 3-D facial model.

REFERENCES

[1] A. Risberg, J. L. Lubker, "Prosody and speechreading", *Speech Transmission Laboratory*, Quarterly Progress and Status Report 4, KTH, Stockholm, 1978.

[2] C. Schwippert, C. Benoit, "Audiovisual intelligibility of an androgynous speaker", in C. Benoit, R. Campbell, eds., *Proceedings of the ESCA Workshop on Audio-Visual Speech Processing (AVSP'97)*, Rhodes, Greece, September, 1997.

[3] E. Agelfors, J. Beskow, M. Dahlquist, B. Granström, M. Lundeberg, G. Salvi, K. E. Spens, T. Öhman, "Synthetic visual speech driven from auditory speech", in D. W. Massoro, ed., *Proceedings from AVSP'99*, University of Santa Cruz, Calif., 1999.

[4] E. Magno-Caldognetto, I. Poggi, "Micro- and macro-bimodality", in C. Benoit, R. Campbell, eds., *Proceedings of the ESCA Workshop on Audio-Visual Speech Processing (AVSP'97)*, Rhodes, Greece, September, 1997.

[5] U. Hadar, T. J. Steiner, E. C. Grant, F. Clifford Rose, "Kinematics of head movements accompanying speech during conversation", *Human Movement Science*, **2**, 35–46 (1983).

[6] D. Bolinger, *Intonation and its Part*, Stanford University Press, Stanford, Calif., 1986.

[7] P. Ekman, "About brows: emotional and conversational signals", in M. von Cranach, K. Foppa, W. Lepenies, D. Ploog, eds., *Human Ethology: Claims and Limits of a New Discipline: Contributions to the Colloquium*, Cambridge, Mass., 1989, pp. 169–248.

[8] D. Massaro, *Perceiving Talking Faces: From Speech Perception to a Behavioral Principle*, *Bradford Books Series in Cognitive Psychology*, MIT Press, Cambridge, Mass., 1997.

[9] C. Benoit, "Why synthesize talking faces?", *Proceedings of the ESCA Workshop on Speech Synthesis*, Autrans, 1990, pp. 253–256.

[10] A. Q. Summerfield, "Lipreading and Audio-Visual Speech Perception", *Philosophical Transactions of the Royal Society of London. Series B: Biological Sciences*, **335**(1273), 71–78 (1992).

[11] C. G. Fisher, "Confusions among visually perceived consonants", *Journal Speech Hearing Research*, **15**, 474–482 (1968).

[12] C. Benoit, T. Lallouache, T. Mohamadi, C. Abry, "A set of French visemes for visual speech synthesis", *Talking Machines: Theories, Models, and Designs*, Elsevier North-Holland, Amsterdam, 1992, pp. 485–504.

[13] V. Fromkin, "Lip positions in American-English vowels", *Language and Speech*, **7**(3), 215–225 (1964).

[14] P. Cosi, E. Magno-Caldognetto, "Lips and jaws movements for vowels and consonants: spatio-temporal characteristics", in D. G. Stork, M. E. Hennecke, eds., *Speechreading by Humans and Machines: Models, Systems, and Applications*, *NATO ASI Series. Series F: Computer and Systems Sciences*, Vol. 150, Springer-Verlag, Berlin, 1996.

[15] E. Magno-Caldognetto, C. Zmarich, P. Cosi, "Statistical definition of visual information for Italian vowels and consonants", in D. Burnham, J. Robert-Ribes, E. Vatikiotis-Bateson, eds., *International Conference on Auditory-Visual Speech Processing AVSP'98*, Terrigal, Australia, 1998, pp. 135–140.

[16] E. Magno-Caldognetto, K. Vagges, C. Zmarich, "Visible articulatory characteristics of the Italian stressed and unstressed vowels", *Proceedings of the XIIIth International Congress of Phonetic Sciences*, Vol. 1, Stockholm, 1995, pp. 366–369.

[17] E. Magno-Caldognetto, C. Zmarich, P. Cosi, F. Ferrero, "Italian consonantal visemes: relationships between spatial/temporal articulatory characteristics and co-produced acoustic signal", in C. Benoit, R. Campbell, eds., *Proceedings of the ESCA Workshop on Audio-Visual Speech Processing (AVSP'97)*, Rhodes, Greece, September, 1997, pp. 5–8.

[18] C. A. Fowler, E. L. Saltzman, "Coordination and coarticulation in speech production", *Language and Speech*, **36**, 171–195 (1993).

[19] A. Löfqvist, "Speech as audible gestures", in W. J. Hardcastle, A. Marchal, eds., *Speech Production and Speech Modeling*, *NATO ASI Series*, Vol. 55, Kluwer Academic Publishers, Dordrecht, 1990, pp. 289–322.

[20] M. M. Cohen, D. W. Massaro, "Modeling coarticulation in synthetic visual speech", in M. Magnenat-Thalmann, D. Thalmann, eds., *Models and Techniques in Computer Animation*, Springer-Verlag, Tokyo, 1993, pp. 139–156.

[21] M. M. Cohen, R. L. Walker, D. W. Massaro, "Perception of synthetic visual speech", in D. G. Stork, M. E. Hennecke, eds., *Speechreading by Humans and Machines, Models, Systems, and Applications*, *NATO ASI Series. Series F: Computer and Systems Sciences*, Vol. 150, Springer-Verlag, Berlin, 1996.

[22] B. LeGoff, C. Benoît, "A French speaking synthetic head", in C. Benoît, R. Campbell, eds., *Proceedings of the ESCA Workshop on Audio-Visual Speech Processing*, Rhodes, Greece, September, 1997, pp. 145–148.

[23] S. A. King, A facial model and animation techniques for animated speech, Ph.D. thesis, Ohio State University, 2001.

[24] B. LeGoff, Synthèse à partir du texte de visage 3D parlant français, Ph.D. thesis, Institut National Polytechnique, Grenoble, France, 1997.

[25] L. Reveret, G. Bailly, P. Badin, "MOTHER: a new generation of talking heads providing a flexible articulatory control for video-realistic speech animation", in B. Yuan, T. Huang, X. Tang, eds., *International Conference on Spoken Language Processing, ICSLP'00*, Vol. II, Beijing, China, 2000, pp. 755–758.

[26] S. E. G. Öhman, "Numerical model of coarticulation", *Journal of Acoustical Society of America*, **41**(2), 311–321 (1967).

[27] F. Elisei, M. Odisio, G. Bailly, P. Badin, "Creating and controlling video-realistic talking heads", *International Conference on Auditory-Visual Speech Proceedings (AVSP'01)*, Aalborg, Denmark, September, 2001, pp. 90–97.

[28] C. Pelachaud, N. I. Badler, M. Steedman, "Generating facial expressions for speech", *Cognitive Science*, **20**(1), 1–46 (1996).

[29] M. Stone, "A three-dimensional model of tongue movement based on ultrasound and x-ray microbean data", *Journal of Acoustical Society of America*, **87**(5), 2207–2217 (1990).

[30] J. P. Lewis, F. I. Parke, "Automated lipsynch and speech synthesis for character animation", in J. H. Caroll, P. Tanner, eds., *Proceedings Human Factors in Computing Systems and Graphics Interface '87*, 1987, pp. 143–147.

[31] P. Kalra, A. Mangili, N. Magnenat-Thalmann, D. Thalmann, "SMILE: a multilayered facial animation system", in T. L. Kunii, ed., *Modeling in Computer Graphics*, Springer-Verlag, Tokyo, 1991, pp. 189–198.

[32] C. Pelachaud, C. W. A. M. van Overveld, C. Seah, "Modeling and animating the human tongue during speech production", *Proceedings of Computer Animation '94*, Geneva, May, 1994, pp. 40–47.

[33] M. M. Cohen, J. Beskow, D. W. Massaro, "Recent developments in facial animation: an inside view", in D. Burnham, J. Robert-Ribes, E. Vatikiotis-Bateson, *International Conference on Auditory-Visual Speech Processing (AVSP'98)*, Terrigal, Australia, 1998, pp. 135–140.

[34] O. Engwall, Using linguopalatal contact patterns to tune 3D tongue model, *Proceedings Eurospeech*, Aalborg, Denmark, September, 2001.

[35] P. Doenges, T. K. Capin, F. Lavagetto, J. Ostermann, I. S. Pandzic, E. Petajan, "MPEG-4: audio/video and synthetic graphics/audio for mixed media", *Signal Processing: Image Communication*, **9**, 433–463 (1997).

[36] J. Ostermann, "Animation of synthetic faces in MPEG-4", *Computer Animation'98*, Philadelphia, Pa., June, 1998, pp. 49–51.

[37] J. N. De Carolis, C. Pelachaud, I. Poggi, F. de Rosis, "Behavior planning for a reflexive agent", *IJCAI'01*, Seattle, Washington, August, 2001, in press.

[38] S. Pasquariello, Modello per l'animazione facciale in MPEG-4, M.S. thesis, University of Rome, Italy, 2000.

[39] F. I. Parke, Computer Generated Animation of Faces, Ph.D. thesis, University of Utah, Salt Lake City, Utah, 1972, UTEC-CSc-72-120.

9

Emotion Recognition and Synthesis Based on MPEG-4 FAPs

Nicolas Tsapatsoulis[1], Amaryllis Raouzaiou[1], Stefanos Kollias[1], Roddy Cowie[2] and Ellen Douglas-Cowie[2]

[1]*National Technical University of Athens, Athens, Greece,* [2]*Queen's University of Belfast, Belfast, Ireland*

ABSTRACT

In the framework of MPEG-4 hybrid coding of natural and synthetic data streams, one can include teleconferencing and telepresence applications, in which a synthetic proxy or a virtual agent is capable of substituting the actual user. Such agents can interact with each other, analyzing input textual data entered by the user and multisensory data, including human emotions, facial expressions and nonverbal speech. This not only enhances interactivity, by replacing single media representations with dynamic multimedia renderings, but also assists human–computer interaction issues, letting the system become accustomed to the current needs and feelings of the user. Actual application of this technology [1] is expected in educational environments, 3-D videoconferencing and collaborative workplaces, online shopping and gaming, virtual communities and interactive entertainment. Facial expression synthesis and animation has gained much interest within the MPEG-4 framework; explicit facial animation parameters (FAPs) have been dedicated to this purpose. However, FAP implementation is an open research area [2]. In this chapter we describe a method for generating emotionally enriched human–computer interaction, focusing on analysis and synthesis of primary [3] and intermediate facial expressions [4]. To achieve this goal we utilize both MPEG-4 facial definition parameters (FDPs) and FAPs. The contribution of the work is twofold: it proposes a way of modeling primary expressions using FAPs and it describes a rule-based technique for analyzing both archetypal and intermediate expressions; for the latter we propose an innovative model generation framework. In particular, a relation between FAPs and the *activation* parameter proposed in classical psychological studies is established, extending the archetypal expression studies that the computer society has concentrated on. The overall scheme leads to a parameterized approach to facial expression synthesis that is compatible with the MPEG-4 standard and can be used for emotion understanding.

9.1 INTRODUCTION

Research on facial expression analysis and synthesis has tended to concentrate on primary or archetypal emotions. The categories that have attracted most interest in human–computer interaction environments in particular are sadness, anger, joy, fear, disgust and surprise. Very few studies that explore nonarchetypal emotions have appeared in computer science literature [4]. This trend may reflect the influence of work by Ekman [5], Friesen [6] and others, who proposed that the archetypal emotions correspond to distinct facial expressions that are supposed to be universally recognizable across cultures. However, psychological researchers working in different traditions [7–9] have investigated a broader variety of emotions. An extensive survey on emotion analysis can be found in Reference 10.

MPEG-4 indicates an alternative way of modeling facial expressions and the underlying emotions that are strongly influenced by neurophysiological and psychological studies. The FAPs that are utilized in the framework of MPEG-4 for facial animation purposes are strongly related to the action units (AUs) that constitute the core of the facial action coding system (FACS) [3].

Psychology contains various ideas that may help researchers in the area of computer graphics and machine vision to exploit the flexibility of MPEG FAPs. One of the best known is the idea that emotions are points in a space with a relatively small number of dimensions. Two dimensions, *activation* and *evaluation*, are sufficient for a first approximation. *Evaluation* summarizes how positive or negative the subject feels; *activation* indicates how energetically he or she is disposed to act. The scheme is useful partly because research such as Whissel's [8] has provided coordinates corresponding to a wide range of emotions.

In this chapter we present a methodology for analyzing both primary and intermediate expressions, taking into account the results of Whissel's study and in particular the *activation* parameter. The proposed methodology consists of four steps:

1. *Description of the archetypal expressions through particular FAPs*: In order to do this, we translate facial muscle movements – describing expressions through muscle actions–into FAPs and create a vocabulary of FAPs for each archetypal expression. FAPs required for the description of the archetypal expressions are also experimentally verified through analysis of prototype datasets. In order to make comparisons with real expression sequences, we model FAPs employed in the facial expression formation through the movement of particular FDP points – the selected FDP points correspond to facial area points that can automatically be detected from real images or video sequences. The derived models serve as a bridge between expression synthesis and analysis [11].
2. *Estimation of the range of variation of FAPs that are involved in each of the archetypal expressions*: This is achieved by analyzing real images and video sequences in a semiautomatic manner and by animating synthesized examples.

3. *Modeling of intermediate expressions*: This is achieved through combination, in the framework of a rule base system, of the *activation* parameter – known from Whissel's work – with the description of the archetypal expressions by FAPs.
4. *Understanding emotions*: Emotion models, created in steps (1) to (4), form the basis of a fuzzy rule system that recognizes the underlying (if any) emotion in facial video sequences.

Figure 9.1 illustrates the way the overall analysis and synthesis system functions. A facial video stream feeds a detection system, whose purpose is to recover the motion of prominent points lying in the facial area and corresponding to specific FDP points. Access to an appropriate facial video sequence is necessary for estimating, with acceptable accuracy, the movement of the facial points (in nonteleconferencing video sequences the solution to this problem is close to impossible). The motion of facial points is mapped to FAPs. A vector is produced consisting of FAP values, which is then compared to a predefined set of emotion profiles (models), by a fuzzy inference system. The output of the system is a decision about the specific emotion conveyed by the real subject or the belief values of the best matching emotions.

It should be noted that the system can be used for animation purposes in very low bit rate environments [12]. The emotion analysis system provides either information about the specific emotion expressed by the subject or simply the movement of particular points within the facial area. In the former case, the system provides the

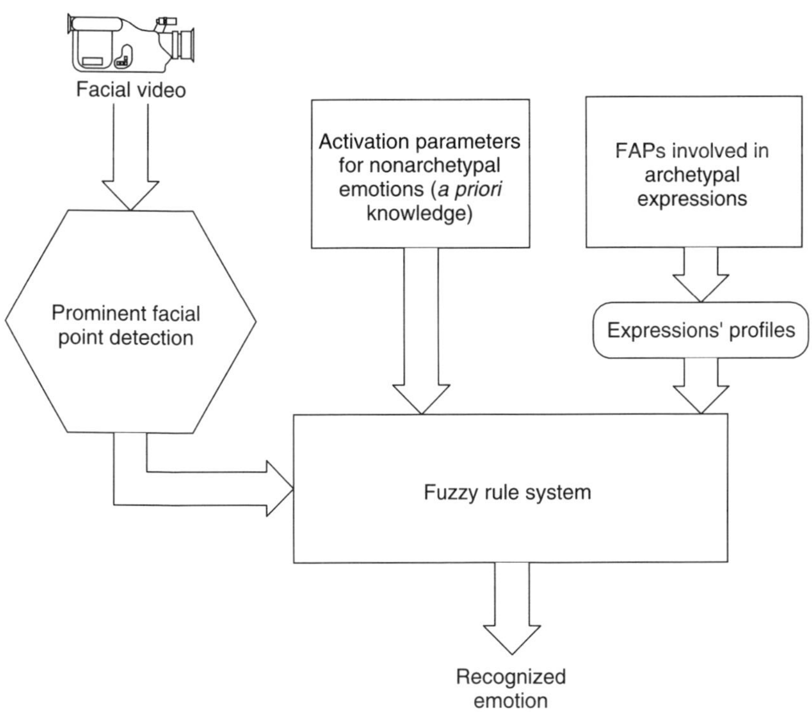

Figure 9.1 Block diagram of the proposed scheme

modification parameters – the FAPs and their appropriate range of variation that are required by the client side application to animate the emotion. In the latter case that corresponds to a failure to recognize a particular emotion, the system simply provides the estimated FAP values (animating the estimated FAP values does not guarantee the creation of a recognizable emotion).

This chapter is organized as follows: Sections 9.2 to 9.4 present the first three parts of the proposed methodology. Section 9.5 describes a way of utilizing the proposed scheme for emotion analysis purposes. Experimental results that illustrate the performance of the presented approach are given in Section 9.6. Finally, conclusions are presented in Section 9.7.

9.2 DESCRIPTION OF THE ARCHETYPAL EXPRESSIONS USING FAPs

In the framework of MPEG-4 standard, one can describe both the anatomy of a human face – basically through FDPs – and the animation parameters, with groups of distinct tokens eliminating the need to specify the topology of the underlying geometry. These tokens can then be mapped to automatically detected measurements and indications of motion on a video sequence; thus, they can help estimate a real expression conveyed by the subject and, if required, approximate it by means of a synthetic one.

Modeling facial expressions and underlying emotions through FAPs serves several purposes:

1. Given the FAP values describing the activation of a face one can form estimates of the emotion expressed by the subject.
2. The methodology ensures that the synthetic sequences created with it are compatible with the MPEG-4 standard.
3. Archetypal expressions occur rather infrequently: in most cases, emotions are expressed through variation of a few discrete facial features that are directly related to particular FAPs. Moreover, distinct FAPs can be utilized for communication between humans and computers in a paralinguistic form, expressed by facial signs.
4. Because FAPs do not correspond to specific models or topologies, synthetic expressions can be overlaid on models or characters other than the subject who originally made the gestures.

Two basic issues should be addressed when modeling archetypal expression: (1) estimation of FAPs that are involved in their formation, (2) definition of the FAP intensities. The former is examined in the current section, while the latter is explained in Section 9.5.

It is clear that the FACS has had a profound influence on research into the analysis of expression. The FACS is a system that tries to extract visually distinguishable facial movements using knowledge of facial anatomy. FACS uses AU as measurement units. Note that an Action Unit does not correspond to a single muscle. It could combine the movement of two muscles or work in the reverse way, that is, split into several muscle movements.

Table 9.1 FAP to AU mapping

Action Units	FAPs
AU1	raise_l_i_eyebrow + raise_r_i_eyebrow
AU2	raise_l_o_eyebrow + raise_r_o_eyebrow
AU3	
AU4	raise_l_o_eyebrow + raise_r_o_eyebrow + raise_l_m_eyebrow + raise_r_m_eyebrow + raise_l_I_eyebrow + raise_r_i_eyebrow + squeeze_l_eyebrow + squeeze_r_eyebrow
AU5	close_t_l_eyelid + close_t_r_eyelid
AU6	lift_l_cheek + lift_r_cheek
AU7	close_b_l_eyelid + close_b_r_eyelid
AU8	
AU9	lower_t_midlip + raise_nose + stretch_l_nose + stretch_r_nose
AU10	raise_nose (+stretch_l_nose + stetch_r_nose) + lower_t_midlip
AU11	
AU12	push_t_lip + push_b_lip(+lower_lowerlip + lower_t_midlip + raise_b_midlip)
AU13	
AU14	
AU15	lower_l_cornerlip + lower_r_cornerlip
AU16	
AU17	depress_chin
AU18	
AU19	
AU20	raise_b_midlip + lower_l_cornerlip + lower_r_cornerlip + stretch_l_cornerlip + stretch_r_cornerlip + lower_t_lip_lm + raise_b_lip_lm + lower_t_lip_lm_o + raise_b_lip_lm_o + raise_l_cornerlip_o + lower_t_lip_rm + raise_b_lip_rm + lower_t_lip_rm_o + raise_b_lip_rm_o + raise_r_cornerlip_o

MPEG-4 FAPs are also strongly related to AUs, as shown in Table 9.1. Description of archetypal expressions by means of muscle movements and AUs has been the starting point for setting the archetypal expression description through FAPs.

Hints for this mapping were obtained from psychological studies [5, 13, 14] that refer to face formation during expression generation, as well as from experimental data provided by classic databases such as Ekman's and MediaLab's (see also Section 9.3). Table 9.2 illustrates the description of archetypal expressions and some variations of them using the MPEG-4 FAPs terminology. It should be noted that the sets shown in Table 9.2 consist of the vocabulary of FAPs to be used for each archetypal expression and not a particular profile for synthesizing–analyzing expressions. This means that, if animated, they would not necessarily produce the corresponding expression. In the following table we define an *expression profile* to be a subset of the FAPs vocabulary, corresponding to a particular expression, accompanied with FAP intensities, that is, the actual ranges of variation that if animated, creates the required expression. Several expression profiles based on the FAPs vocabulary proposed in Table 9.2 are shown in the experimental results section.

Table 9.2 FAPs vocabulary for archetypal expression description

Joy	open_jaw (F_3), lower_t_midlip (F_4), raise_b_midlip (F_5), stretch_l_cornerlip (F_6), stretch_r_cornerlip (F_7), raise_l_cornerlip (F_{12}), raise_r_cornerlip (F_{13}), close_t_l_eyelid (F_{19}), close_t_r_eyelid (F_{20}), close_b_l_eyelid (F_{21}), close_b_r_eyelid (F_{22}), raise_l_m_eyebrow (F_{33}), raise_r_m_eyebrow (F_{34}), lift_l_cheek (F_{41}), lift_r_cheek (F_{42}), stretch_l_cornerlip_o (F_{53}), stretch_r_cornerlip_o (F_{54})
Sadness	close_t_l_eyelid (F_{19}), close_t_r_eyelid (F_{20}), close_b_l_eyelid (F_{21}), close_b_r_eyelid (F_{22}), raise_l_i_eyebrow (F_{31}), raise_r_i_eyebrow (F_{32}), raise_l_m_eyebrow (F_{33}), raise_r_m_eyebrow (F_{34}), raise_l_o_eyebrow (F_{35}), raise_r_o_eyebrow (F_{36})
Anger	lower_t_midlip (F_4), raise_b_midlip (F_5), push_b_lip (F_{16}), depress_chin (F_{18}), close_t_l_eyelid (F_{19}), close_t_r_eyelid (F_{20}), close_b_l_eyelid (F_{21}), close_b_r_eyelid (F_{22}), raise_l_i_eyebrow (F_{31}), raise_r_i_eyebrow (F_{32}), raise_l_m_eyebrow (F_{33}), raise_r_m_eyebrow (F_{34}), raise_l_o_eyebrow (F_{35}), raise_r_o_eyebrow (F_{36}), squeeze_l_eyebrow (F_{37}), squeeze_r_eyebrow (F_{38})
Fear	open_jaw (F_3), lower_t_midlip (F_4), raise_b_midlip (F_5), lower_t_lip_lm (F_8), lower_t_lip_rm (F_9), raise_b_lip_lm (F_{10}), raise_b_lip_rm (F_{11}), close_t_l_eyelid (F_{19}), close_t_r_eyelid (F_{20}), close_b_l_eyelid (F_{21}), close_b_r_eyelid (F_{22}), raise_l_i_eyebrow (F_{31}), raise_r_i_eyebrow (F_{32}), raise_l_m_eyebrow (F_{33}), raise_r_m_eyebrow (F_{34}), raise_l_o_eyebrow (F_{35}), raise_r_o_eyebrow (F_{36}), squeeze_l_eyebrow (F_{37}), squeeze_r_eyebrow (F_{38})
Disgust	open_jaw (F_3), lower_t_midlip (F_4), raise_b_midlip (F_5), lower_t_lip_lm (F_8), lower_t_lip_rm (F_9), raise_b_lip_lm (F_{10}), raise_b_lip_rm (F_{11}), close_t_l_eyelid (F_{19}), close_t_r_eyelid (F_{20}), close_b_l_eyelid (F_{21}), close_b_r_eyelid (F_{22}), raise_l_m_eyebrow (F_{33}), raise_r_m_eyebrow (F_{34}), lower_t_lip_lm_o (F_{55}), lower_t_lip_rm_o (F_{56}), raise_b_lip_lm_o (F_{57}), raise_b_lip_rm_o (F_{58}), raise_l_cornerlip_o (F_{59}), raise_r_cornerlip_o (F_{60})
Surprise	open_jaw (F_3), raise_b_midlip (F_5), stretch_l_cornerlip (F_6), stretch_r_cornerlip (F_7), raise_b_lip_lm (F_{10}), raise_b_lip_rm (F_{11}), close_t_l_eyelid (F_{19}), close_t_r_eyelid (F_{20}), close_b_l_eyelid (F_{21}), close_b_r_eyelid (F_{22}), raise_l_i_eyebrow (F_{31}), raise_r_i_eyebrow (F_{32}), raise_l_m_eyebrow (F_{33}), raise_r_m_eyebrow (F_{34}), raise_l_o_eyebrow (F_{35}), raise_r_o_eyebrow (F_{36}), squeeze_l_eyebrow (F_{37}), squeeze_r_eyebrow (F_{38}), stretch_l_cornerlip_o (F_{53}), stretch_r_cornerlip_o (F_{54})

9.3 THE RANGE OF VARIATION OF FAPs IN REAL VIDEO SEQUENCES

An important issue, useful to both emotion analysis and synthesis systems, is the range of variation of the FAPs that are involved in facial expression formation. From the synthesis point of view, a study has been carried out [2] that refers to FAPs range definition. However, the suggested ranges of variation are rather loose and cannot be used for analysis purposes. In order to have clear cues about FAPs range of variation in real video sequences, we analyzed two well-known datasets showing archetypal expressions, Ekman's (static) [5] and MediaLab's (dynamic) [15], and computed statistics about the involved FAPs. Both sets show extreme cases of expressions, rather than everyday ones. However, they can be used for setting limits to the variance of the respective FAPs [16, 17]. To achieve this, a way of modeling FAPs through the movement of facial points is required. Analysis of FAPs range of variation in real images and video sequences is used next for two purposes:

1. To verify and complete the proposed vocabulary for each archetypal expression.
2. To define profiles of archetypal expressions.

9.3.1 Modeling FAPs through the Movement of Facial Points

Although FAPs are practical and very useful for animation purposes, they are inadequate for analyzing facial expressions from video scenes or still images. The main reason is the absence of quantitative definitions for FAPs as well as their nonadditive nature. Note that the same problem holds for the FACS AUs. This is quite reasonable, given the strong relationship between AUs and FAPs (see Table 9.1). In order to measure facial-related FAPs in real images and video sequences, it is necessary to define a way of describing them through the movement of points that lie in the facial area and that can be automatically detected. Such a description could gain advantage from the extended research on automatic facial point detection [18, 19]. Quantitative description of FAPs based on particular FDP points, which correspond to movement of protuberant facial points, provides the means of bridging the gap between expression analysis and animation–synthesis. In the expression analysis case, the nonadditive property of the FAPs can be addressed by a fuzzy rule system similar to the one described later in Section 9.5.

Quantitative modeling of FAPs is implemented using the features labeled as f_i ($i = 1, \ldots, 15$) in the third column of Table 9.3 [16]. The feature set employs FDP points that lie in the facial area and under some constraints, can be automatically detected and tracked. It consists of distances, noted as $s(x, y)$ where x and y correspond to FDP points shown in Figure 9.2b, between these protuberant points, some of which are constant during expressions and are used as reference points. Distances between reference points are used for normalization [see Figure 9.2a]. The units for f_i are identical to those corresponding to FAPs, even in cases where no one-to-one relation exists.

It should be noted that not all FAPs included in the vocabularies shown in Table 9.2 can be modeled by distances between facial protuberant points (e.g. *raise_b_lip_lm_o*, *lower_t_lip_lm_o*). In such cases, the corresponding FAPs are retained in the vocabulary and their ranges of variation are experimentally defined on the basis of facial animations. Moreover, some features serve for the estimation of the range of variation of more than one FAP (e.g. features f_{12} to f_{15}).

9.3.2 Vocabulary Verification

To obtain clear cues about the FAPs range of variation in real video sequences as well as to verify the vocabulary of FAPs involved in each archetypal emotion, we analyzed Ekman's and MediaLab's datasets, which show archetypal expressions. The analysis was based on the quantitative modeling of FAPs described in the previous section. Computed statistics are summarized in Table 9.4. Mean values provide typical values that can be used for particular expression profiles, while the standard deviation can

Table 9.3 Quantitative FAPs modeling (1) $s(x, y)$ is the Euclidean distance between the FDP points x and y shown in Figure 9.2b, (2) $D_{\text{i-NEUTRAL}}$ refers to the distance D_i when the face is its in neutral position

FAP name	Feature for the description	Utilized feature	Unit
squeeze_l_eyebrow (F_{37})	$D_1 = s(4.6, 3.8)$	$f_1 = D_{\text{1-NEUTRAL}} - D_1$	ES
squeeze_r_eyebrow (F_{38})	$D_2 = s(4.5, 3.11)$	$f_2 = D_{\text{2-NEUTRAL}} - D_2$	ES
lower_t_midlip (F_4)	$D_3 = s(9.3, 8.1)$	$f_3 = D_3 - D_{\text{3-NEUTRAL}}$	MNS
raise_b_midlip (F_5)	$D_4 = s(9.3, 8.2)$	$f_4 = D_{\text{4-NEUTRAL}} - D_4$	MNS
raise_l_i_eyebrow (F_{31})	$D_5 = s(4.2, 3.8)$	$f_5 = D_5 - D_{\text{5-NEUTRAL}}$	ENS
raise_r_i_eyebrow (F_{32})	$D_6 = s(4.1, 3.11)$	$f_6 = D_6 - D_{\text{6-NEUTRAL}}$	ENS
raise_l_o_eyebrow (F_{35})	$D_7 = s(4.6, 3.12)$	$f_7 = D_7 - D_{\text{7-NEUTRAL}}$	ENS
raise_r_o_eyebrow (F_{36})	$D_8 = s(4.5, 3.7)$	$f_8 = D_8 - D_{\text{8-NEUTRAL}}$	ENS
raise_l_m_eyebrow (F_{33})	$D_9 = s(4.4, 3.12)$	$f_9 = D_9 - D_{\text{9-NEUTRAL}}$	ENS
raise_r_m_eyebrow (F_{34})	$D_{10} = s(4.3, 3.7)$	$f_{10} = D_{10} - D_{\text{10-NEUTRAL}}$	ENS
open_jaw (F_3)	$D_{11} = s(8.1, 8.2)$	$f_{11} = D_{11} - D_{\text{11-NEUTRAL}}$	MNS
close_t_l_eyelid (F_{19}) – close_b_l_eyelid (F_{21})	$D_{12} = s(3.2, 3.4)$	$f_{12} = D_{12} - D_{\text{12-NEUTRAL}}$	IRISD
close_t_r_eyelid (F_{20}) – close_b_r_eyelid (F_{22})	$D_{13} = s(3.1, 3.3)$	$f_{13} = D_{13} - D_{\text{13-NEUTRAL}}$	IRISD
stretch_l_cornerlip (F_6) (stretch_l_cornerlip_o)(F_{53}) – stretch_r_cornerlip (F_7) (stretch_r_cornerlip_o) (F_{54})	$D_{14} = s(8.4, 8.3)$	$f_{14} = D_{14} - D_{\text{14-NEUTRAL}}$	MW
squeeze_l_eyebrow (F_{37}) and squeeze_r_eyebrow (F_{38})	$D_{15} = s(4.6, 4.5)$	$f_{15} = D_{\text{15-NEUTRAL}} - D_{15}$	ES

define the range of variation (see also Section 9.3.3). The units of the values shown are those of the corresponding FAPs [2]. The symbol (*) expresses the absence of the corresponding FAP in the vocabulary of that particular expression, while the symbol (–) shows that although the corresponding FAP is included in the vocabulary, it has not been verified by the statistical analysis. The latter case shows that not all FAPs included in the vocabulary are experimentally verified.

The detection of the facial point subset used to describe the FAPs involved in the archetypal expressions was based on the work presented in Reference 20. To obtain accurate detection, in many cases, human assistance was necessary. The authors are working toward a fully automatic implementation of the FDP points detection procedure.

Figure 9.3 illustrates particular statistics, computed over the previously described datasets, for the expression *joy*. In all diagrams, the horizontal axis shows the indices of the features defined in the third column of Table 9.3, while the vertical axis shows the minimum, maximum and mean values of the corresponding feature. From this figure it is confirmed, for example, that *lower_t_midlip* (feature with index 3) that refers to lowering the middle of the upper lip is employed, because even the maximum value for this FAP is below zero. In the same way, the FAPs *raise_l_m_eyebrow, raise_r_m_eyebrow, close_t_l_eyelid, close_t_r_eyelid,*

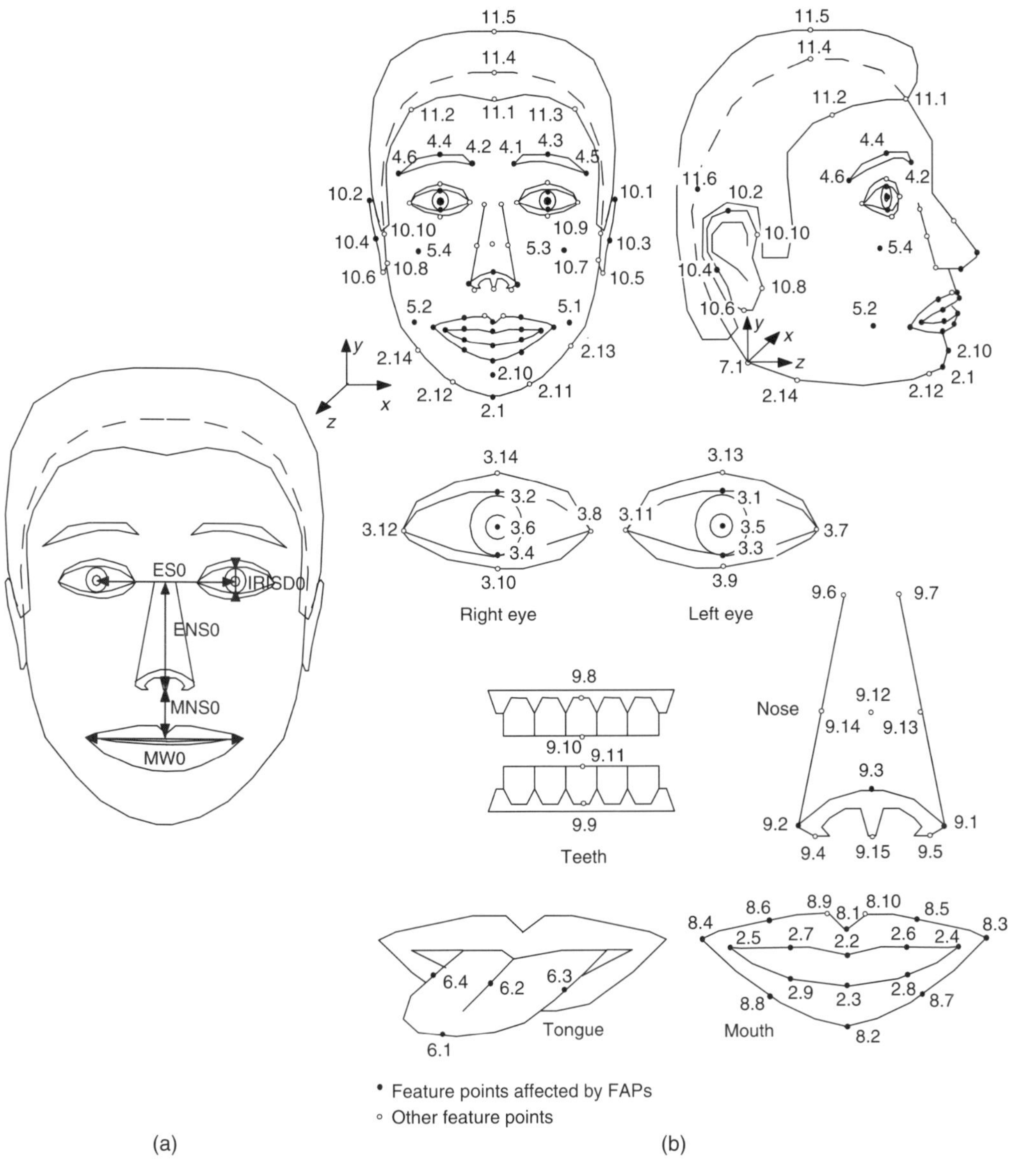

Figure 9.2 (a) Normalization distances; (b) FDP points

close_b_l_eyelid, *close_b_r_eyelid*, *stretch_l_cornerlip*, *stretch_r_cornerlip* (indices 9, 10, 12, 13, 14) are verified. Some of the aforementioned FAPs are described using a single variable. For example, the *stretch_l_cornerlip* and *stretch_r_cornerlip* are both modeled via f_{14} (their values, shown in Table 9.4, are equal to the half value of feature f_{14}). Similar to Figure 9.3, Figure 9.4 illustrates feature statistics for the expression *surprise*.

Table 9.4 Statistics for the vocabulary of FAPs for the archetypal expression. The symbol (*) expresses the absence of the corresponding FAP in the vocabulary of the particular expression while symbol (–) shows that although the corresponding FAP is included in the vocabulary, it has not been verified by the statistical analysis

FAP name (symbol)	Stats	Anger	Sadness	Joy	Disgust	Fear	Surprise
open_jaw (F_3)	Mean	*	*	–	–	291	885
	StD	*	*	–	–	189	316
lower_t_midlip (F_4)	Mean	73	*	−271	−234	–	*
	StD	51	*	110	109	–	*
raise_b_midlip (F_5)	Mean	*	*	–	−177	218	−543
	StD	*	*	–	108	135	203
stretch_l_cornerlip (F_6), stretch_l_cornerlip_o (F_{53}),	Mean	*	*	234	*	*	−82
stretch_ r_cornerlip (F_7), stretch_r_cornerlip_o (F_{54})	StD	*	*	98	*	*	39
lower_t_lip_lm (F_8)	Mean	*	*	*	–	*	*
	StD	*	*	*	–	*	*
lower_t_lip_rm (F_9)	Mean	*	*	*	–	*	*
	StD	*	*	*	–	*	*
raise_b_lip_lm (F_{10})	Mean	*	*	*	–	*	*
	StD	*	*	*	–	*	*
raise_b_lip_rm (F_{11})	Mean	*	*	*	–	*	*
	StD	*	*	*	–	*	*
close_t_l_eyelid (F_{19}),	Mean	–	−153	−254	203	−244	−254
close_b_l_eyelid (F_{21})	StD	–	112	133	148	126	83
close_t_r_eyelid (F_{20}),	Mean	–	−161	−242	211	−249	−252
close_b_r_eyelid (F_{22})	StD		109	122	145	128	81
raise_l_i_eyebrow (F_{31})	Mean	−83	85	*	*	104	224
	StD	48	55	*	*	69	103
raise_r_i_eyebrow (F_{32})	Mean	−85	80	*	*	111	211
	StD	51	54	*	*	72	97
raise_l_m_eyebrow (F_{33})	Mean	−149	–	24	−80	72	144
	StD	40	–	22	53	58	64
raise_r_m_eyebrow (F_{34})	Mean	−144	–	25	−82	75	142
	StD	39	–	22	54	60	62
raise_l_o_eyebrow (F_{35})	Mean	−66	–	*	*	–	54
	StD	35	–	*	*	–	31
raise_r_o_eyebrow (F_{36})	Mean	−70	–	*	*	–	55
	StD	38		*	*	–	31
squeeze_l_eyebrow (F_{37})	Mean	57	*	*	*	–	–
	StD	28	*	*	*	–	–
squeeze_r_eyebrow (F_{38})	Mean	58	*	*	*	–	–
	StD	31	*	*	*	–	–

Table 9.4 *(continued)*

FAP name (symbol)	Stats	Anger	Sadness	Joy	Disgust	Fear	Surprise
lift_l_cheek (F_{41})	Mean	*	*	–	*	*	*
	StD	*	*	–	*	*	*
lift_r_cheek (F_{42})	Mean	*	*	–	*	*	*
	StD	*	*	–	*	*	*
stretch_l_cornerlip_o (F_{53})	Mean	*	*	–	*	*	–
	StD	*	*	–	*	*	–
stretch_r_cornerlip_o (F_{54})	Mean	*	*	*	*	*	–
	StD	*	*	*	*	*	
lower_t_lip_lm_o (F_{55})	Mean	*	*	*	–	*	*
	StD	*	*	*	–	*	*
lower_t_lip_rm_o (F_{56})	Mean	*	*	*	–	*	*
	StD	*	*	*	–	*	*
raise_b_lip_lm_o (F_{57})	Mean	*	*	*	–	*	*
	StD	*	*	*	–	*	*
raise_b_lip_rm_o (F_{58})	Mean	*	*	*	–	*	*
	StD	*	*	*	–	*	*
raise_l_cornerlip_o (F_{59})	Mean	*	*	*	–	*	*
	StD	*	*	*	–	*	*
raise_r_cornerlip_o (F_{60})	Mean	*	*	*	–	*	*
	StD	*	*	*	–	*	*

9.3.3 Creating Archetypal Expression Profiles

An archetypal *expression profile* is a set of FAPs accompanied by the corresponding range of variation, which, if animated, produces a visual representation of the corresponding emotion. Typically, a profile of an archetypal expression consists of vocabulary definitions coupled with the appropriate ranges of variation for the corresponding subset of FAPs. The statistical expression analysis performed on the aforementioned datasets is useful for FAPs vocabulary completion and verification, as well as for a rough estimation of the range of variation of FAPs, but not for profile creation. In order to define exact profiles for the archetypal expressions three steps were followed:

1. Subsets of FAPs that are candidates to form an archetypal expression were defined by translating the face formations proposed by psychological studies [5, 13, 14] to FAPs.
2. Initial ranges of variation were computed on the basis of the statistics shown in Table 9.4 (see the following table for a detailed description).
3. The corresponding profiles were animated to verify the appropriateness of derived representations.

The initial range of variation for the FAPs was computed as follows: Let $m_{i,j}$ and $\sigma_{i,j}$ be the mean value and standard deviation of FAP F_j for the archetypal

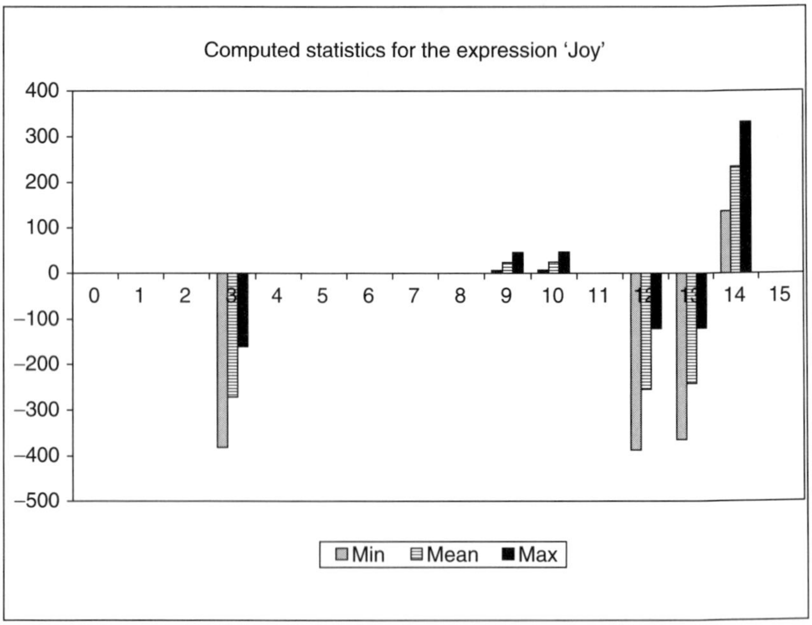

Figure 9.3 Computed statistics for the expression 'Joy'. Horizontal axis shows the indices of the features defined in the third column of Table 9.3 while vertical axis shows the value of the corresponding feature

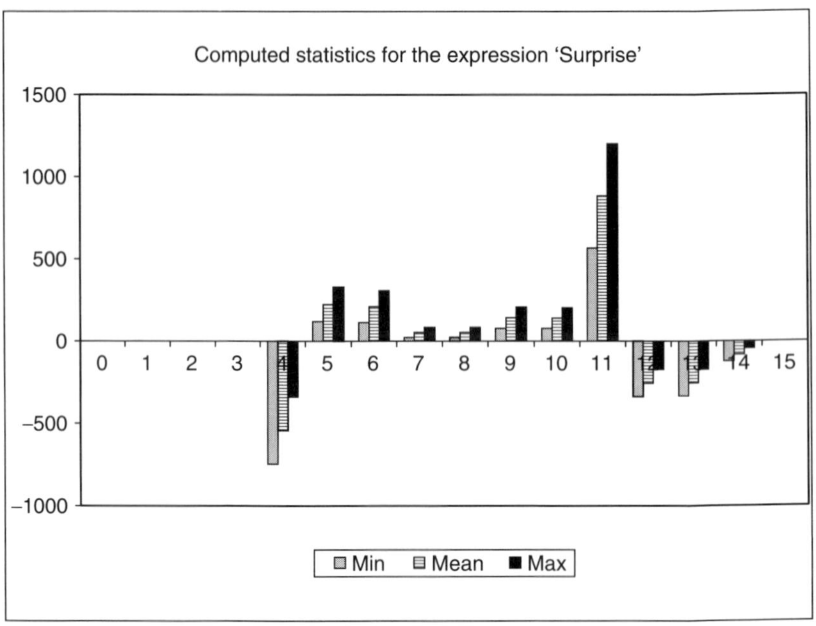

Figure 9.4 Computed statistics for the expression 'Surprise'. The horizontal axis shows the indices of the features defined in the third column of Table 9.3 while the vertical axis shows the value of the corresponding feature

expression i (where $I = \{1 \Rightarrow \text{Anger}, 2 \Rightarrow \text{Sadness}, 3 \Rightarrow \text{Joy}, 4 \Rightarrow \text{Disgust}, 5 \Rightarrow \text{Fear}, 6 \Rightarrow \text{Surprise}\}$), as estimated in Table 9.4. The initial range of variation $X_{i,j}$ of FAP F_j for the archetypal expression i is defined as:

$$X_{i,j} = [m_{i,j} - \sigma_{i,j}, m_{i,j} + \sigma_{i,j}] \tag{9.1}$$

for bidirectional, and

$$X_{i,j} = [\max(0, m_{i,j} - \sigma_{i,j}), m_{i,j} + \sigma_{i,j}] \text{ or } X_{i,j} = [m_{i,j} - \sigma_{i,j}, \min(0, m_{i,j} + \sigma_{i,j})] \tag{9.2}$$

for unidirectional FAPs [2].

Following the procedure described in the preceding text, Table 9.5 was produced showing examples of archetypal expression profiles:

Table 9.5 Profiles for the archetypal emotions

Profiles	FAPs and range of variation
Anger ($P_A^{(0)}$)	$F_4 \in [22, 124]$, $F_{31} \in [-131, -25]$, $F_{32} \in [-136, -34]$, $F_{33} \in [-189, -109]$, $F_{34} \in [-183, -105]$, $F_{35} \in [-101, -31]$, $F_{36} \in [-108, -32]$, $F_{37} \in [29, 85]$, $F_{38} \in [27, 89]$
$P_A^{(1)}$	$F_{19} \in [-330, -200]$, $F_{20} \in [-335, -205]$, $F_{21} \in [200, 330]$, $F_{22} \in [205, 335]$, $F_{31} \in [-200, -80]$, $F_{32} \in [-194, -74]$, $F_{33} \in [-190, -70]$, $F_{34} =\in [-190, -70]$
$P_A^{(2)}$	$F_{19}[-330, -200]$, $F_{20} \in [-335, -205]$, $F_{21} \in [200, 330]$, $F_{22} \in [205, 335]$, $F_{31} \in [-200, -80]$, $F_{32} \in [-194, -74]$, $F_{33} \in [70, 190]$, $F_{34} \in [70, 190]$
$P_A^{(3)}$	$F_{16} \in [45, 155]$, $F_{18} \in [45, 155]$, $F_{19} \in [-330, -200]$, $F_{20} \in [-330, -200]$, $F_{31} \in [-200, -80]$, $F_{32} \in [-194, -74]$, $F_{33} \in [-190, -70]$, $F_{34} \in [-190, -70]$, $F_{37} \in [65, 135]$, $F_{38} \in [65, 135]$
$P_A^{(4)}$	$F_{16} \in [-355, -245]$, $F_{18} \in [145, 255]$, $F_{19} \in [-330, -200]$, $F_{20} \in [-330, -200]$, $F_{31} \in [-200, -80]$, $F_{32} \in [-194, -74]$, $F_{33} \in [-190, -70]$, $F_{34} \in [-190, -70]$, $F_{37} \in [65, 135]$, $F_{38} \in [65, 135]$
Sadness ($P_S^{(0)}$)	$F_{19} \in [-265, -41]$, $F_{20} \in [-270, -52]$, $F_{21} \in [-265, -41]$, $F_{22} \in [-270, -52]$, $F_{31} \in [30, 140]$, $F_{32} \in [26, 134]$
Joy ($P_J^{(0)}$)	$F_4 \in [-381, -161]$, $F_6 \in [136, 332]$, $F_7 \in [136, 332]$, $F_{19} \in [-387, -121]$, $F_{20} \in [-364, -120]$, $F_{21} \in [-387, -121]$, $F_{22} \in [-364, -120]$, $F_{33} \in [2, 46]$, $F_{34} \in [3, 47]$, $F_{53} \in [136, 332]$, $F_{54} \in [136, 332]$
$P_J^{(1)}$	$F_6 \in [160, 240]$, $F_7 \in [160, 240]$, $F_{12} \in [260, 340]$, $F_{13} \in [260, 340]$, $F_{19} \in [-449, -325]$, $F_{20} \in [-426, -302]$, $F_{21} \in [325, 449]$, $F_{22} \in [302, 426]$, $F_{33} \in [70, 130]$, $F_{34} \in [70, 130]$, $F_{41} \in [130, 170]$, $F_{42} \in [130, 170]$, $F_{53} \in [160, 240]$, $F_{54} \in [160, 240]$
$P_J^{(2)}$	$F_6 \in [160, 240]$, $F_7 \in [160, 240]$, $F_{12} \in [260, 340]$, $F_{13} \in [260, 340]$, $F_{19} \in [-449, -325]$, $F_{20} \in [-426, -302]$, $F_{21} \in [-312, -188]$, $F_{22} \in [-289, -165]$, $F_{33} \in [70, 130]$, $F_{34} \in [70, 130]$, $F_{41} \in [130, 170]$, $F_{42} \in [130, 170]$, $F_{53} \in [160, 240]$, $F_{54} \in [160, 240]$

(*continued overleaf*)

Table 9.5 (*continued*)

Profiles	FAPs and range of variation
$P_J^{(3)}$	$F_6 \in [160, 240]$, $F_7 \in [160, 240]$, $F_{12} \in [260, 340]$, $F_{13} \in [260, 340]$, $F_{19} \in [-449, -325]$, $F_{20} \in [-426, -302]$, $F_{21} \in [61, 185]$, $F_{22} \in [38, 162]$, $F_{33} \in [70, 130]$, $F_{34} \in [70, 130]$, $F_{41} \in [130, 170]$, $F_{42} \in [130, 170]$, $F_{53} \in [160, 240]$, $F_{54} \in [160, 240]$
Disgust ($P_D^{(0)}$)	$F_4 \in [-343, -125]$, $F_5 \in [-285, -69]$, $F_{19} \in [55, 351]$, $F_{20} \in [66, 356]$, $F_{21} \in [55, 351]$, $F_{22} \in [66, 356]$, $F_{33} \in [-123, -27]$, $F_{34} \in [-126, -28]$
Fear ($P_F^{(0)}$)	$F_3 \in [102, 480]$, $F_5 \in [83, 353]$, $F_{19} \in [-370, -118]$, $F_{20} \in [-377, -121]$, $F_{21} \in [-370, -118]$, $F_{22} \in [-377, -121]$, $F_{31} \in [35, 173]$, $F_{32} \in [39, 183]$, $F_{33} \in [14, 130]$, $F_{34} \in [15, 135]$
$P_F^{(1)}$	$F_3 \in [400, 560]$, $F_5 \in [333, 373]$, $F_{19} \in [-400, -340]$, $F_{20} \in [-407, -347]$, $F_{21} \in [-400, -340]$, $F_{22} \in [-407, -347]$
$P_F^{(2)}$	$F_3 \in [400, 560]$, $F_5 \in [307, 399]$, $F_{19} \in [-530, -470]$, $F_{20} \in [-523, -463]$, $F_{21} \in [-530, -470]$, $F_{22} \in [-523, -463]$, $F_{31} \in [460, 540]$, $F_{32} \in [460, 540]$, $F_{33} \in [460, 540]$, $F_{34} \in [460, 540]$, $F_{35} \in [460, 540]$, $F_{36} \in [460, 540]$
$P_F^{(3)}$	$F_3 \in [400, 560]$, $F_5 \in [-240, -160]$, $F_{19} \in [-630, -570]$, $F_{20} \in [-630, -570]$, $F_{21} \in [-630, -570]$, $F_{22} \in [-630, -570]$, $F_{31} \in [460, 540]$, $F_{32} \in [460, 540]$, $F_{37} \in [60, 140]$, $F_{38} \in [60, 140]$
$P_F^{(4)}$	$F_3 \in [400, 560]$, $F_5 \in [-240, -160]$, $F_{19} \in [-630, -570]$, $F_{20} \in [-630, -570]$, $F_{21} \in [-630, -570]$, $F_{22} \in [-630, -570]$, $F_{31} \in [460, 540]$, $F_{32} \in [460, 540]$, $F_{33} \in [360, 440]$, $F_{34} \in [360, 440]$, $F_{35} \in [260, 340]$, $F_{36} \in [260, 340]$, $F_{37} \in [60, 140]$, $F_{38} \in [60, 140]$
$P_F^{(5)}$	$F_3 \in [400, 560]$, $F_5 \in [-240, -160]$, $F_{19} \in [-630, -570]$, $F_{20} \in [-630, -570]$, $F_{21} \in [-630, -570]$, $F_{22} \in [-630, -570]$, $F_{31} \in [460, 540]$, $F_{32} \in [460, 540]$, $F_{33} \in [360, 440]$, $F_{34} \in [360, 440]$, $F_{35} \in [260, 340]$, $F_{36} \in [260, 340]$, $F_{37} \in 0$, $F_{38} \in 0$
$P_F^{(6)}$	$F_3 \in [400, 560]$, $F_5 \in [-240, -160]$, $F_8 \in [-120, -80]$, $F_9 \in [-120, -80]$, $F_{10} \in [-120, -80]$, $F_{11} \in [-120, -80]$, $F_{19} \in [-630, -570]$, $F_{20} \in [-630, -570]$, $F_{21} \in [-630, -570]$, $F_{22} \in [-630, -570]$, $F_{31} \in [460, 540]$, $F_{32} \in [460, 540]$, $F_{33} \in [360, 440]$, $F_{34} \in [360, 440]$, $F_{35} \in [260, 340]$, $F_{36} \in [260, 340]$, $F_{37} \in 0$, $F_{38} \in 0$
$P_F^{(7)}$	$F_3 \in [400, 560]$, $F_5 \in [-240, -160]$, $F_{19} \in [-630, -570]$, $F_{20} \in [-630, -570]$, $F_{21} \in [-630, -570]$, $F_{22} \in [-630, -570]$, $F_{31} \in [360, 440]$, $F_{32} \in [360, 440]$, $F_{33} \in [260, 340]$, $F_{34} \in [260, 340]$, $F_{35} \in [160, 240]$, $F_{36} \in [160, 240]$
$P_F^{(8)}$	$F_3 \in [400, 560]$, $F_5 \in [-240, -160]$, $F_{19} \in [-630, -570]$, $F_{20} \in [-630, -570]$, $F_{21} \in [-630, -570]$, $F_{22} \in [-630, -570]$, $F_{31} \in [260, 340]$, $F_{32} \in [260, 340]$, $F_{33} \in [160, 240]$, $F_{34} \in [160, 240]$, $F_{35} \in [60, 140]$, $F_{36} \in [60, 140]$
$P_F^{(9)}$	$F_3 \in [400, 560]$, $F_5 \in [307, 399]$, $F_{19} \in [-630, -570]$, $F_{20} \in [-623, -563]$, $F_{21} \in [-630, -570]$, $F_{22} \in [-623, -563]$, $F_{31} \in [460, 540]$, $F_{32} \in [460, 540]$, $F_{33} \in [460, 540]$, $F_{34} \in [460, 540]$, $F_{35} \in [460, 540]$, $F_{36} \in [460, 540]$
Surprise ($P_{Su}^{(0)}$)	$F_3 \in [569, 1201]$, $F_5 \in [-746, -340]$, $F_6 \in [-121, -43]$, $F_7 \in [-121, -43]$, $F_{19} \in [-337, -170]$, $F_{20} \in [-333, -171]$, $F_{21} \in [-337, -170]$, $F_{22} \in [-333, -171]$, $F_{31} \in [121, 327]$, $F_{32} \in [114, 308]$, $F_{33} \in [80, 208]$, $F_{34} \in [80, 204]$, $F_{35} \in [23, 85]$, $F_{36} \in [24, 86]$, $F_{53} \in [-121, -43]$, $F_{54} \in [-121, -43]$

Table 9.5 *(continued)*

Profiles	FAPs and range of variation
$P_{Su}^{(1)}$	$F_3 \in [1150, 1252]$, $F_5 \in [-792, -700]$, $F_6 \in [-141, -101]$, $F_7 \in [-141, -101]$, $F_{10} \in [-530, -470]$, $F_{11} \in [-530, -470]$, $F_{19} \in [-350, -324]$, $F_{20} \in [-346, -320]$, $F_{21} \in [-350, -324]$, $F_{22} \in [-346, -320]$, $F_{31} \in [314, 340]$, $F_{32} \in [295, 321]$, $F_{33} \in [195, 221]$, $F_{34} \in [191, 217]$, $F_{35} \in [72, 98]$, $F_{36} \in [73, 99]$, $F_{53} \in [-141, -101]$, $F_{54} \in [-141, -101]$
$P_{Su}^{(2)}$	$F_3 \in [834, 936]$, $F_5 \in [-589, -497]$, $F_6 \in [-102, -62]$, $F_7 \in [-102, -62]$, $F_{10} \in [-380, -320]$, $F_{11} \in [-380, -320]$, $F_{19} \in [-267, -241]$, $F_{20} \in [-265, -239]$, $F_{21} \in [-267, -241]$, $F_{22} \in [-265, -239]$, $F_{31} \in [211, 237]$, $F_{32} \in [198, 224]$, $F_{33} \in [131, 157]$, $F_{34} \in [129, 155]$, $F_{35} \in [41, 67]$, $F_{36} \in [42, 68]$
$P_{Su}^{(3)}$	$F_3 \in [523, 615]$, $F_5 \in [-386, -294]$, $F_6 \in [-63, -23]$, $F_7 \in [-63, -23]$, $F_{10} \in [-230, -170]$, $F_{11} \in [-230, -170]$, $F_{19} \in [-158, -184]$, $F_{20} \in [-158, -184]$, $F_{21} \in [-158, -184]$, $F_{22} \in [-158, -184]$, $F_{31} \in [108, 134]$, $F_{32} \in [101, 127]$, $F_{33} \in [67, 93]$, $F_{34} \in [67, 93]$, $F_{35} \in [10, 36]$, $F_{36} \in [11, 37]$

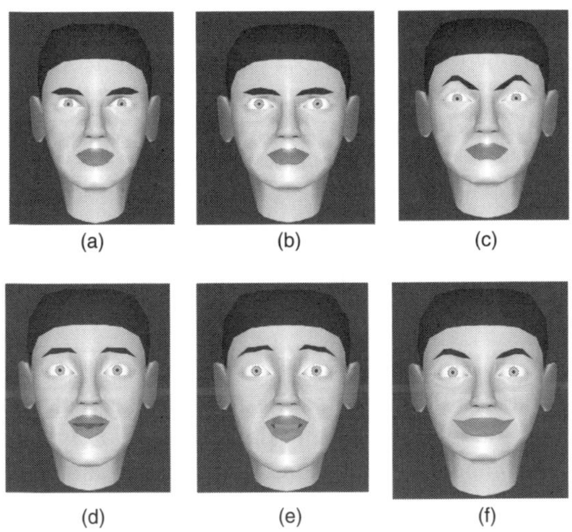

Figure 9.5 Examples of animated profiles (a)–(c) anger; (d)–(e) surprise; (f) joy

Generally, for animation purposes, every MPEG-4 decoder has to provide and use an MPEG-4-compliant face model whose geometry can be defined using FDP points or it should define the animation rules based on face animation tables (FAT). Using FATs, we can specify which model vertices should be moved for each FAP, and how. We can also define the transformed nodes of the face as well the kind of transformations. For our experiments on setting the archetypal expression profiles we used the face model developed in the context of the European Project *ACTS MoMuSys* [21]. This is freely available at the website *http://www.iso.ch/ittf*.

Figure 9.5 shows some examples of animated profiles. Figure 9.5a shows a particular profile for the archetypal expression *anger*, while Figures 9.5b,c show alternative profiles of the same expression. The difference between them is due to FAP intensities. Difference in FAP intensities is also shown in Figure 9.5d and e, both illustrating profiles of expression *surprise*. Finally, Figure 9.5f shows an example of a profile of the expression *joy*.

9.4 CREATING PROFILES FOR NONARCHETYPAL EXPRESSIONS

In this section we propose a method for creating profiles for nonarchetypal expressions. Since computer scientists and engineers have carried out a limited number of studies dealing with emotions other than the archetypal ones [10], it is necessary to search in other subject–discipline bibliographies. Psychologists have examined a broader set of emotions [17], but very few of the studies provide results that can be exploited in computer graphics and machine vision fields. One of these studies carried out by Whissel [8], suggests that emotions are points in a space spanning a relatively small number of dimensions that seem to occupy two axes: *activation* and *evaluation*, as shown in Table 9.6. *Activation* is the degree of arousal associated with terms such as *patient* (at 3.3) representing a midpoint, *surprised* (over 6) representing high activation and *bashful* (around 2) representing low *activation*. *Evaluation* is the degree of pleasantness associated with the terms *guilty* (at 1.1) representing the negative extreme and *delighted* (at 6.4) representing the positive extreme [8]. In our experience, the estimation of *evaluation* through gross facial attributes is difficult (even intractable). On the other hand, it does appear possible to estimate *activation* on the basis of facial points' movement.

The third column in Table 9.6 represents Plutchik's [7] observation that emotion terms are unevenly distributed through the space defined by dimensions such as Whissell's. Instead, they tend to form an approximately circular pattern called *emotion wheel*. Shown values refer to an angular measure that runs from *Acceptance* (0) to *Disgust* (180).

For the creation of profiles for intermediate emotions we consider two cases:

1. Emotions that are similar in nature to an archetypal one; for example, they may differ only in the intensity of muscle actions.
2. Emotions that cannot be considered as related to any of the archetypal ones.

In both cases we proceed by following the steps enumerated in the following text:

1. Utilize either the *activation* parameter or Plutchik's *angular* measure as *a priori* knowledge about the intensity of facial actions for several emotions. This knowledge is combined with the profiles of archetypal expressions through a rule-based system to create profiles for intermediate emotions.
2. Animate the produced profiles for testing/correcting their appropriateness in terms of the visual similarity with the requested emotion.

9.4.1 Universal Emotion Categories

As a general rule, one can define six broad categories, each one characterized by an archetypal emotion. Within each of these categories, intermediate expressions are described by different emotional and optical intensities, as well as minor variation in expression details. From the synthetic point of view, emotions that belong to the same category can be rendered by animating the same FAPs using different intensities. For example, the emotion group *fear* also contains *worry* and *terror* [14]; reducing or increasing the intensities of the relevant FAPs allows these two emotions to be

Table 9.6 Emotion words from Whissel's [8] study

	Activation	Evaluation	Angle		Activation	Evaluation	Angle
Accepting			0	Disgusted	5	3.2	161.3
Adventurous	4.2	5.9	270.7	Disinterested	2.1	2.4	127.3
Affectionate	4.7	5.4	52.3	Disobedient			242.7
Afraid	4.9	3.4	70.3	Displeased			181.5
Aggressive	5.9	2.9	232	Dissatisfied	4.6	2.7	183
Agreeable	4.3	5.2	5	Distrustful	3.8	2.8	185
Amazed	5.9	5.5	152	Eager	5	5.1	311
Ambivalent	3.2	4.2	144.7	Ecstatic	5.2	5.5	286
Amused	4.9	5	321	Elated			311
Angry	4.2	2.7	212	Embarrassed	4.4	3.1	75.3
Annoyed	4.4	2.5	200.6	Empty	3.1	3.8	120.3
Antagonistic	5.3	2.5	220	Enthusiastic	5.1	4.8	313.7
Anticipatory	3.9	4.7	257	Envious	5.3	2	160.3
Anxious	6	2.3	78.3	Exasperated			239.7
Apathetic	3	4.3	90	Expectant			257.3
Apprehensive			83.3	Forlorn			85
Ashamed	3.2	2.3	83.3	Furious	5.6	3.7	221.3
Astonished	5.9	4.7	148	Generous			328
Attentive	5.3	4.3	322.4	Gleeful	5.3	4.8	307
Awed			156.7	Gloomy	2.4	3.2	132.7
Bashful	2	2.7	74.7	Greedy	4.9	3.4	249
Bewildered	3.1	2.3	140.3	Grief-stricken			127.3
Bitter	6.6	4	186	Grouchy	4.4	2.9	230
Boastful	3.7	3	257.3	Guilty	4	1.1	102.3
Bored	2.7	3.2	136	Happy	5.3	5.3	323.7
Calm	2.5	5.5	37	Helpless	3.5	2.8	80
Cautious	3.3	4.9	77.7	Hesitant			134
Cheerful	5.2	5	25.7	Hopeful	4.7	5.2	298
Confused	4.8	3	141.3	Hopeless	4	3.1	124.7
Contemptuous	3.8	2.4	192	Hostile	4	1.7	222
Content	4.8	5.5	338.3	Humiliated			84
Contrary	2.9	3.7	184.3	Impatient	3.4	3.2	230.3
Cooperative	3.1	5.1	340.7	Impulsive	3.1	4.8	255
Critical	4.9	2.8	193.7	Indecisive	3.4	2.7	134
Curious	5.2	4.2	261	Indignant			175
Daring	5.3	4.4	260.1	Inquisitive			267.7
Defiant	4.4	2.8	230.7	Interested			315.7
Delighted	4.2	6.4	318.6	Intolerant	3.1	2.7	185
Demanding	5.3	4	244	Irritated	5.5	3.3	202.3
Depressed	4.2	3.1	125.3	Jealous	6.1	3.4	184.7
Despairing	4.1	2	133	Joyful	5.4	6.1	323.4
Disagreeable	5	3.7	176.4	Loath	3.5	2.9	193
Disappointed	5.2	2.4	136.7	Lonely	3.9	3.3	88.3
Discouraged	4.2	2.9	138	Meek	3	4.3	91

(*continued overleaf*)

Table 9.6 (*continued*)

	Activation	Evaluation	Angle		Activation	Evaluation	Angle
Nervous	5.9	3.1	86	Self-conscious			83.3
Obedient	3.1	4.7	57.7	Self-controlled	4.4	5.5	326.3
Obliging	2.7	3	43.3	Serene	4.3	4.4	12.3
Outraged	4.3	3.2	225.3	Shy			72
Panicky	5.4	3.6	67.7	Sociable	4.8	5.3	296.7
Patient	3.3	3.8	39.7	Sorrowful	4.5	3.1	112.7
Pensive	3.2	5	76.7	Stubborn	4.9	3.1	190.4
Perplexed			142.3	Submissive	3.4	3.1	73
Playful			269.7	Surprised	6.5	5.2	146.7
Pleased	5.3	5.1	328	Suspicious	4.4	3	182.7
Possessive	4.7	2.8	247.7	Sympathetic	3.6	3.2	331.3
Proud	4.7	5.3	262	Terrified	6.3	3.4	75.7
Puzzled	2.6	3.8	138	Timid			65
Quarrelsome	4.6	2.6	229.7	Tolerant			350.7
Ready			329.3	Trusting	3.4	5.2	345.3
Receptive			32.3	Unaffectionate	3.6	2.1	227.3
Reckless			261	Uncertain			139.3
Rebellious	5.2	4	237	Uncooperative			191.7
Rejected	5	2.9	136	Unfriendly	4.3	1.6	188
Remorseful	3.1	2.2	123.3	Unhappy			129
Resentful	5.1	3	176.7	Unreceptive			170
Revolted			181.3	Unsympathetic			165.6
Sad	3.8	2.4	108.5	Vacillating			137.3
Sarcastic	4.8	2.7	235.3	Vengeful			186
Satisfied	4.1	4.9	326.7	Watchful			133.3
Scared			66.7	Wondering	3.3	5.2	249.7
Scornful	5.4	4.9	227	Worried	3.9	2.9	126

synthesized or discriminated. In the case of expression profiles, this affects the range of variation of the corresponding FAPs that is appropriately translated. The fuzziness that is introduced by the varying scale of the change of FAP intensity, also provides assistance in achieving some differentiation between outputs associated with similar situations. This ensures on the one hand, that synthesis will not render 'robotlike' animation, but noticeably more realistic results; and on the other hand, that analysis systems could in principle discriminate 'neighboring' emotions.

Let $P_i^{(k)}$ be the kth profile of emotion i and $X_{i,j}^{(k)}$ be the range of variation of FAP F_j involved in $P_i^{(k)}$. If A, I are emotions belonging to the same universal emotion category, A being the archetypal and I the intermediate one, then the following rules are applied:

Rule 1:	$P_A^{(k)}$ and $P_I^{(k)}$ employ the same FAPs.
Rule 2:	The range of variation $X_{I,j}^{(k)}$ is computed by $X_{I,j}^{(k)} = \frac{a_I}{a_A} X_{A,j}^{(k)}$
Rule 3:	a_A and a_I are the values of the *activation* parameter for emotion words A and I obtained from Whissel's study [8].

9.4.2 Intermediate Emotions

Creating profiles for emotions that do not clearly belong to a universal category is not straightforward. Apart from estimating the range of variations for FAPs, one should also define the vocabulary of FAPs for the particular emotion. In order to proceed we utilize both the *emotion wheel* of Plutchik [7], especially the *angular* measure (shown also in Table 9.6); and the *activation* parameter. Let I be an intermediate emotion lying between archetypal emotions A_1 and A_2 (that are supposed to be the nearest, with respect to the two sides of emotion I) according to their *angular* measure. Let also V_{A_1} and V_{A_2} be the vocabularies (sets of FAPs) corresponding to A_1 and A_2, respectively. The vocabulary V_I of emotion I emerges as the union of vocabularies V_{A_1} and V_{A_2}, that is, $V_I = V_{A_1} \cup V_{A_2}$.

As already stated in Section 9.2, defining a vocabulary is not enough for modeling expressions–profiles should be created for this purpose. This poses a number of interesting issues, such as: (1) what happens if an FAP is included in both V_{A_1} and V_{A_2}, but, with contradictory motion directions? (2) What happens if an FAP is included in only one of the vocabularies? In our approach, FAPs included in both V_{A_1} and V_{A_2}, that also have a common motion direction are retained in the new profile (their range of variation emerges as a weighted average of the consisting ones). FAPs included in only one of the vocabularies are averaged with the respective neutral position. The same applies in the case of contradictory FAPs (FAPs included in both V_{A_1} and V_{A_2}, but that have, however, contradictory motion directions). Averaging of the intensities usually favors the most exaggerated of the emotions that are combined, whereas FAPs with contradicting intensities cancel out. In practice, this approach works successfully, as shown in the actual results that follow. In the following table we describe the way to merge profiles of archetypal emotions and create profiles of intermediate ones:

Let $P_{A_1}^{(k)}$ be the kth profile of emotion A_1 and $P_{A_2}^{(l)}$ the lth profile of emotion A_2, then the following rules are applied so as to create a profile $P_I^{(m)}$ for the intermediate emotion I:

Rule 1: $P_I^{(m)}$ includes FAPs that are involved either in $P_{A_1}^{(k)}$ or $P_{A_2}^{(l)}$.

Rule 2: If F_j is an FAP involved in both $P_{A_1}^{(k)}$ and $P_{A_2}^{(l)}$ with the same sign (direction of movement), then the range of variation $X_{I,j}^{(k)}$ is computed as a weighted translation of $X_{A_1,j}^{(k)}$ and $X_{A_2,j}^{(l)}$ (where $X_{A_1,j}^{(k)}$ and $X_{A_2,j}^{(l)}$ are the ranges of variation of FAP F_j involved in $P_{A_1}^{(k)}$ and $P_{A_2}^{(l)}$, respectively) in the following way: (1) the translated range of variations $t(X_{A_1,j}^{(k)}) = \frac{a_I}{a_{A_1}} X_{A_1,j}^{(k)}$ and $t(X_{A_2,j}^{(k)}) = \frac{a_I}{a_{A_2}} X_{A_2,j}^{(k)}$ of $X_{A_1,j}^{(k)}$ and $X_{A_2,j}^{(l)}$ are computed, (2) the center and length $c_{A_1,j}^{(k)}, s_{A_1,j}^{(k)}$ of $t(X_{A_1,j}^{(k)})$ and $c_{A_2,j}^{(k)}, s_{A_2,j}^{(k)}$ of $t(X_{A_2,j}^{(k)})$ are also computed, (3) the length of $X_{I,j}^{(k)}$ is $s_{I,j}^{(m)} = \frac{\omega_I - \omega_{A_1}}{\omega_{A_2} - \omega_{A_1}} s_{A_1,j}^{(k)} + \frac{\omega_{A_2} - \omega_I}{\omega_{A_2} - \omega_{A_1}} s_{A_2,j}^{(l)}$ and its midpoint is

$$c_{I,j}^{(m)} = \frac{\omega_I - \omega_{A_1}}{\omega_{A_2} - \omega_{A_1}} c_{A_1,j}^{(k)} + \frac{\omega_{A_2} - \omega_I}{\omega_{A_2} - \omega_{A_1}} c_{A_2,j}^{(l)}$$

Rule 3: If F_j is involved in both $P_{A_1}^{(k)}$ and $P_{A_2}^{(l)}$ but with contradictory signs (opposite directions of movement), then the range of variation $X_{I,j}^{(k)}$ is computed by $X_{I,j}^{(m)} = \frac{a_I}{a_{A_1}} X_{A_1,j}^{(k)} \cap \frac{a_I}{a_{A_2}} X_{A_2,j}^{(l)}$. In case $X_{I,j}^{(k)}$ is eliminated (which is the most possible situation) then F_j is excluded from the profile.

Rule 4: If F_j is involved only in one of $P_{A_1}^{(k)}$ and $P_{A_2}^{(l)}$, then the range of variation $X_{I,j}^{(k)}$ will be averaged with the corresponding of the neutral face position, that is, $X_{I,j}^{(m)} = \frac{a_I}{2 * a_{A_1}} X_{A_1,j}^{(k)}$ or $X_{I,j}^{(m)} = \frac{a_I}{2 * a_{A_2}} X_{A_2,j}^{(l)}$

Rule 5: a_{A_1}, a_{A_2} and a_I are the values of the *activation* parameter for emotion words A_1, A_2 and I, obtained from Whissel's study [8].

Rule 6: $\omega_{A_1}, \omega_{A_2}$ and $\omega_I, \omega_{A_1} < \omega_I < \omega_{A_2}$ are the *angular* parameters for emotion words A_1, A_2 and I, obtained from Plutchik's study [7].

It should be noted that the profiles, created using the aforementioned rules have to be animated for testing and correction purposes. The final profiles are those that present an acceptable visual similarity with the requested real emotion.

9.5 THE EMOTION ANALYSIS SYSTEM

In this section we present a way of utilizing profile-based emotion modeling for emotion understanding purposes. By doing so, we show that modeling emotions serves purposes related to both synthesis and analysis.

Figures 9.6 and 9.7 show the way the emotion analysis system functions. Let us consider as input to the emotion analysis system a 15-element length feature vector $\underline{f}$ that corresponds to the 15 features f_i shown in the third column of Table 9.3. The particular values of $\underline{f}$ can be rendered to FAP values as shown in the first column of the same table (see also Section 9.3.1) resulting in an input vector $\underline{G}$. The elements of $\underline{G}$ express the observed values of the corresponding FAPs (for example G_1 refers to the value of F_{37}).

Let $X_{i,j}^{(k)}$ be the range of variation of FAP F_j involved in the kth profile $P_i^{(k)}$ of emotion i. If $c_{i,j}^{(k)}$ and $s_{i,j}^{(k)}$ are the middle point and length of interval of $X_{i,j}^{(k)}$, respectively, then we describe a fuzzy class $A_{i,j}^{(k)}$ for F_j, using the membership function $\mu_{i,j}^{(k)}$ shown in Figure 9.8. Let also $\Delta_i^{(k)}$ be the set of classes $A_{i,j}^{(k)}$ that correspond to profile $P_i^{(k)}$; the beliefs $p_i^{(k)}$ and b_i that are observed through the vector $\underline{G}$ facial state corresponds to profile $P_i^{(k)}$ and emotion i respectively and are computed through the following equations:

$$p_i^{(k)} = \prod_{A_{i,j}^{(k)} \in \Delta_i^{(k)}} r_{i,j}^{(k)} \tag{9.3}$$

$$b_i = \max_k (p_i^{(k)}) \tag{9.4}$$

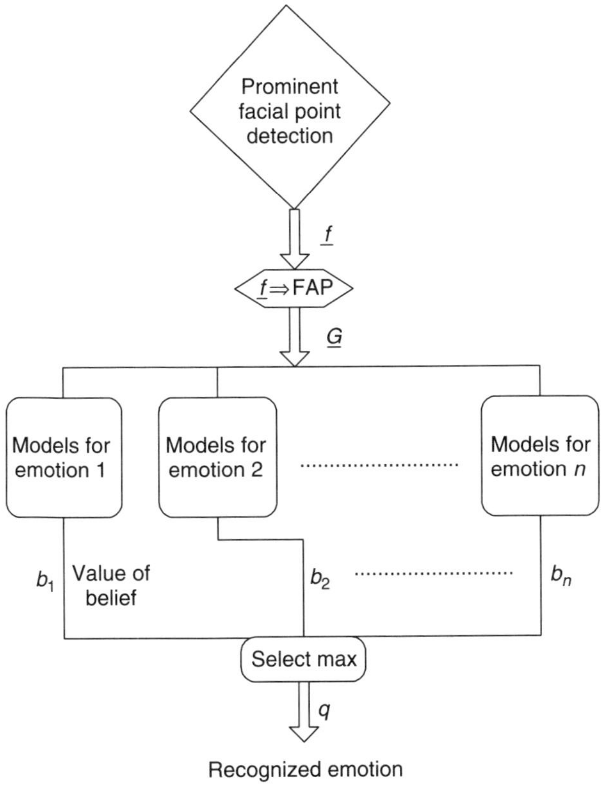

Figure 9.6 The emotion analysis system

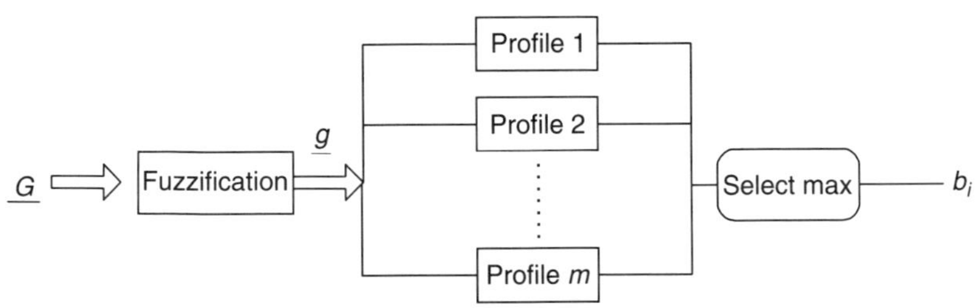

Figure 9.7 The fuzzy inference subsystem

where

$$r_{i,j}^{(k)} = \max\{g_i \cap A_{i,j}^{(k)}\} \tag{9.5}$$

expresses the *relevance* $r_{i,j}^{(k)}$ of the ith element of the input feature vector with respect to class $A_{i,j}^{(k)}$. Actually $\underline{g} = A'(\underline{G}) = \{g_1, g_2, \ldots\}$ is the fuzzified input vector resulting from a *singleton* fuzzification procedure [22].

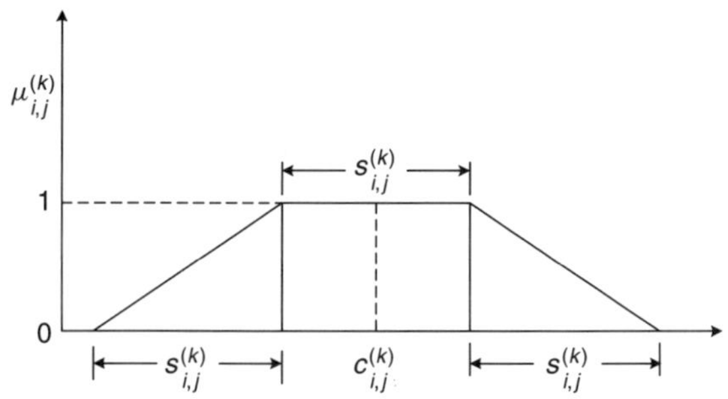

Figure 9.8 The form of membership functions

If a final decision about what is the observed emotion has to be made then the following equation is used:

$$q = \arg\max_i b_i \tag{9.6}$$

It is observed through Equation (9.3) that the various emotion profiles correspond to the fuzzy intersection of several sets and are implemented through a τ-*norm* of the form $t(a, b) = a \cdot b$. Similarly, the belief that an observed feature vector corresponds to a particular emotion results from a fuzzy union of several sets (see Equation 9.4) through an σ-*norm* that is implemented as $u(a, b) = \max(a, b)$.

It should be noted that in the previously described emotion analysis system no hypothesis has been made about the number of recognizable emotions. This number is limited only by the number of profiles that have been modeled. Thus, the system can be used for analyzing either a few of the archetypal emotions or many more, using the methodology described in Section 9.4 to create profiles for nonarchetypal emotions.

9.6 EXPERIMENTAL RESULTS

In this section we show the efficiency of the proposed scheme in modeling archetypal and intermediate emotions according to the methodology described in the previous sections. Animated profiles were created using the face model developed in the context of the European Project *ACTS MoMuSys* [21], as well as the 3-D model of the software package *Poser*, Edition 4 of Curious Labs Company. This model has separate parts for each moving face part. The Poser model interacts with the controls in Poser and has joints that move naturally, as in a real person. Poser mirrors real-face movements by adding joint parameters to each face part. This allows us to manipulate the figure based on those parameters. We can control the eyes, the eyebrows and the mouth of the model by filling the appropriate parameters. To achieve this, a mapping from FAPs to Poser parameters is necessary. We did this mapping mainly experimentally. The relationship between FAPs and Poser parameters is more or less straightforward.

The first set of experiments shows synthesized archetypal expressions (see Figure 9.9) created by using the Poser software package. The 3-D nature of the face

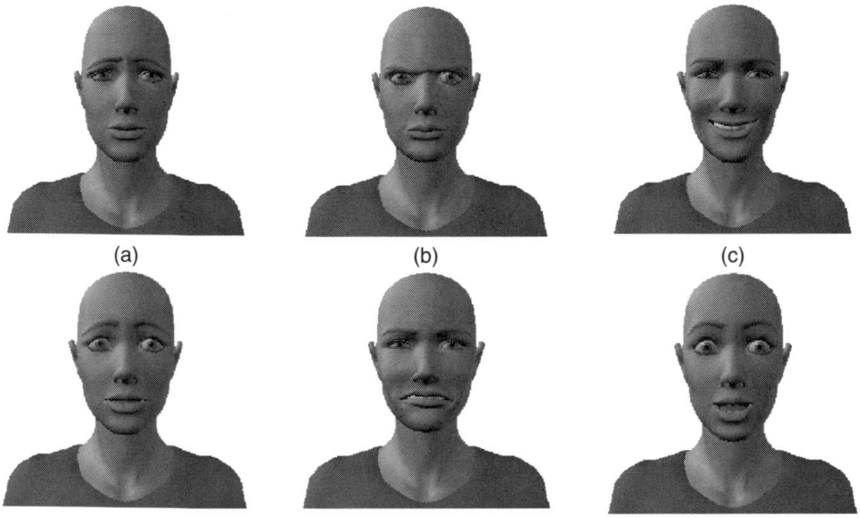

Figure 9.9 Synthesized archetypal expressions created using the 3-D model of the POSER software package (a) sadness; (b) anger; (c) joy; (d) fear; (e) disgust and (f) surprise

model renders the underlying emotions in a more natural way than the MPEG-4 compatible face model (compare Figures 9.5e and f for the emotions *surprise* and *joy* with the Figures 9.9f and 9.9c respectively). However, in both cases the synthesized examples are rather convincing.

The second set of experiments show particular examples in creating nonarchetypal expressions based on our proposed method. More details are given in the following sections.

9.6.1 Creating Profiles for Emotions Belonging to a Universal Category

In this section we illustrate the proposed methodology for creating profiles for emotions that belong to the same universal category as an archetypal one. Emotion terms *afraid*, *terrified* and *worried* are considered to belong to the emotion category *fear* [13] whose modeling base is the term *afraid* (see Figures 9.10 and 9.11). Table 9.7 shows profiles for the terms *terrified* and *worried* that have been generated from the basic profile of *afraid* (in particular $P_{\mathrm{F}}^{(8)}$). The range of variation $X_{\mathrm{T},j}^{(8)}$ of FAP F_j belonging to the eighth profile of the emotion term *terrified* is computed by the equation $X_{\mathrm{T},j}^{(8)} = (6.3/4.9)X_{\mathrm{F},j}^{(8)}$, where $X_{\mathrm{F},j}^{(8)}$ is the range of variation of FAP F_j belonging to the eighth profile of the emotion term *afraid*. Similarly $X_{\mathrm{W},j}^{(8)} = (3.9/4.9)X_{\mathrm{F},j}^{(8)}$ is the range of variation of FAP F_j belonging to the eighth profile of the emotion term *worried*. Figures 9.10 (a) to (c) and 9.11 (a) to (c) show the animated profiles for the emotion terms *afraid*, *terrified* and *worried* respectively. The FAP values that we used are the median ones of the corresponding ranges of variation.

Table 9.7 Created profiles for the emotions terror and worry

Emotion term	Activation	Profile
Afraid	4.9	$F_3 \in [400, 560]$, $F_5 \in [-240, -160]$, $F_{19} \in [-630, -570]$, $F_{20} \in [-630, -570]$, $F_{21} \in [-630, -570]$, $F_{22} \in [-630, -570]$, $F_{31} \in [260, 340]$, $F_{32} \in [260, 340]$, $F_{33} \in [160, 240]$, $F_{34} \in [160, 240]$, $F_{35} \in [60, 140]$, $F_{36} \in [60, 140]$
Terrified	6.3	$F_3 \in [520, 730]$, $F_5 \in [-310, -210]$, $F_{19} \in [-820, -740]$, $F_{20} \in [-820, -740]$, $F_{21} \in [-820, -740]$, $F_{22} \in [-820, -740]$, $F_{31} \in [340, 440]$, $F_{32} \in [340, 440]$, $F_{33} \in [210, 310]$, $F_{34} \in [210, 310]$, $F_{35} \in [80, 180]$, $F_{36} \in [80, 180]$
Worried	3.9	$F_3 \in [320, 450]$, $F_5 \in [-190, -130]$, $F_{19} \in [-500, -450]$, $F_{20} \in [-500, -450]$, $F_{21} \in [-500, -450]$, $F_{22} \in [-500, -450]$, $F_{31} \in [210, 270]$, $F_{32} \in [210, 270]$, $F_{33} \in [130, 190]$, $F_{34} \in [130, 190]$, $F_{35} \in [50, 110]$, $F_{36} \in [50, 110]$

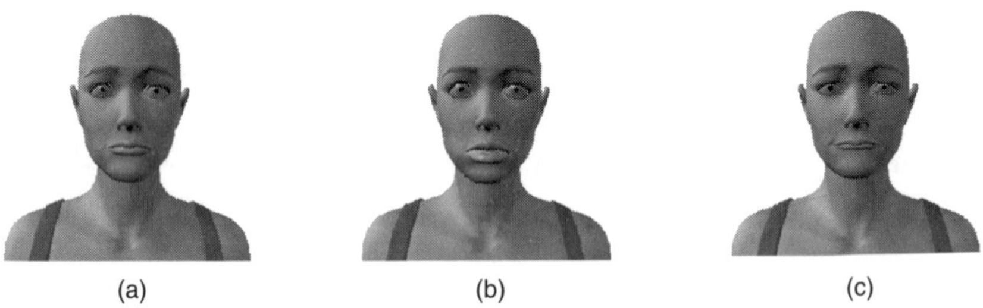

Figure 9.10 Poser face model. Animated profiles for emotion terms (a) afraid; (b) terrified and (c) worried

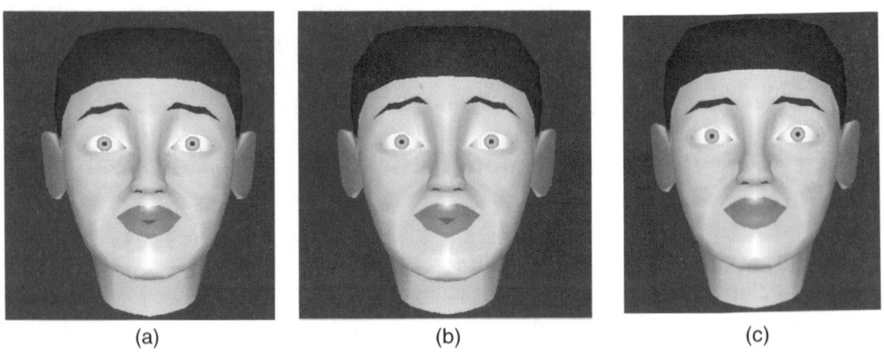

Figure 9.11 MPEG-4 face model. Animated profiles for emotion terms (a) afraid; (b) terrified and (c) worried

9.6.2 Creating Profiles for Nonarchetypal Emotions

In this section we describe a method for creating a profile for the emotion *guilt*. According to Plutchik's *angular* measure (see Table 9.6), the emotion term *guilty* (angular measure 102.3 degrees) lies between the archetypal emotion terms *afraid* (angular measure 70.3 degrees) and *sad* (angular measure 108.5 degrees), being closer to the latter (see Figures 9.12 and 9.13). According to Section 9.4.2 the vocabulary V_G of emotion

Table 9.8 Created profile for the emotion guilt

Emotion term	Activation	Angular measure	Profile
Afraid	4.9	70.3	$F_3 \in [400, 560]$, $F_5 \in [-240, -160]$, $F_{19} \in [-630, -570]$, $F_{20} \in [-630, -570]$, $F_{21} \in [-630, -570]$, $F_{22} \in [-630, -570]$, $F_{31} \in [260, 340]$, $F_{32} \in [260, 340]$, $F_{33} \in [160, 240]$, $F_{34} \in [160, 240]$, $F_{35} \in [60, 140]$, $F_{36} \in [60, 140]$
Guilty	4	102.3	$F_3 \in [160, 230]$, $F_5 \in [-100, -65]$, $F_{19} \in [-110, -310]$, $F_{20} \in [-120, -315]$, $F_{21} \in [-110, -310]$, $F_{22} \in [-120, -315]$, $F_{31} \in [61, 167]$, $F_{32} \in [57, 160]$, $F_{33} \in [65, 100]$, $F_{34} \in [65, 100]$, $F_{35} \in [25, 60]$, $F_{36} \in [25, 60]$
Sad	3.9	108.5	$F_{19} \in [-265, -41]$, $F_{20} \in [-270, -52]$, $F_{21} \in [-265, -41]$, $F_{22} \in [-270, -52]$, $F_{31} \in [30, 140]$, $F_{32} \in [26, 134]$

guilt emerges as the union of vocabularies V_F and V_S, that is, $V_G = V_F \cup V_S$, where V_F and V_S are the vocabularies corresponding to emotions *fear* and *sad* respectively. Table 9.8 shows a profile for the term *guilty* generated from an underlying profile of the term *afraid* (in particular $P_F^{(8)}$) and *sad* ($P_S^{(0)}$). FAPs F_3, F_5, F_{33} to F_{36} are included only in the $P_F^{(8)}$ and therefore the corresponding ranges of variation in the emerging *guilty* profile $P_G^{(m)}$ (*m*th *guilty* profile) are computed by averaging the ranges of variation of $P_F^{(8)}$ with the neutral face, according to *Rule 4* (see Section 9.4.2); for example $X_{G,3}^{(m)} = (4/2 * 4.9)X_{F,4}^{(8)}$. FAPs F_{19} to F_{22}, F_{31}, F_{32} are included in both $P_F^{(8)}$ and $P_S^{(0)}$, with the same direction of movement, thus *Rule 2* is followed. For example, the range of variation $X_{G,19}^{(m)}$ for FAP F_{29} term is computed as follows:

$$t(X_{F,19}^{(8)}) = \frac{4}{4.9} X_{F,19}^{(8)} => [-510, -460], c_{F,19}^{(8)} = -485, s_{F,19}^{(8)} = 50,$$

$$t(X_{S,19}^{(0)}) = \frac{4}{3.9} X_{S,19}^{(0)} => [-270, -42], c_{S,19}^{(0)} = -156, s_{S,19}^{(9)} = 228,$$

since $\omega_F = 70.3°$, $\omega_S = 108.5°$, $\omega_G = 102.3°$, $c_{G,19}^{(m)} = [(102.3 - 70.3)/(108.5 - 70.3)] * (-156) + [(108.5 - 102.3)/(108.5 - 70.3)] * (-485) = -209 s_{G,19}^{(m)} = [(102.3 - 70.3)/(108.5 - 70.3)] * 228 + [(108.5 - 102.3)/(108.5 - 70.3)] * 50 = 199$, and $X_{G,19}^{(m)}$ corresponds to the range $[-110, -310]$.

9.7 CONCLUSION–DISCUSSION

In this chapter we have proposed a complete framework for creating visual profiles based on FAPs for intermediate (not primary) emotions. Emotion profiles can serve either the vision part of an emotion recognition system or a client side application that creates synthetic expressions. The main advantage of the proposed system is its flexibility:

- No hypothesis needs to be made about what the facial points detection system is (see Figure 9.1); it is enough to detect the movement of a predefined set of FDP points.

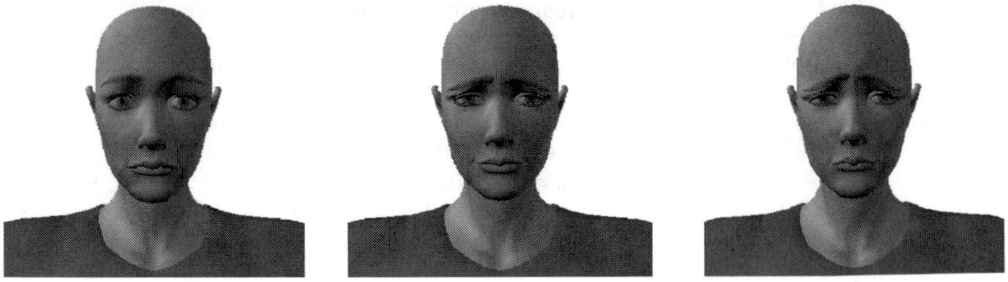

(a) (b) (c)

Figure 9.12 Poser face model: Animated profiles for emotion terms (a) afraid; (b) guilty and (c) sad

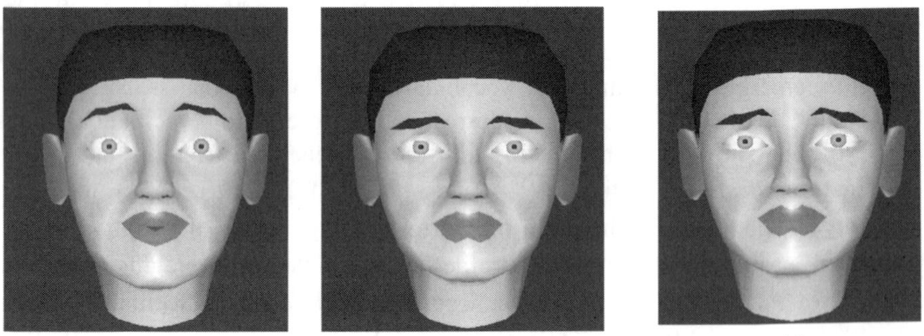

(a) (b) (c)

Figure 9.13 MPEG-4 face model. Animated profiles for emotion terms (a) afraid; (b) guilty and (c) sad

- The system is extensible with respect to completing (or modifying) the proposed vocabulary of FAPs for the archetypal expressions
- The range of variation of FAPs that are involved in the archetypal expression profiles can be modified. Note, however, that this modification affects the profiles of the nonarchetypal emotions.
- The system is extensible with respect to the number of nonarchetypal expressions that can be modeled.
- The system can be used either for expression synthesis or for expression analysis. In the former case, a rule-based procedure serves as an agent for synthesizing expressions, while in the latter case a fuzzy inference system provides the means of an autonomous emotion analysis system.

Exploitation of the results obtained by psychological studies related with emotion recognition from computer scientists is possible, although not straightforward. We have shown that terms such as the *emotion wheel* and *activation* are suitable for extending the emotions that can be visually modeled. Extension of these results combining audio and visual emotion analysis systems is currently under investigation in the framework – an EC-funded project called ERMIS [23].

REFERENCES

[1] ISO/IEC JTC1/SC29/WG11 N3205, Multi-Users Technology (Requirements and Applications), Maui, December, 1999.

[2] M. Tekalp, J. Ostermann, "Face and 2-D mesh animation in MPEG-4", Tutorial Issue on the MPEG-4 Standard, *Signal Processing: Image Communication*, **15**(4–5), 387–421 (2000).

[3] P. Ekman, W. Friesen, *The Facial Action Coding System*, Consulting Psychologists Press, San Francisco, Calif., 1978, *http://www.paulekman.com*.

[4] EC TMR Project PHYSTA Report, Development of Feature Representation from Facial Signals and Speech, January, 1999.

[5] P. Ekman, "Facial expression and emotion", *American Psychologist*, **48**, 384–392 (1993).

[6] P. Ekman, W. V. Friesen, *Pictures of Facial Affect*, Consulting Psychologists Press, Palo Alto, Calif., 1978.

[7] R. Plutchik, *Emotion: A Psychoevolutionary Synthesis*, Harper and Row, New York, 1980.

[8] C. M. Whissel, "The dictionary of affect in language", in R. Plutchik, H. Kellerman, eds., *Emotion: Theory, Research and Experience: vol. 4, The Measurement of Emotions*, Academic Press, New York, 1989.

[9] J. Russell, L. Barrett-Feldman "Core affect, prototypical emotional episodes, and other things called emotion: dissecting the elephant", *Journal of Personality and Social Psychology*, **76**, 805–819 (1999).

[10] R. Cowie, E. Douglas-Cowie, N. Tsapatsoulis, G. Votsis, S. Kollias, W. Fellenz, J. Taylor, "Emotion recognition in human–computer interaction", *IEEE Signal Processing Magazine*, **18**(1), 32–80 (2001).

[11] N. Tsapatsoulis, K. Karpouzis, G. Stamou, F. Piat, S. Kollias, A fuzzy system for emotion classification based on the MPEG-4 facial definition parameter set, *Proceedings of the EUSIPCO-2000*, Tampere, Finland, September, 2000.

[12] A. Raouzaiou, N. Tsapatsoulis, S. Kollias, Parameterised facial expression synthesis for videoconferencing applications, *Proceedings of International Workshop on Very Low Bitrate Video Coding (VLBV01)*, Athens, Greece, October, 2001.

[13] F. Parke, K. Waters, *Computer Facial Animation*, A K Peters, Wellesley, Mass., 1996.

[14] G. Faigin, *The Artist's Complete Guide to Facial Expressions*, Watson-Guptill, New York, 1990.

[15] *http://www.media.mit.edu/*.

[16] K. Karpouzis, N. Tsapatsoulis, S. Kollias, Moving to continuous facial expression space using the MPEG-4 facial definition parameter (FDP) set, *Proceedings of the Electronic Imaging 2000 Conference of SPIE*, San Jose, Calif., January, 2000.

[17] EC TMR Project PHYSTA Report, Review of Existing Techniques for Human Emotion Understanding and Applications in Human–Computer Interaction, October, 1998.

[18] P. Chellapa, C. Wilson, S. Sirohey, "Human and machine recognition of faces: a survey", *Proceedings of IEEE*, **83**(5), 705–740 (1995).

[19] H. A. Rowley, S. Baluja, T. Kanade, "Neural network-based face detection", *IEEE Transactions on PAMI*, **20**(1), 23–28 (1998).

[20] K.-M. Lam, H. Yan, "An analytic-to-holistic approach for face recognition based on a single frontal view", *IEEE Transactions on PAMI*, **20**(7), 1998, *http://www.computer.org/tpami/tp1998/i7toc.htm*.

[21] ISTFACE – MPEG-4 Facial Animation System – Version 3.3.1 Gabriel Abrantes, (Developed in the Context of the European Project ACTS MoMuSys © 97–98 Instituto Superior Tecnico).

[22] G. J. Klir, B. Yuan, *Fuzzy Sets and Fuzzy Logic, Theory and Applications*, Prentice Hall, Upper Saddle River, N. J., 1995.

[23] IST Project: Emotionally Rich Man-Machine Interaction Systems (ERMIS), 2001–2003.

10

The InterFace Software Platform for Interactive Virtual Characters

Igor S. Pandzic[1], Michele Cannella[2], Franck Davoine[3], Robert Forchheimer[1], Fabio Lavagetto[4], Haibo Li[5], Andrew Marriott[6], Sotiris Malassiotis[7], Montse Pardas[8], Roberto Pockaj[3] and Gael Sannier[9]

[1]*Linköping University, Linköping, Sweden,* [2]*TAU SARL, Italy,* [3]*Université de Technologie de Compiegne, Compiegne, France,* [4]*University of Genova, Genova, Italy,* [5]*Umeå University, Umeå, Sweden,* [6]*Curtin University, Perth, Australia,* [7]*ITI, Thessaloniki, Greece,* [8]*Universitat Politècnica de Catalunya, Barcelona, Spain,* [9]*W Interactive SARL, Gaillard, France*

ABSTRACT

We describe the Common Software Platform (CSP) for interactive virtual characters developed within the IST project InterFace. Collaborative research effort has put numerous demands on the development of a common platform, and these demands had to be reconciled with the most important requirement to have a working system with optimal performance integrated as a result of the project. The initial effort put into designing this common platform is already paying off in terms of easier collaboration within the project. More interesting from an external point of view is the Integrated Common Software Platform I-CSP, which we regard as a practical implementation platform for interactive virtual characters on the Internet. After presenting the initial requirements, analysis and design choices, this platform is described in detail.

10.1 INTRODUCTION

The objective of the InterFace project [1] is to define new models and implement advanced tools for audio–video analysis, synthesis and representation in order to make man–machine interaction as natural as possible on the basis of everyday human communication means such as speech, facial expressions and body gestures from both sides.

Man to machine action applies through audio–video channels, which is integrated and analyzed coherently not only to perform low-level tasks, such as word recognition or eye movement tracking, but also high-level interpretation and data fusion, such as speech emotion understanding or facial expression classification. Machine to man action, on the other hand, is based on humanlike audio–video feedback simulating a 'person in the machine'. The human machine is also a humane machine, capable of providing interaction consistent with the multimodal stimuli, which have been received and processed. The human machine will have a face and a voice and will have a body and gestures. The human machine will use all its artificial senses to 'understand' high-level messages coming from the interacting man and will use all its virtual actuators to provide back acoustic–visual responses. It has been shown that even a primitive incarnation of this future vision has benefits for the end user [2], and applications for this kind of technology are outlined in References 3 to 6.

The project is working on a number of technologies for bidirectional human communication through an extended library of audio–video analysis/synthesis tools. The InterFace project was oriented from the very beginning to provide concrete demonstration of its application potentialities. Owing to this orientation, it was important to plan for the integration of all technologies implemented within the project, as well as existing ones, into a CSP used to implement applications based on the Internet.

One of the foundations of the InterFace project is that the sharing of developed tools should be simple and that all partners should be able to run all the tools and also be able to try out new applications based on the developed tools. Thus, a special task was defined from the very start, that is, to agree upon a CSP. All partners have been allocated resources to participate in this task with the understanding that a working proposition is to be available during the first quarter of the project. This resulted in the development of the software architecture called the CSP consisting of two parts: the Network CSP (N-CSP), suitable for easy tool exchange between distant partners during the development phase, and the I-CSP, intended for final integration of the tools into a working application.

This paper describes the InterFace CSP. In the first section, the requirements on the CSP, possible solutions and trade-offs are presented. The chosen solution is introduced. In the following sections, the two distinct architectures forming the CSP are introduced and one of them, the I-CSP, is presented in detail.

10.2 REASONING BEHIND THE INTERFACE PLATFORM

In this section, we introduce the requirements that were set on the design of the CSP and present several possible solutions that were proposed and evaluated during the early decision process. We discuss the relative merits of the proposed solutions within the context of given requirements and finally introduce a suitable solution resulting from these considerations.

10.2.1 Requirements

In order to allow assessment of different solutions, a set of requirements on the platform were formulated. The discussed requirements were as follows:

- Any partner in the consortium should be able to easily run and incorporate software developed by any other partner. In this way, all partners would have access to the full set of tools and demos as they are developed.
- The above-mentioned requirement should not mean that the partners need to have access to all source codes. Thus, it should be possible to use proprietary codes in the project.
- The Common Software Platform should furthermore allow a strategy for generating PC-specific versions of selected tools for real-time and/or consumer-oriented demonstration purposes.

The following justifications were given for these requirements. First, by allowing all partners to run the tools, malfunctioning or poorly defined functions are detected early after a tool has been supplied to the project. Secondly, all partners are made aware about new tools (preferably through the project Web page). This gives rise to ideas for new applications and also gives feedback to the partners responsible for the integration.

The second requirement *is not* an encouragement to prevent the exchange of source codes between the partners. The idea is instead to allow a technical solution that makes it possible to use proprietary codes for experimental purposes. Such codes might come from other projects or be confidential for commercial reasons.

Finally, the third requirement states that the agreed-upon concept should allow for high-performance demonstrations. If these cannot be built directly onto the development platform, there should be a strategy on how to incorporate the tools so that computing or communication overhead is minimized.

10.2.2 Possible Solutions

There exist several possible solutions that fulfill the requirements. Each solution has its advantages and drawbacks. Below, we indicate the main candidates that were proposed.

The common hardware/OS platform

A straightforward and simple solution is to specify exactly a specific hardware, operating system and possibly a language. Examples are the PC hardware platform running either Windows or Linux. If C/C++ is used and the object format is agreed upon, compiled modules can be easily shared and used by all partners. The exchange format would be static or dynamic link libraries (DLLs).

The main drawback of this solution is the low flexibility associated with allowing only one specific hardware/OS platform. Partners who are used to other machines such as Sun, SGI and so on would have to leave their preferred programming environment together with possibly a wealth of earlier derived tools.

A common language

Alternatively, the consortium could decide upon a common language such as Java or C++. The exchange format could be DLLs or Java Bytecode. This would allow people to write and debug codes on different hardware/OS platforms. However, for compatibility across platforms without having to share source codes, only Java is an alternative. A main drawback of this solution is the loss in performance, appearing when running Bytecode programs.

The network computing solution

A third concept is based on the assumption that the computers used by the partners can communicate with each other over a network such as the Internet. This opens up the possibility to actually access any tool or function without having to port it to another system.

For such a concept to work, it is necessary to define protocols that enable cross-platform communication over the network. Fortunately, there are several possible solutions along these lines. One could, for example, use sockets that are low-level interfaces between processes running on different machines, implementations of the Microsoft Distributed Component Object Model (DCOM) or the JavaSoft's Java/Remote Method Invocation (Java/RMI). Another interesting solution is to use the OMG CORBA (Object Management Group Common Object Request Broker Architecture) [7]. This is a proven technique for doing function calls over the network. Besides being a tool for distributing functions over a network, CORBA also offers a standardized way of declaring the input and output variables. Thus, all tools developed within the project have a clear definition of what is sent to them and what will be returned.

Thus, it is seen that the network-oriented solution offers multiple platforms and multiple language environments. A drawback is that there is a communication delay each time a function is accessed.

10.2.3 The Chosen Solution

Choosing the right solution for the InterFace CSP is then a decision that needs to take into account and balance the level of flexibility in terms of hardware and operating systems versus efficiency. The three main approaches mentioned above are summarized in Figure 10.1.

It was soon realized that any single proposal would not fulfill all the requirements alone. Thus, a structure was discussed in which several of the above-mentioned techniques could be combined. In fact, it was finally identified that all three proposals had a place in the Common Platform but in different contexts.

CORBA has obvious advantages in the research phase of tools development. It allows each partner to use his/her favorite system for development and evaluation. By 'wrapping' a developed tool into a CORBA wrap (examples are shown in Appendix A), the tool immediately becomes available for testing by the other partners. An additional advantage is that CORBA forces the developer to generate well-defined descriptions of the input/output variables for his tools.

Considering the strategic goals of InterFace, the benefits of a shared and world-wide accessible platform were anticipated. For this reason, CORBA was selected as

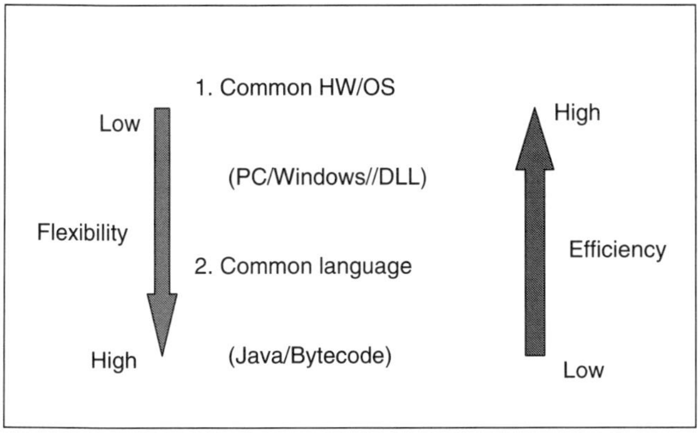

Figure 10.1 Summary of software platforms

the basis for the platform. All synthesis and analysis tools for video and audio must be available in such a *network access* form.

However, the InterFace project is not only meant for tools development from a research point of view but also to show that the produced tools can achieve sufficiently good performance to allow new applications on the Internet. Example applications are the demonstrators to be assembled as part of the project. For such applications, it may be necessary that the whole application be compiled to run on the same machine. In particular, this holds for applications that process real-time video.

For this reason, it was decided that the subset of tools aimed at the real-time demonstrations should be available also for the PC/Windows platform, preferably as DLLs. Since many of the partners already use PC/Windows machines, it is a simple task for them to generate DLLs. Tools that are developed on other machines need to be recompiled. This has to be taken into consideration by those partners who use non-PC systems and who want the system integrators to include their tools into the demonstrators. As an example, it is necessary to clearly separate machine-dependent parts from machine-independent parts.

Finally, many applications, if not all, from the InterFace project are meant to run interactively over the Internet. Typically, a client-server concept is used. Also, in many cases, the client will be a Web browser such as Netscape. For this reason, it was agreed that Java would be used when applicable, particularly for the synthesis tools in which the processing demand is expected to be lower than for the analysis. This is not in conflict with either the CORBA or the DLL requirements since Java programs can run in a CORBA environment.

On the basis of all these considerations, two distinct concepts are used within the InterFace project: the N-CSP and the I-CSP. Both of them integrate the tools developed within the project into two distinct platforms:

- N-CSP is based on CORBA and offers the highest flexibility for the interchange of tools. It is used for testing the integration of the tools while they are being developed and improved. Integration of tools in the N-CSP guarantees their compatibility. It is not used for real-time interaction within the components. This architecture is mainly dedicated for developers.

- I-CSP is defined on a fixed platform (Windows + Netscape), a client-server architecture and a combination of C/C++ and Java implementation options [using Java Native Interface(JNI)], which offers optimal efficiency for demonstrations and applications, while providing a solid architecture for the integration of tools. The goal is to provide a system dedicated to users.

In both platforms, the MPEG-4 International Standard [8] has been used wherever appropriate for communication between various modules.

10.3 NETWORK COMMON SOFTWARE PLATFORM (N-CSP)

N-CSP is based on CORBA and offers the highest flexibility for the interchange of tools among all partners. It allows to verify the compliance of the provided tools with the specifications and their correct functioning, demonstrate them on a Web page, collect comments based on trials and thus make sure that the tools are ready for integration into the I-CSP. The network-based architecture is illustrated in Figure 10.2. An INTERFACE application consists of the application core and various modules. Each module may be resident in a different hardware platform than the application core itself. The communication between the different modules and the core application program is taken care of by CORBA.

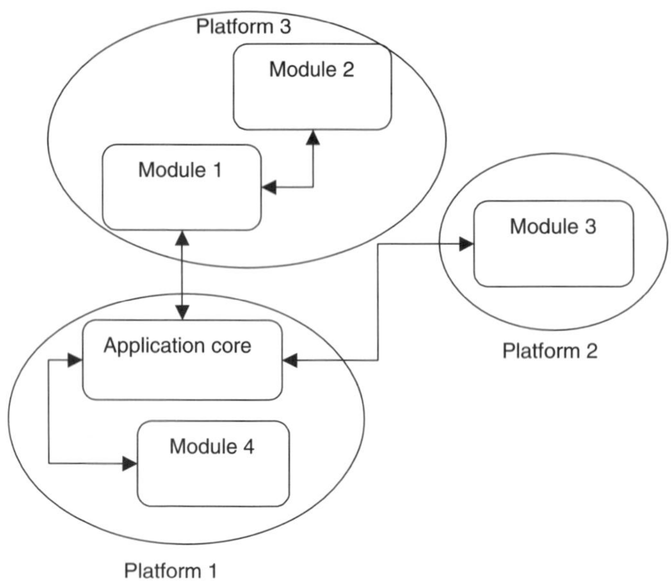

Figure 10.2 Overview of the network common software platform (N-CSP)

10.4 INTEGRATED COMMON SOFTWARE PLATFORM (I-CSP)

The I-CSP defines the set of hardware platforms, software platforms, communication protocols and interfaces as well as the architecture putting them together in a final system used for the demonstration of the InterFace project and the applications.

As shown in Figure 10.3, the I-CSP consists of a server and a client.

The *Server* consists of a standard HTTP server and the InterFace Extensions, which are additional processes needed to achieve particular InterFace functions. The platform for the Server is Windows NT or an equivalent. However, to a largest possible extent, care is taken to write software for the InterFace Extensions so that it can be ported easily to other platforms such as Linux, Solaris and other flavors of Unix. Also, as long as it does not interfere with main design goals, a reasonable effort is made to design the server in such a way that it can run without modifications on a different Windows platform, that is, the same platform as the client, allowing the client and the server to be run on the same host for one-computer demonstrations.

The *Client* is fully embedded in a Web browser. For the purposes of demonstration, the I-CSP is limited to the Netscape browser running on a Windows 98 or on an equivalent platform. The reason for this limitation is the need to use components written in native code that cannot be easily rewritten in a platform-independent way. However, wherever possible, components are written in a fully platform-independent way (i.e. Java applet running in a Web browser). The advantages of this approach is that parts of the system may be fully platform-independent from the beginning and the possibility is left open to replace more components with platform-independent ones to finally reach a fully platform-independent system.

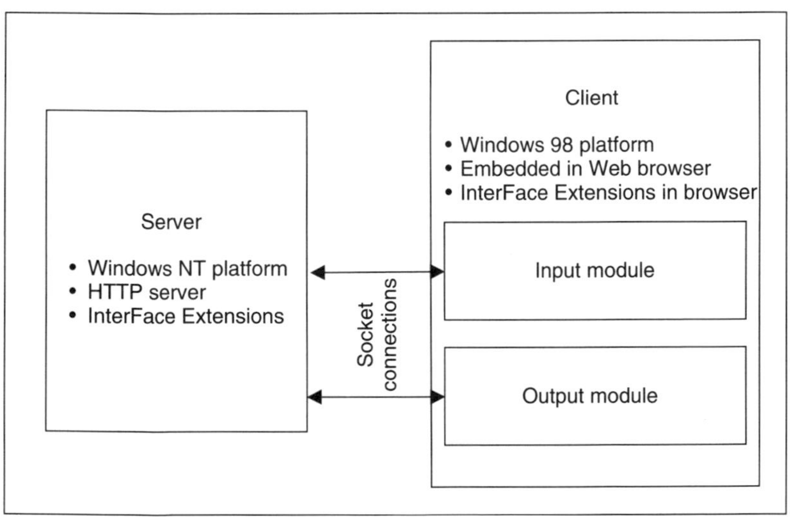

Figure 10.3 Overview of the interface integrated common software platform (I-CSP)

The Client consists of the *Input and Output Modules*. The Input Module deals with all user inputs. The Output Module deals with all outputs to the user. Each of these modules is integrated independently with the browser and has an independent socket connection to the server. This allows easier integration process because work can be done on each module separately; another advantage is the independent layout of the two modules in a Web page; finally, this leaves a possibility for certain applications to use just one of the modules independently.

10.4.1 The Server

The overview of the Server is shown in Figure 10.4. The Server consists of a standard HTTP server and the InterFace Extensions. The InterFace Extensions consist of

- the Session Manager
- the Decoder
- the Dialogue Manager
- the Text-to-Speech Module
- the Phoneme/Bookmark to FAP Converter
- the Face and Body Animation (FBA) Encoder
- the Audio Encoder
- the Multiplexer

The modules are implemented as libraries (DLLs) and integrated into a single application running multiple threads as necessary.

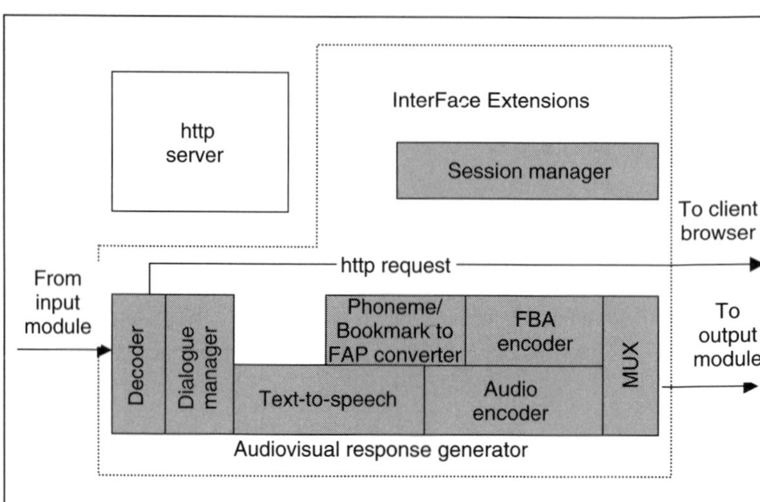

Figure 10.4 The server

The Session Manager

The Session Manager is a process that manages the networking of new clients, their connection and disconnection. The Session Manager spawns new threads dedicated to the audiovisual response generator for each new connecting client and establishes connections. The session manager will also be used to collect some information about the user (if available) such as his IP address, the country he is coming from and so on.

The Data Decoding Module

This module decodes the data coming through a socket from the Client Input Module and passes the decoded data to the Dialogue Manager (DM). It will also manage some http requests in order to be able to synchronize the content of the client Web pages with the virtual character presentation.

The Dialogue Manager

The DM provides the 'intelligence' to the system. On the basis of the inputs in the form of text annotated with emotional tags, it produces answers that are based on a knowledge base and a set of rules determining its behavior. The output is in a form of text annotated with emotions, which are encapsulated in FAP bookmarks.

The DM is based on ALICE [9], a freeware artificial intelligence tool.

The Text-to-Speech Module

The Text-to-Speech (TTS) module produces speech on the basis of text input and emotion annotation, which it receives from the DM encoded in an MPEG-4 TTS bit stream. It is highly coupled with the audio and FBA encoders in order to produce a streaming audio and FBA output. It also produces a time-stamped phoneme list, together with time-stamped bookmarks containing emotions/expressions in the form of MPEG-4 Facial Animation Parameters (FAPs).

The Phoneme/Bookmark to FAP Converter

The converter takes as input the time-stamped phonemes and bookmarks produced by the TTS module. It produces a stream of FAP frames ready for encoding. There are different implementations of the Phoneme/Bookmark to FAP Converter – it can either use the high-level FAPs to encode visemes and expressions or it can generate corresponding facial motion using the low-level FAPs directly. The former approach obviously saves a lot of bandwidth, while the later approach potentially offers higher control over the facial expressions. However, whether there is an actual improvement in quality by doing so has not been studied.

The FBA Encoder

As the TTS module produces speech, a corresponding MPEG-4 FBA bit stream containing FBA actions (visemes, expressions, gestures) is encoded by the FBA Encoder.

The Audio Encoder

As the TTS module produces speech, it is encoded on the fly into an audio bit stream.

The Multiplexer

The MUX multiplexes the FBA and Audio bit streams into a single bit stream with corresponding time stamp information (alternatively, a simpler solution may be to keep separate FBA and Audio bit streams and stream them through separate channels).

10.4.2 The Input Module of the Client

The Input Module is responsible for managing inputs from the user (from a keyboard and a mouse and also from a microphone and a camera), doing a preliminary analysis on collected data and sending them to the InterFace server.

We can distinguish three main sources of data:

- keyboard and mouse are used to control the client functions and to interact with the application;
- microphone is connected to an audio card and produces an audio stream, which is passed to an audio analysis tool;
- camera is connected to a video capturing card and produces a video stream used to feed one or more video analysis tools.

The output of the module is an encoded data stream sent over the Internet to the InterFace server. The data stream contains the outputs of analysis tools, other audio and video data (if needed) and text data provided by the user to the DM.

A summary of module inputs and outputs can be found in Figure 10.5.

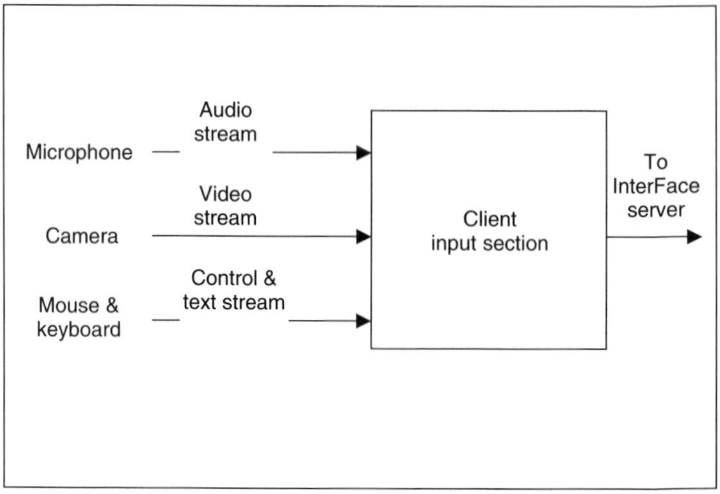

Figure 10.5 The input module of the client

The client input module consists of a set of modules:

- Audio–Video Capturing Module
- Audio Analysis Module
- Video Analysis Modules (both low-level and high-level)
- Data Encoding Module
- Graphical User Interface (GUI)

All the modules are embedded inside a Netscape Navigator browser running on an MS Windows 98/2000 platform. Wherever possible, components are written in Java 2 (jdk 1.2) to get full advantage of the new capabilities of the language such as security, Corba integration, audio/video management and JNI. Unfortunately, to use Java 2, the current version of Netscape Navigator browser must be enhanced with a plug-in (provided for free by Sun). In Figure 10.6, module connections are shown.

The Audio–Video Capturing Module

The Audio–Video Capturing Module produces audio and video streams using capturing devices installed in the PC. The streams can be forwarded to the analysis tools or temporarily saved to a local file (if analysis is not performed in real time). The data format of audio–video streams is to be defined together with tool providers.

The Audio Analysis Module

The Audio Analysis Module comprehends a set of tools used to extract information from an audio stream containing captured speech. The analysis is to be speaker-independent and multilingual.

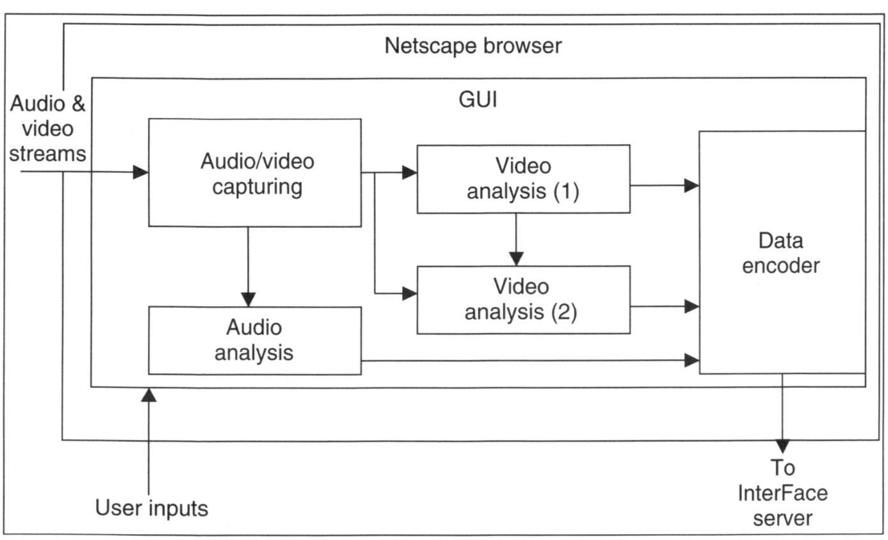

Figure 10.6 Details of the input module of the client

It is possible to identify two main tasks:

- *speech recognition* to identify phonemes and their duration (to drive synthesis) and to interpret sentences from detected keywords (input to DM)
- *emotion analysis* to extract a set of parameters that would characterize the emotions of the speaker.

The outputs of the module are:

- parameters (to be defined) to represent detected emotions in speech
- phonemes extracted along with duration
- sentences recognized from the speech to create a text stream for server DM.

The Video Analysis Modules

Video Analysis Modules are tools used to extract different levels of information from a video stream. We can classify them into low-level analysis and high-level analysis functions (marked with 1 and 2 in Figure 10.6).

Low-level analysis takes a video stream as input and is capable of detecting

- presence and position of a human face in the scene;
- geometric characteristics, structure, texture and feature points of the face;
- motion (both global and local) of the face [10–12].

The output can be an MPEG-4 FAP and Facial Definition Parameter (FDP) stream.

High-level analysis [13, 14] can use the outputs of low-level analysis and directly use the video stream (to improve robustness). It can detect

- facial expressions with intensities: joy, sadness, anger, fear, disgust, surprise or neutral;
- emotions from facial features;
- other visual behaviors (gaze, nods, shakes).

The output can be MPEG-4 high-level FAPs.

The Data Encoding Module

The Data Encoding Module manages the encoding of extracted data produced by other modules (and possibly some raw audio and video) and the transmission (through a private socket) of data to the Server.

The Graphical User Interface

The GUI manages the interaction between the user and the functions provided by the other modules. The GUI is written in Java (Java applet) and is embedded in a Netscape browser window.

10.4.3 The Output Module of the Client

The Output Module provides the audiovisual output comprising speech, Facial Animation and Body Animation.

The Output Module is a single, multithreaded Java Applet running in a Netscape browser on a Windows platform. The various components within the Output Module, illustrated in Figure 10.7, are integrated through Java APIs. Within each component, it is possible to use JNI to call functions in native code. This allows a reasonably easy integration of modules written in C/C++, while leaving the opportunity to integrate, at a later stage, pure Java components and obtain a platform-independent system.

The session control

The Session Control component communicates with the Session Manager on the Server. Its task is to establish the connection with the server, open a session, provide necessary information to DEMUX, enabling it to open the streaming connection with the DM, and close the session at the end.

The demultiplexer

The DEMUX demultiplexes the main bit stream into audio and FBA bit streams.

The audio decoder/renderer

The Audio Decoder/Renderer decodes the audio bit stream and renders (plays) the sound through the local audio system.

The FBA decoder

The FBA Decoder decodes the FBA bit stream and provides the FBA parameters at appropriate times to FBA Modules by calling their APIs. It performs the necessary frame-skipping for synchronization.

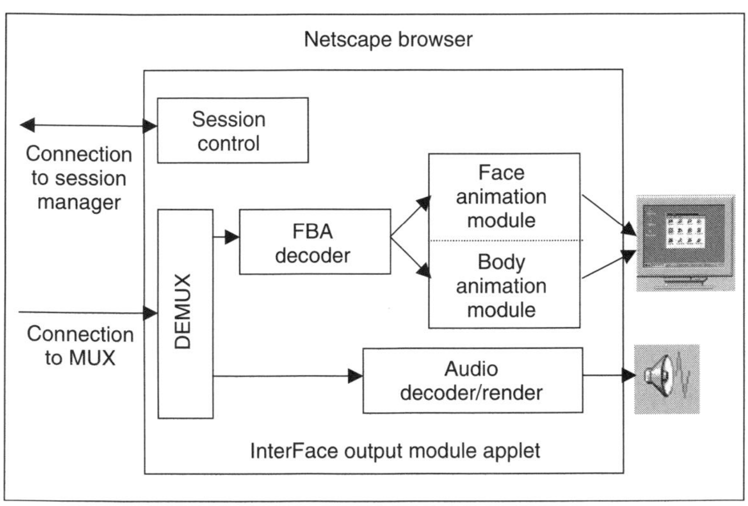

Figure 10.7 The output module of the client

The face and body animation modules

The FBA Modules load, animate and render to the screen (within the Applet surface in the browser window) the face and the body of the Virtual Character. Where necessary, they use HTTP protocol to obtain from the Server any configuration data that they need (generic face model etc.).

In order to obtain a resulting Virtual Character as a unit, the FBA Modules need to be closely integrated. This can be done by exporting the face geometry from the Face Animation Module in such a way that the Body Animation Module can place it correctly, thus obtaining the full representation of the virtual character.

Alternatively, the Face Animation Module can be used without the Body Animation to render just the face; in this case, the Face Animation Module is a fully independent unit with no particular requirements for integration with the Body Animation Module.

The FBA Modules load, animate and render to the screen (within the Applet surface in the browser window) the Face and the Body of the Virtual Character. Their APIs allow to load and delete the face and body, to pass MPEG-4 FAPs to the face, to pass MPEG-4 Body Animation Parameters to the body, perform the appropriate deformation corresponding to those parameters and render the face and the body.

Examples of possible face animation module implementations are found in References 15 and 16.

10.5 CONCLUSION

We have presented the software platform for interactive virtual characters and the design choices leading to such a platform on the basis of our experience in a joint, international research effort. We believe that the presented experience and architecture can be helpful in similar research efforts. More importantly, we believe that our I-CSP is a good basis for various applications for interactive virtual characters in the Internet environment.

ACKNOWLEDGMENT

This research is supported by the European Commission through the InterFace project contract, IST-1999-10036 of the fifth framework program.

REFERENCES

[1] The InterFace Project, IST-1999-10036, *www.ist-interface.org*.

[2] I. S. Pandzic, J. Ostermann, D. Millen, "Synthetic faces: What are they good for?", *The Visual Computer Journal*, **15**(7–8), 330–340 (1999).

[3] S. Beard, A. Marriott, R. Pockaj, A humane interface, *OZCHI 2000 Conference on Human-Computer Interaction: Interfacing Reality in the New Millennium*, Sydney, Australia, December, 2000.

[4] C. Bonamico, F. Lavagetto, Virtual talking heads for tele-education applications, *SSGRR2001 International Conference Advances in Infrastructure for Electronic Business, Science, and Education on the Internet L'Aquila*, Italy, August, 2001.

[5] A. Marriott, R. Pockaj, C. Parker, "The face of E-commerce", in S. Rahman, R. Bignall, eds., *Internet Commerce and Software Agents – Cases, Technologies and Opportunities*, Idea Group Publishing, 2000.

[6] I. S. Pandzic, "Life on the web", *Software Focus Journal*, **2**(2), 52–59 (2001).

[7] *www.corba.org*.

[8] Moving Picture Experts Group, ISO/IEC 14496 – MPEG-4 International Standard, *www.cselt.it/mpeg*.

[9] *www.alicebot.org*.

[10] J. Ahlberg, "Using the active appearance algorithm for face and facial feature tracking", *2nd International Workshop on Recognition, Analysis and Tracking of Faces and Gestures in Realtime Systems (RATFFG-RTS)*, Vancouver, Canada, July, 2001, pp. 68–72.

[11] J. Ahlberg, "Facial feature tracking using an active appearance-driven model", *International Conference on Augmented, Virtual Environments and 3D Imaging (ICAV3D)*, Mykonos, Greece, May, 2001, pp. 116–199.

[12] S. Dubuisson, F. Davoine, J.-P. Cocquerez, Automatic facial feature extraction and facial expression recognition, *3rd International Conference on AVBPA*, Halmstad, Sweden, June, 2001.

[13] M. Pardàs, M. Losada, Facial parameter extraction system based on active contours, *Proceedings of the IEEE International Conference on Image Processing, ICIP01*, Athens, Greece, October, 2001.

[14] M. Pardàs, A. Bonafonte, "Facial animation parameters extraction and expression detection using HMM", *Proceedings Euroimage ICAV3D 2001*, May, 2001, pp. 120–123.

[15] C. Bonamico, R. Pockaj, C. Braccini, "A Java-based MPEG-4 facial animation player", *ICAV3D'01 International Conference on Augmented, Virtual Environments and 3D Imaging*, Mykonos, Greece, May, June, 2001, pp. 335–338.

[16] I. S. Pandzic, A web-based MPEG-4 facial animation system, *Euroimage ICAV3D 2001*, Mykonos, Greece, 2001; demonstration at *www.icg.isy.liu.se/~igor/MpegWeb*.

PART 4
APPLICATIONS

Although three decades old, computer facial animation is still a very young and exciting field. In its classical application field of computer animation for film, there is a lot of room for innovation. Potentially even more exciting are newly emerging applications in the fields of communications and human–computer interface (HCI).

An overview of FA for HCI is provided by Mariott in Chapter 12, wherein he also introduces the Virtual Humans Markup Language (VHML). In Chapter 11, Li and Forchheimer extend the description of model-based coding and its relation to the MPEG-4 standared.

The other chapters introduce several concrete applications proposed by companies. EPTAMEDIA is proposing virtual guides and other similar applications in Chapter 16. AT&T is offering a talking e-mail (Chapter 13) and experimenting with an electronic agent (Chapter 14). face2face proposes MPEG-4 FA tools for computer animation professionals (Chapter 15).

This part gives a snapshot of emerging and potential applications of MPEG-4 FA.

11

Model-based Coding: The Complete System

Haibo Li[1] and Robert Forchheimer[2]

[1]*Umeå University, Umeå, Sweden,* [2]*Linköping University, Linköping, Sweden*

11.1 HISTORY

The basic idea of using model-based coding for communication of human faces has been around for a long time [1–12]. In the early 1980s a detailed concept was worked out for the first time and preliminary results were shown [13, 14]. The idea behind model-based (face) coding is to parameterize a talking face, and then to extract and transmit the parameters describing the facial movements such as the facial expressions and the head motion. In the receiver, the parameters are used to control the face animation in order to reconstruct the talking head. Since high-level animation parameters are enough to describe a talking face, very high compression can be achieved with this scheme. The idea to employ face animation as a video decoder brought about a totally new way of thinking into how to achieve very low bit rate video coding. Recall that at that time the popular models for image coding were statistical ones, for example, Markov models. It was rare to use deterministic models, especially high-level semantic models for image coding. As mentioned above, model-based coding was not an isolated idea, there were earlier precursors that had laid down a solid basis [1]. Some of the important works that have had an impact on model-based coding are

1. Parke's work on a parameterized face [15]
2. Phong's shading model [16]
3. Ekman's facial action coding system (FACS) [17]
4. The tremendous growth in interest in image understanding and computer vision [18, 19].

Although most components of model-based coding had been identified and solutions to implement each component were known in the early stage, the quality of the reconstructed image was not at all satisfactory. Artificial facial texture generated from the Phong's shading model made it very difficult for the image-coding community to believe that this could become a feasible coding scheme. This impression remained until texture mapping techniques were introduced into model-based coding [20, 21]. With the help of texture mapping techniques, the quality of image synthesis was dramatically enhanced. The improved image quality of the artificial faces showed the potential of model-based coding and eventually overcame the prejudice against model-based coding.

In the beginning of the 1990s, research in model-based coding started to spread. Since the framework of this scheme had been made public [22], and its attractive future was also demonstrated, a new round of research was focusing on implementation of the individual components. It was realized that the main problems were how to extract the model parameters, such as motion, shape, illuminance and so on and that these problems were quite hard to solve. Out of these problems the most difficult and interesting one is how to estimate the facial movement, including global head motion and local facial expressions. This problem was very challenging indeed! Available knowledge from the computer vision field was primarily focused on how to estimate 3-D rigid motion [23]. Very little work had been done on nonrigid motion, particularly facial motion! For several years image-coding groups played a leading role in facial motion estimation research [20, 24].

An important event in the development of model-based coding was the start of the MPEG-4 standardization effort in 1995. The original objective of MPEG-4 was to develop a new standard for very low bit rate video coding. At that time there was no clear idea on how to achieve substantial improvements in bit rate but it was widely accepted that model-based coding was a promising candidate. Today it can be argued that it was the development of model-based coding that paved the way for MPEG-4 to include face modeling. After this decision was taken, animation researchers started to become very active in the standardization work and made major contributions. Their philosophy and vision have deeply influenced the face animation standard.

11.2 CODING PRINCIPLE AND ARCHITECTURES

11.2.1 The MDL Principle

Today, very realistic images can be created through computer graphics technology. A real-world image can be described with the *scene*, the *objects*, the *light-sources* and the *camera*. For each object in the scene, its shape, location, orientation, motion and optical properties of the object's surface can be described by graphics data, called *scene descriptors*. Every scene descriptor contained in the graphics data has certain impact on the resulting synthetic image. The graphics data can be viewed as a very compact code for the synthetic image. The scene descriptors can be extracted through computer vision techniques that correspond to the image encoder. The computer graphic algorithm is used to render the image from the graphics data, which forms the image decoder (Figure 11.1). In theory, ultimate image compression can be achieved by means of a database that stores descriptions of predefined well-known objects, instead of describing the objects from scratch [25]. The graphic data can then contain references to this database.

Obviously, this kind of image-compression method would be lossy since the synthetic image would presumably differ slightly from the real image. This is why even from the very beginning model-based coding was of lossy nature. Model-based coding can be viewed as an implementation of the ultimate image-compression idea.

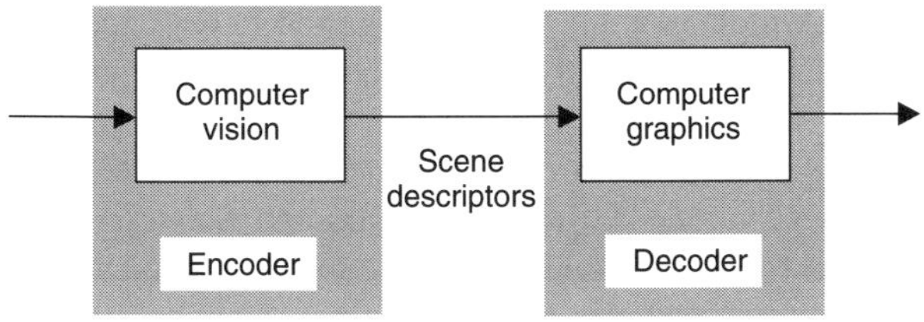

Figure 11.1 Computer graphics as decoder

Realizing that it is very hard to extract scene descriptors from real-world scenes in general, coding of human faces became the natural target. Around human faces, a set of physical models has been utilized to help in the extraction of scene descriptors. With the help of the models, extraction of scene descriptors is reduced to a simpler problem, that is, *to estimate the parameters associated with the models.*

In model-based coding, all technical problems such as how to choose suitable models, how many parameters to use and how to estimate the parameters are centered around three key factors: **I**, the image to be compressed, **M**, the physical models used to describe the image and **p**, the associated parameters. The basic concept of model-based coding can be described through the principle of minimum description length (MDL) [25]:

$$\mathbf{C}(\mathbf{I}, \mathbf{p}) = \mathbf{C}(\mathbf{p}) + \min_{\mathrm{M}}(\mathbf{C}(\mathbf{M}|\mathbf{p}) + \mathbf{C}(\mathbf{I}|\mathbf{M}))$$

where $\mathbf{C}(\cdot)$ is the coding cost.

The total cost $\mathbf{C}(\mathbf{I}, \mathbf{p})$ consists of three terms, that is, $\mathbf{C}(\mathbf{p})$ the cost to describe all parameters, $\mathbf{C}(\mathbf{M}|\mathbf{p})$ the cost to describe the models when the parameters are known and $\mathbf{C}(\mathbf{I}|\mathbf{M})$ the cost to measure how well the image can be described by the models.

The MDL principle tells us that it is not a good idea to use very detailed models to reduce the cost $\mathbf{C}(\mathbf{I}|\mathbf{M})$, since we have to take the descriptive complexity of the models into account. Similarly, it is also a bad idea to adopt a large number of parameters, which will lead to a high cost of $\mathbf{C}(\mathbf{p})$. In real situations, we have to learn how to make a compromise between image quality, models and parameters. This principle has been used in the following analysis.

11.2.2 Coding Architectures

A complete coding chain consists of an encoder and a decoder. Normally, image analysis resides in the encoder to extract scene parameters, and image synthesis is used in the decoder to reconstruct the image from the scene parameters. In terms of the coupling relationship between the encoder and decoder and how image analysis and synthesis interact, the coding architecture can be divided into two types: open-loop and closed-loop architectures.

Open-loop architecture

In an open-loop structure, image analysis and image synthesis are two totally independent modules. They do not even need to share the same physical models as shown in

Figure 11.2. in which an analysis model is used in the encoder while a synthesis model is used in the decoder. The only connection between image analysis and image synthesis is given by the scene parameters. Since there is no constraint on the definition of analysis models and synthesis models, a protocol is needed for the scene parameters to make the synthesis understand the scene parameters extracted from the analysis module. In fact MPEG-4 took this strategy to solve the communication problem between image analysis and image synthesis by setting up a standard for face animation. 86 facial features are selected and defined in MPEG-4 as shown in Figure 11.3. An important consideration behind this choice is based on the fact that MPEG-4 was not supposed to standardize any particular face model. Instead, only animation protocols, face animation parameters (FAP) and face definition parameters (FDP) were standardized. It is these protocols that make it possible for different animators to use the same parameter set. Obviously, this gives animators the freedom to render their talking heads with their favorite models since the decoder is separated from the encoder, and image analysis is separated from synthesis. While enjoying this freedom, some serious problems arise from an image communication point of view. The most obvious one is that it is

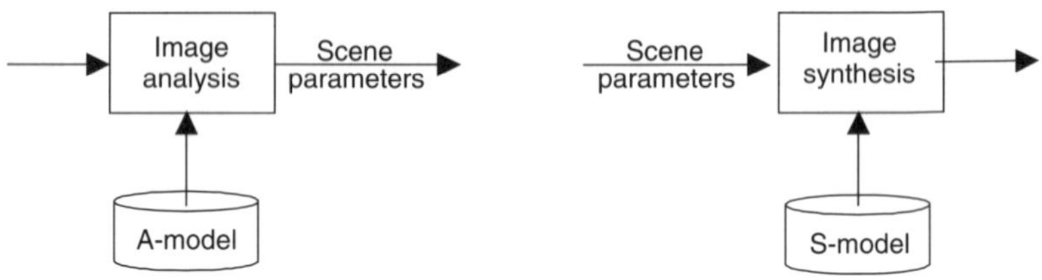

Figure 11.2 Open-loop architecture

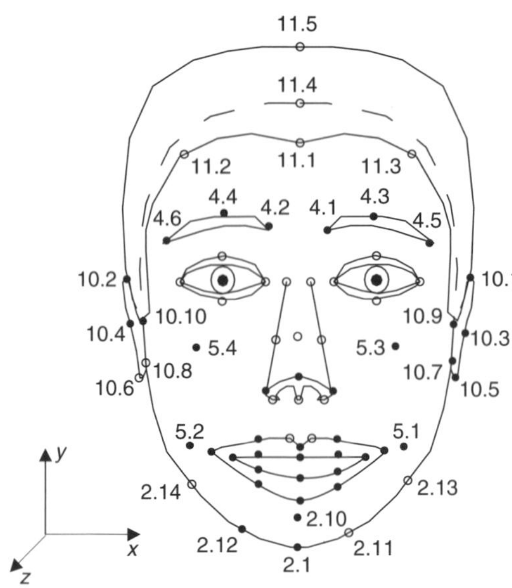

Figure 11.3 An MPEG-4 face

impossible for the encoder to control the quality of the image synthesis. That is, when scene parameters are extracted by an image analysis module, no information will be fed back into the encoder about the quality of the synthesized image. Of course, for face animation it is not necessary to assume that the animation parameters come from a corresponding image analysis module, these parameters can just as well have been generated from speech or even from text! However, for model-based coding, adopting an open-loop architecture where animation parameters have to be extracted from a video sequence and analysis needs help from synthesis is not a good strategy. This naturally leads to the closed-loop structure that has been adopted in model-based coding since the early 1990s [26–28].

Closed-loop architecture

In the closed-loop architecture, image synthesis is embedded into the image analysis loop as shown in Figure 11.4. Thus, the encoder contains both image analysis and synthesis, while only image synthesis is contained in the decoder, just as in an open-loop architecture. With such a closed-loop architecture, it is possible to check the quality of the extracted parameters directly through an image synthesis module. Note that the image synthesis module employed in the image analysis loop is identical to the one in the targeted decoder. The difference between the original image and its rendered version can be used to further improve the operation of image analysis. In this way, the encoder is fully aware of the quality of the rendered image at the decoder and knows if updated parameters make a real improvement. In fact, this kind of strategy has been used for a long time in video and audio coding under the name of predictive coding or Differential Pulse Code Modulation (DPCM). Why model-based coding adopts such an architecture lies in its strategy to make use of synthesis to simplify the image analysis problem. The ultimate goal of the image analysis is to construct 3-D models for the scenes and objects such as human faces and shoulders, and eventually to extract the parameters associated with these models. This corresponds to a

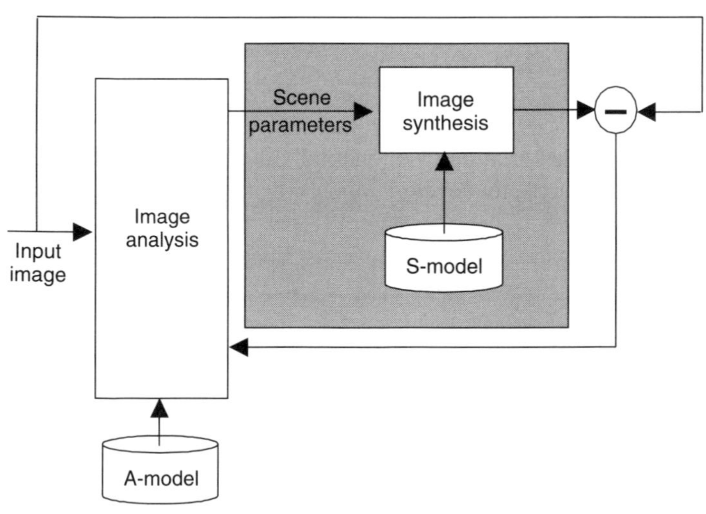

Figure 11.4 A closed-loop architecture

type of inverse problems, which are known to be ill-posed and ill-conditioned [29–32]. Generally, no simple and robust solutions exist for this type of inverse problems. The general image analysis tasks defined in model-based coding are thus rather difficult. Fortunately, for model-based coding, the good news is that we only need to handle specific objects, like human faces, thus prior 3-D models can be made and utilized. 3-D models impose very strong constraints on the solution space, which can transfer an ill-conditioned problem to a robust problem as nonlinear constraints may become linear. This naturally suggests the usage of synthesis to help analysis. Such a strategy is called ABS (analysis-by-synthesis). The closed-loop scheme is a way to implement this. Although being a powerful architecture it is not always devoid of difficulties. One problem is that the encoder will require all decoders to employ identical models, for example, 3-D face models. In reality, there are some ways, like transcoding, to release this strong constraint. This is addressed further in Section 11.5.

11.3 MODELING

The topic of this section is to review how to build parametric models for human faces. Such a model includes facial motion, shape and texture, as well as the camera, and light models. Modeling plays a key role in model-based coding. It is also interesting to examine how many parameters are needed to describe a talking head scene. The number of parameters is directly related to model complexity (MC). It is not easy to answer this question, which is essentially due to the fact that we have no knowledge about object complexity (OC) when we build a model for an object like a face. However, in reality we know whether the chosen model is good or bad. This can be learnt from the modeling performance. With this we mean, for example, the signal-to-noise ratio, given a certain coding cost. If the modeling complexity matches the OC, good performance can be expected. On the contrary, if the MC does not match the OC, the performance will be reduced regardless of whether the MC is higher or lower than the OC (Figure 11.5). In this section, we review the modeling works used in model-based coding.

11.3.1 Facial Shape Modeling

Although most faces have similar structure and the same feature set, there is obviously considerable variation from one individual face to another. One of the challenges of face animation is to develop models that support and allow these variations in

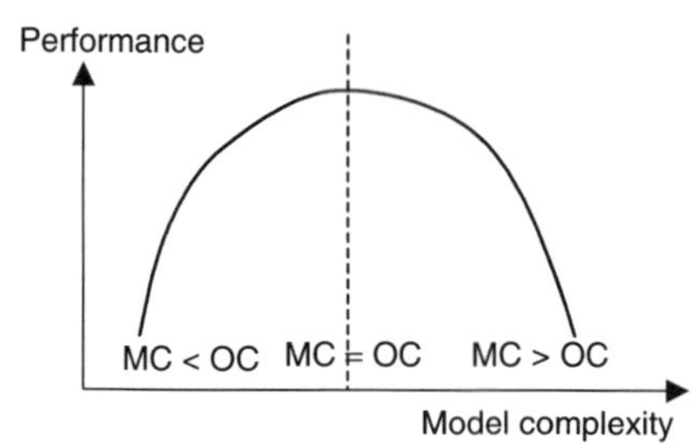

Figure 11.5 Relation between system performance and model complexity

facial geometry. There are several ways to model facial geometry, such as volume representation and surface representation [33]. In model-based coding, the most popular representation is based on polygonal surfaces, typically, a triangular mesh. To develop a triangular mesh face model the key issue is the choice of vertices of the mesh, which raises two important questions:

1. How many vertices of a triangular mesh are needed to accurately represent a face?
2. How does one choose the location of the vertices?

To answer the first question we can use signal-sampling theory [34]. A face geometry can be viewed as a 2-D signal $z(x, y)$. If we could compute the bandwidth of the 2-D geometry function $z(x, y)$, $\mathbf{B}$, then the 2-D sampling theorem will tell us that the number of samples needed per unit area to specify their shape is given by $\mathbf{f} > 2\mathbf{B}$.

Knowing the sampling rate, the minimum number of vertices is known in theory. Note that the bandwidth, $\mathbf{B}$, is the measure of OC and the number of vertices is a measure of MC. In the face animation literature, model density is another name for MC. It is worthwhile to point out that the number of vertices required from the sampling theorem may become quite large. In practice, a rather small number of vertices are often chosen on the basis of the following considerations:

1. for most applications, accurate representation is not absolutely necessary, a good representation with less vertices may be good enough,
2. visual distortion caused by an inaccurate geometry model can go unnoticed, particularly after texture mapping is applied,
3. having a large number of vertices will give rise to computational difficulties.

Now to the second question about how to choose the vertices. Naturally, if it is affordable to employ the number of vertices computed from the Signal Sampling Theory, a regular and even two-dimensional sampling mesh can be used and it is not necessary to worry about what kind of sampling pattern one should employ. The problem is when only a small number of vertices is allowed in reality. Then the sampling pattern has to be chosen carefully. A reasonable criterion in the choice of the vertices is that the density of the triangular mesh should be distributed according to the surface curvature, which can be measured with the Extended Gaussian Image concept [35, 36]. Areas of high surface curvature, such as the nose, the mouth, around the eyes and so on need a high density of triangles, whereas areas of lower curvature (the forehead, cheeks and neck) need less triangles. Furthermore, other factors under consideration from an animation point of view are [33]:

1. the mesh must approximate the face well for each expression;
2. the mesh edges must coincide with the creases of the face;
3. the mesh edges must coincide with color boundaries.

In addition, from a perspective of image analysis, the vertices should be easily identified and measured visually.

The birth of the CANDIDE face model [37] was based on these considerations. One important observation shown in Figure 11.6 is that only about one hundred

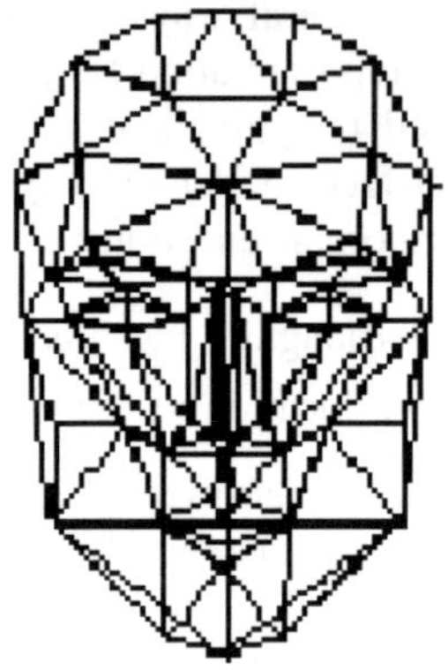

Figure 11.6 CANDIDE, a 3-D facial model

vertices are used to specify the facial geometry. These vertices correspond to almost all the feature points defined in MPEG-4. Because of the low number of vertices, this model allows fast image rendering also on low-performance platforms such as handheld devices.

11.3.2 Facial Expressions

Research by Paul Ekman and others [17] has identified a small number of basic emotions that people can identify from facial expressions with a high degree of accuracy. These emotions are *sadness, anger, joy, disgust* and *surprise*. However, within each of these categories there are a wide range of facial expression intensities and variations in expression details. Obviously, the basic step to handle emotions is to model facial expressions. Let us first look at the expressive structure of the face.

The various facial expressions are driven by the muscular activities (Figure 11.7) that are the direct result of the mental condition and the emotional state of the individual [17]. Facial expressions represent changes in neuromuscular activity that lead to visually detectable changes in the facial appearance (Figure 11.8). Bassili [38] observed and verified that facial motion dominates the facial expressions. He showed that facial

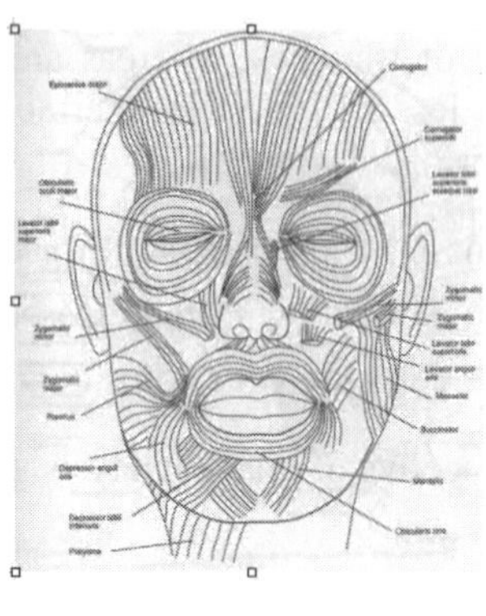

Figure 11.7 Muscular distribution in the face [From Parke and Waters (1996)] [33]

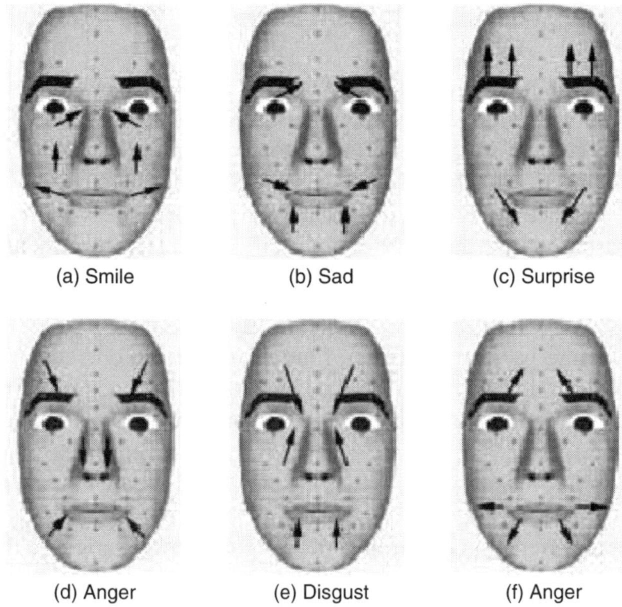

Figure 11.8 Facial expressions are dominated by facial motions [From Bassili (1978)] [38]

expressions could be identified by facial motion cues even without any facial texture information. This observation has been explicitly or implicitly employed in the study of facial expression recognition [39]. Similarly, in both face animation and model-based coding, facial motion has been used to characterize facial expressions [17]. Now let us look at how to model facial expressions based on facial motion.

The most direct way to characterize facial expressions is to develop a language to describe how each point on the face moves. In MPEG-4, a set of facial feature points are defined through the facial animation parameters (FAPs) and each point can move independently in one of three dimensions. Obviously, it is quite flexible to use the 84 MPEG-4 feature points to describe facial expressions. The price paid for this great flexibility is the high cost to specify the 66 low-level FAPs. Unlike the general cases under consideration by MPEG-4, in model-based coding, natural facial expressions are extracted from a real expressive face. The feature points on the face cannot move independently. They are physically constrained by facial muscles and skin. To exploit the implicit constraints, a more efficient way to characterize natural facial expressions, namely action units (AU), has traditionally been used in model-based coding [28, 24].

The facial action coding system, FACS, was introduced by Ekman and Friesen [17], to describe all visually distinguishable facial movements. In FACS, AU are defined to account for changes in facial expressions relative to a neutral face. Each action unit is a minimal action that cannot be divided into smaller actions. There are a total of 46 AU. Although a facial movement might involve only one action unit, most natural facial movements are modeled as a linear combination of these AU.

$$E(x, y, z) = \sum_i a_i \phi_i(x, y, z)$$

where $\phi_i(x, y, z)$ are the AU and $E(x, y, z)$ represents the total motion of a point in the face.

The combination of these AU results in a large set of possible facial expressions. For example, an expression of happiness can be synthesized by a linear combination of pulling lip corners (AU12 + 13) and mouth opening (AU25 + 27) with upper-lip raiser (AU10). Unfortunately, AU cannot be used directly since they are model-dependent. The AU defined in FACS are verbal and pictorial descriptions of facial motion. In model-based coding the AU for a specific model is implemented. Alternatively, one can use the AU descriptions to group FAPs activated by the same muscle and estimate the FAP coefficient values for a fully activated AU [40].

Coming back to the issue of human emotions, one could ask if the AU can also be used to characterize them? We all have the experience that we can read the emotion by looking at someone's face. Emotion can be viewed as dynamic facial expressions. It should be noted that the AU are spatially localized patterns giving expressions that are static. But does this imply that the AU are difficult to use to characterize dynamic facial expressions? In fact, the difficulty does not lie in the AU themselves but in how to employ them. According to our common experience, most facial expressions can be classified properly by human beings from static face pictures. A reasonable interpretation about our ability to read emotion from static images is that a neutral face is always implicitly defaulted in our minds. When we watch a picture containing an expressed face, the difference between the expressed face and the neutral face implies dynamic information. To employ the AU to characterize emotion, the key is to exploit a temporal combination of AU specifying individual facial expressions. In this way, modeling an emotion becomes a problem of how to model and learn the temporal rules that govern the facial expression variations. The temporal behavior of emotions can be extracted on the basis of the observation that the measured AU at each frame look apparently random. However, they are fully controlled by invisible internal states. Therefore, it is natural that hidden markov models (HMM) have become useful to model and specify dynamic facial expressions [41, 42]. In this type of approach the AU are chosen as observations of the underlying HMMs. Through the HMM framework, the AU are probabilistically coupled to dynamic facial expressions. This is a very suitable approach for practical applications in which real facial motion is almost never completely localized and where the detection of a unique set of AU for a specific facial expression is not guaranteed [42].

11.3.3 Facial Motion Modeling

It is very important to build a flexible motion model that can characterize the nonrigid nature of the 3-D motion of a human face. Assume a point p in the face is represented by a vector $\mathbf{s} = (x, y, z)^{\mathrm{T}}$. After facial motion it moves to a new position $\mathbf{s}' = (x', y', z')^{\mathrm{T}}$; the change in position of this point can be written as follows according to the Helmholtz Theory on nonrigid motion [27].

$$\mathbf{s}' = \mathbf{Rs} + \mathbf{T} + \mathbf{Ds}$$

where $\mathbf{R}$ is a rotation matrix, $\mathbf{T}$ is a translation vector, and $\mathbf{D}$ is a deformation matrix that represents the facial skin deformation caused by the facial expressions. Although the mathematical form of the above motion model is simple, it is not possible to use the model directly to estimate 3-D motion of a human face. This is simply because **R**, **T** and **D** are in general point-dependent! Fortunately, face motion is a special type of nonrigid motion. It can be modeled as a *dominant rigid* motion and a *slight nonrigid motion*. Thus, **R** and **T** can be considered as global motion parameters, which are point-independent. However, the deformation matrix also contains nine unknown parameters that are point-dependent. We should keep in mind that the local nonrigid motion is caused by the facial expression. This implies that the motion of any point in a face is not free or independent but constrained by muscle and skin. If these inherent constraints are utilized, the number of unknown motion parameters will be dramatically reduced. This strategy has been widely used in model-based coding.

Before analyzing the three motion components in more detail, let us first make a comment about the difference between interframe motion and global motion. Figure 11.9 is used to illustrate the difference, where $\mathbf{P} = [\mathbf{R}, \mathbf{T}, \mathbf{D}]$. The interframe motion refers to the relative rotation, translation and deformation between frames I_{t-1} and I_t. For the frame t, the global motion is the relative motion between frame I_t and the reference frame. Unfortunately, in most cases $\mathbf{P}_t$ will not be equal to $\mathbf{P}_{t-1} + \delta\mathbf{P}_{t-1}$!

(a) Translation

The translational motion component is represented as the 3-D location of the object reference frame relative to the current camera reference frame using the vector

$$\mathbf{T} = (T_x, T_y, T_z)^{\mathrm{T}}$$

The T_x and T_y correspond to directions parallel to the image plane, while the T_z corresponds to the depth of the object along the optical axis. For typical video camera focal lengths, even with wide-angle lenses, there is already much less sensitivity to T_z motion than there is to (T_x, T_y) motion. Therefore, in real applications, either T_z will not be used at all, or T_z/f will be used instead [43].

(b) Rotation

The 3-D rotation is defined as the relative rotation between the object reference frame and the current camera reference frame. There are several ways to represent a rotation

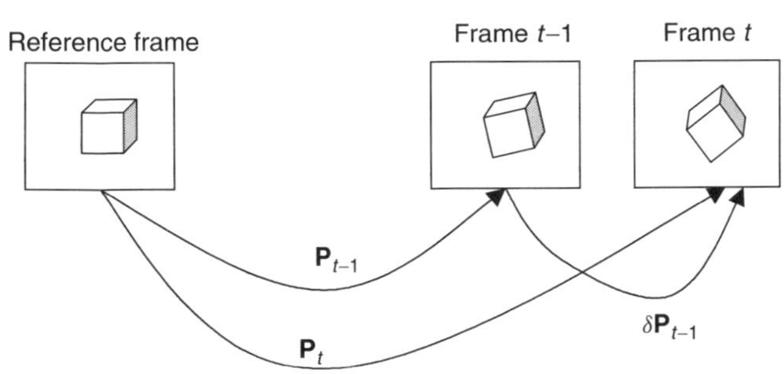

Figure 11.9 Motion estimation problem

matrix **R**. The most popular representation is in terms of Euler angles, in which the rotation is composed of three rotations, $(\omega_x, \omega_y, \omega_z)$ around orthogonal coordinate axes fixed in space. The major advantage to use Eular angles is its intuitive and simple interpretation when the rotation angles are small. Euler angles have been widely used in estimating relative motion between frames in model-based coding [18, 20, 27].

$$\begin{pmatrix} 1 & -\delta\omega_z & \delta\omega_y \\ \delta\omega_z & 1 & -\delta\omega_x \\ -\delta\omega_y & \delta\omega_x & 1 \end{pmatrix}$$

Thus, three parameters $(\delta\omega_x, \delta\omega_y, \delta\omega_z)$ can be used to represent the interframe rotation.

Unfortunately, the representation in terms of Eular angles is not a good choice for global rotation. The best way to represent global rotation is to use the unit quaternion, $\mathbf{q} = (q_1, q_2, q_3, q_4)$, in which a general 3-D rotation is modeled as a rotation about a specified axis as shown in Figure 11.10. The major advantage lies in the simple relationship between the rotation quaternion and the interframe motion [44–46]. Thus, three quaternion parameters are often used in recursive motion estimation (the four elements of the unit quaternion only have three degrees of freedom due to the normality constraint).

(c) Deformation

Here are three different ways to characterize facial deformation.

(1) Using MPEG-4 FAPs: The most direct way to characterize the local facial motion is to employ the facial feature points specified in MPEG-4. Each low-level FAP denotes a movement of one facial feature point.

(2) Using the action units: Since deformation parameters will be extracted from a live face, the representation based on the AU will be very compact. We can write

$$\mathbf{Ds} = \mathbf{\Phi A}$$

where m facial expression movement parameters have been collected into the vector **A**. The *3xm* matrix $\mathbf{\Phi}$ determines how a certain point **s** is affected by **A** [24].

(3) Using motion units: Different from the AU, the motion units (MUs) are learnt directly from a training set that contains a large enough number of expressive face sequences. This is done in the following way: a 3-D face model is first fitted to the

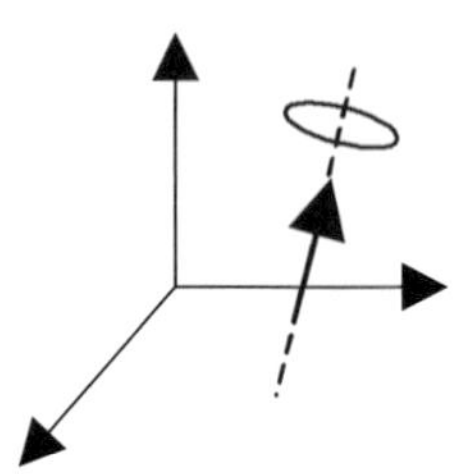

Figure 11.10 3-D rotation specified by a quaternion

expressive face in the video, the motion traces of the vertices of the model are then recoded and analyzed. Principal facial motion patterns, extracted from the motion traces, are defined as the MUs, $\mathbf{\Psi}$. Thus facial deformation can also be represented as

$$\mathbf{Ds} = \mathbf{\Psi A}$$

where m motion parameters have been collected in the vector $\mathbf{A}$. The matrix $\mathbf{\Psi}$ contains principal facial motion patterns [46].

11.3.4 Facial Texture Modeling

The facial texture is needed on the decoder side to synthesize the face image. Principally, only one single facial texture is needed for synthesizing all future frames. However, if a highly realistic face is the objective, the texture will have to be sent more than once to cope with temporal changes such as wrinkles and changing lighting conditions. Therefore, a highly compact representation of facial texture is needed. In addition, a facial texture model could also be used to provide a good criterion of 'faceness' (distance to the face space as described below), to the parameter estimation [48, 49].

On the basis of the fact that all human faces are similar to each other, a highly efficient way to represent face textures is the eigenface model [48, 49]

$$I = \sum_{k=1}^{M} \alpha_k \phi_k$$

where ϕ_k are eigenfaces. This formula tells us that a human face can be represented by a linear combination of other face images (eigenfaces). In this representation the key is how to calculate eigenfaces. Eigenfaces are extracted from a given face image database. Although an eigenface is also a linear combination of faces in the database, it is not straightforward to extract eigenfaces. This is due to the fact that large translations or scale-changes make the face image set highly nonconvex [51, 52]. This is illustrated in Figure 11.11 where **C**, a linear combination of **A** and **B**, is clearly not part of the face space.

The nonconvexity implies that a linear combination cannot represent the data satisfactorily. To handle this problem, an alignment procedure has to be called for. An efficient alignment is the geometrical normalization introduced in Reference 52 and illustrated in Figure 11.12. It has been shown that the face image set is more convex after geometrical normalization than before the normalization. Figure 11.13 gives some resulting examples.

The top row in Figure 11.13 shows five images from the training set after they have been geometrically normalized. The bottom row shows the average image followed by the first four eigenfaces.

An example of a representation using eigenfaces is shown in Figure 11.14. The original face to the very left (not part of the training set) is geometrically normalized (second image) and then projected onto a database of M = 40 eigenfaces calculated

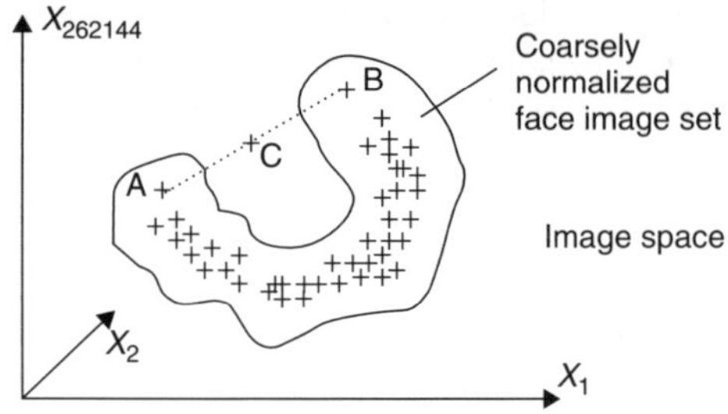

Figure 11.11 A nonconvex face space [Image courtesy of Jacob Ström (1998)] [52]

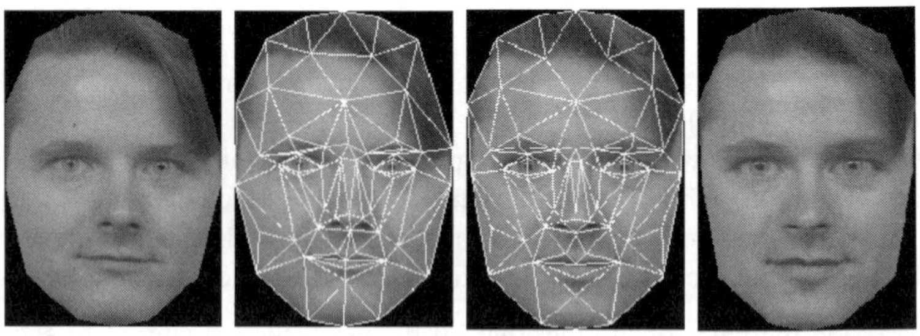

Figure 11.12 Geometrical normalization [Images courtesy of Jacob Ström (1998)] [52]

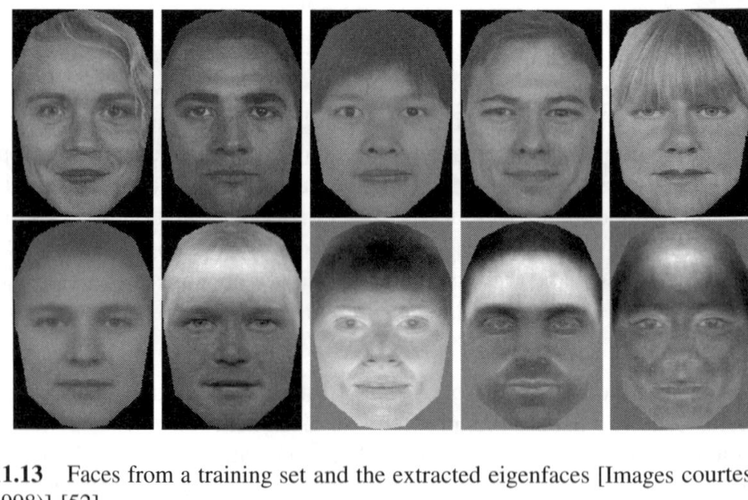

Figure 11.13 Faces from a training set and the extracted eigenfaces [Images courtesy of Jacob Ström (1998)] [52]

from a training set of N = 200 images. The reconstructed face is shown in the third image, and the extreme right image results after inverse geometrical normalization.

For the generic face texture, roughly fifty eigenfaces are enough to represent any normalized face texture. For model-based coding purpose, the same technique can be used to extract personal eigenfaces, which are expected to cover diverse facial expressions. Since it is just for personal purpose, there is reason to believe that a smaller number of eigenfaces is needed to span the expressive face space.

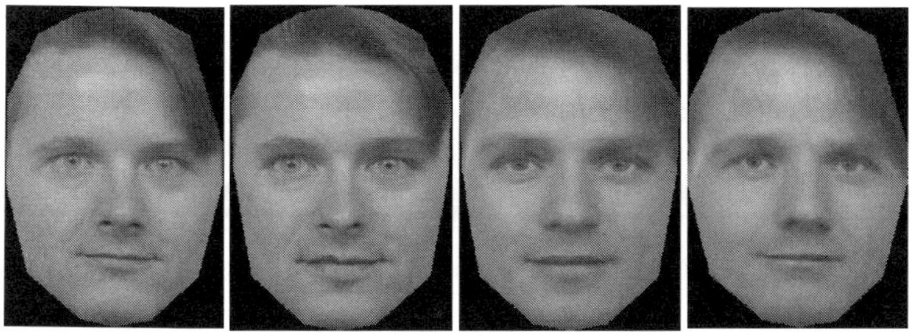

Figure 11.14 An illustration of a face represented by eigenfaces [Images courtesy of Jacob Ström (1998)] [52]

11.3.5 Camera Model

The mapping from a 3-D coordinate onto the image plane can be modeled as the *perspective* projection

$$\begin{pmatrix} X \\ Y \end{pmatrix} = \begin{pmatrix} x \\ y \end{pmatrix} \frac{f}{z}$$

where f is the focal length, (X, Y) is the image coordinate, and (x, y, z) is the 3-D coordinate. In this model, f only varies the scaling of the image. This perspective projection is often referred to as a Pinhole camera (illustrated in Figure 11.5).

When the distance between the 3-D object and the camera is large, the perspective projection degenerates into the *orthogonal* projection (Figure 11.16)

$$\begin{pmatrix} X \\ Y \end{pmatrix} = f \begin{pmatrix} x \\ y \end{pmatrix}$$

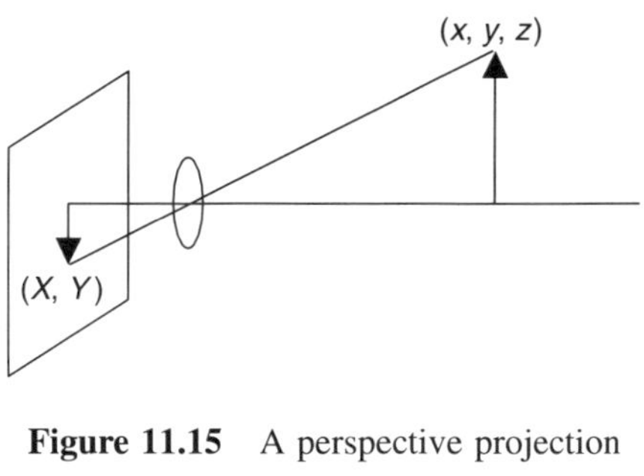

Figure 11.15 A perspective projection

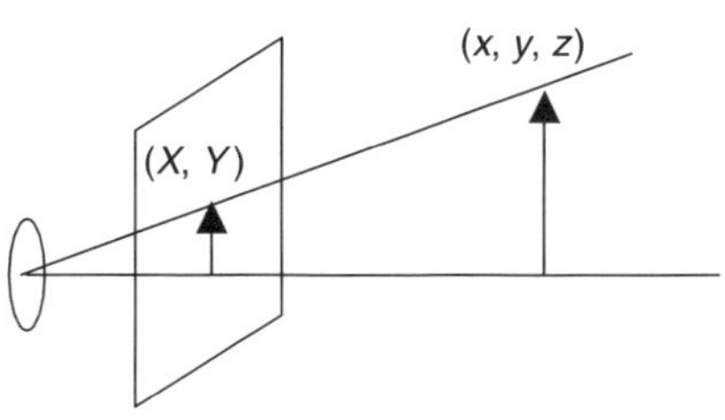

Figure 11.16 A center projection

Both *perspective* and *orthogonal* projection models are popular in model-based coding [27, 43].

Recently, a different projection model, the *center projection*, has been receiving much attention. This projection has long been used in the field of photogrammetry and has been adopted in recursive 3-D motion estimation [43, 45]. The projection is defined as

$$\begin{pmatrix} X \\ Y \end{pmatrix} = \begin{pmatrix} x \\ y \end{pmatrix} \frac{1}{1 + z\beta}$$

The focal length f is parameterized by its inverse, $\beta = 1/f$. It is claimed that this projection decouples the camera focal length from the depth of the point, and using β instead of f does not exhibit numerical ill conditioning. Interestingly, the *center* projection can unify both *perspective* and *orthogonal* projection. This is easy to see since all rays project orthogonally onto the image plane when $\beta = 0$.

11.3.6 Illuminance Modeling

It has been shown [18] that the brightness of a point in the image plane can be modeled by

$$I = L\frac{\pi}{4}\left(\frac{d}{f}\right)^2 \cos^4 \alpha$$

where L is the scene radiance, d/f is the camera parameter, and α is the angle between the optical axis and the viewing direction (Figure 11.17). This formula clearly shows that image brightness is proportional to scene radiance [18]. Furthermore, the scene radiance of a Lambertian surface illuminated by a light source of radiance E is

$$L = \frac{1}{\pi} E\rho \cos\theta$$ where ρ is the so-called *albedo*, a reflectance factor.

From the formulas we see that if the lighting in the scene changes, the brightness of corresponding points on the image plane may differ significantly. If the orientation of the object surface relative to a light source changes because of object motion, brightness will not remain constant. It is very vital to handle the illuminance problem

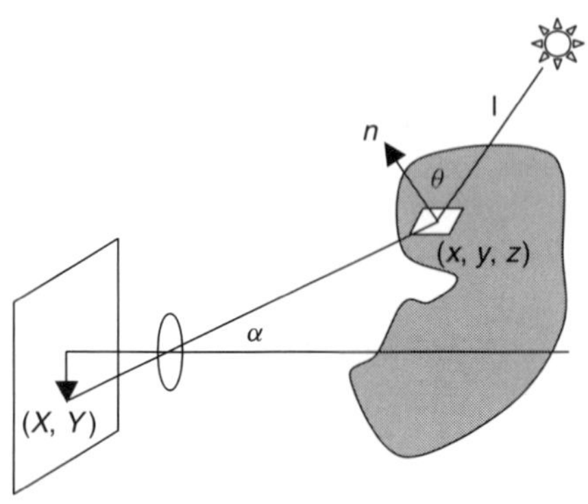

Figure 11.17 An illuminated scene

in model-based coding for at least two reasons:the first is that the brightness constancy assumption is the key in most algorithms to estimate the motion of objects between two images; the second is that the knowledge of illuminance can be used to manipulate and enhance the synthesized objects.

There are two ways to model the varying illuminance in model-based coding.

(a) Implicit approach

In this approach the image intensity is modeled as a product of irradiance $\mathbf{F}(x, y)$ and reflectance $\mathbf{R}(x, y)$ [18]

$$\mathbf{I}(x, y) = \mathbf{F}(x, y)\mathbf{R}(x, y)$$

$\mathbf{R}(x, y)$ is the intrinsic feature of the image. If the illuminance factor, $\mathbf{F}(x, y)$ can be removed from the image intensity then $\mathbf{R}(x, y)$ can be considered as an illuminance-independent factor, which can be used for parameter estimation.

(b) Explicit approach

Rather than removing the disturbing effects, explicit information about the illuminance changes can be estimated [53, 54]. Assume that in the original scene the incident light is composed of ambient light and a directional light source with illuminant direction $\mathbf{l}$, and the texture map $\mathbf{I}_{\text{ref}}$ of the 3-D object model is acquired under ambient light. Then we have the relation [53]

$$\mathbf{I}_t = \mathbf{I}_{\text{ref}}(\text{C}_{\text{amb}} + \text{C}_{\text{dir}} \max \{-\mathbf{n.l}, 0\})$$

where C_{amb} and C_{dir} control the relative intensity of ambient and directional light, respectively. The good news for model-based coding is that the surface normal $\mathbf{n}$ can be directly obtained from the employed 3-D model. The equation contains 4 illuminance parameters (or 8 in the color case). A linear solution for the illuminance parameters has been suggested [53]. Experiments show that just considering the simple Lambert reflection model, the reconstruction quality increases by more than 6 dB compared to the case of ambient illuminance where the image is simply scaled! This shows the power of using illuminance models. Experiments also show that if more complicated illuminance models are employed, it seems that no significant improvement can be expected. This shows that a simple Lambertian model matches the head-and-shoulder scene quite well.

11.3.7 Parameter Summary

We have examined the most important modeling issues regarding talking head scenes. In Table 11.1 we tabulate all the parameters suggested for use in a model-based coding system.

It may come as a surprise that 400 parameters are needed to characterize a talking head! Yet, it is reasonable to assume that the system complexity of a talking head is really this large. A more interesting question is how to estimate these 400 parameters from a video sequence, which is the topic of the next section.

Table 11.1 Parameters used in model-based coding

Name of parameters	No. of parameters	Meaning of parameters
Motion parameters	6 (5)	Rotation and translation
Local motion parameters	66 (20)	Face animation parameters (action units or motion units)
Facial geometry	84 × 3	Face definition parameters
Camera parameters	1	Focal length
Facial texture parameters	50	Eigen faces
Illuminance parameters	4 (8)	Gray (color) image
Total	300–400	

11.4 PARAMETER ESTIMATION

The parameters describing a talking head are implicit and cannot be measured directly. In model-based coding, the *analysis-by-synthesis* principle described in Section 11.2 is commonly used to estimate parameters: The parameters $\mathbf{p}$ are estimated through the following steps: search through the space of possible parameter values, then project model instances back into the image $\mathbf{I}_{\text{rec}}(\mathbf{p})$ and compare it with the observed image $\mathbf{I}$. The difference, $\mathbf{E}(\mathbf{p}) = \text{difference}(\mathbf{I}, \mathbf{I}_{\text{rec}})$ is used to guide a new search. This process is repeated until the projected image $\mathbf{I}_{\text{rec}}(\mathbf{p})$ is consistent with the observed image $\mathbf{I}$. The best set of parameters to interpret the object in the image is then the set that optimizes the measure $\mathbf{E}(\mathbf{p})$.

In one approach, all the pixels in the face image are used in the comparison. An alternative approach is to employ feature points [45, 46, 55]. Motion field measured from interframe displacement of these feature points is used as the observation. It is claimed that observation based on a motion field is less sensitive to illuminance effect than that based on image intensity. In this section we review only parameter estimation techniques based on a whole frame. Parameter estimation based on feature points is addressed in Section 11.5.

Within the framework of ABS, parameter estimation is treated as a general optimization problem. When applying optimization techniques to our problem, we have to distinguish between two cases: (1) *Initialization* and (2) *Successive Estimation*. In the initialization case, the objective is to get the initial state of the system such as the location, orientation, size, color of the object and light directions. A typical initialization task in model-based coding is the 'adaptation of a human face model'. Initialization is extremely important if we want to build an automatic model-based coding system. Unfortunately, less attention has been paid to solving the initialization problem. In contrast, successive estimation has always been the center topic in model-based coding research. In successive estimation, finding rough location and approximated parameters of the target object in an image are available from prior processing and the task is then to correct this initial approximation. This can be performed with local optimization techniques [56, 57]. On the contrary, it is much harder to solve the initialization problem. In the remaining part of this section we focus on how to attack this problem.

11.4.1 Parameter Search

In the absence of an initial approximation, an alternative is to perform a blind search through the parameter space. As described above, the key step in this search is the image rendering based on the assumed models. Rendering speed is then a key issue. Let us look at how many search loops a PC can perform per second. As is known a typical PC of today has a graphics card with specialized hardware for 3-D graphics, including texture mapping. It is very fast in rendering an image from a 3-D model. According to measurements [47], the time to render an image, including both texture mapping and image transfer, is around 5 milliseconds. Furthermore, making use of Single Instruction Multiple Data (SIMD) instructions, the rendering time is reduced to less than one millisecond! Thus, with a modern PC, we can perform matches between the observed image and the image generated from the model at a rate of 200 to 1000 times per second. Let us look at the size of the corresponding parameter space to see if it is possible to employ a blind search to find the parameters.

The parameter search space

We have seen that around 4×10^2 parameters are needed to characterize a talking head scene. If the number of levels associated with a parameter is 10^2, then the space of possible solutions is prohibitively huge, 10^{800}. Fortunately, for the concrete problems of model-based coding, the solution space is much smaller than this huge space. Let us take the typical initialization problem as example: adaptation of a human face model. To minimize the number of parameters to be determined, a personal face model is employed. By the personal face model we imply that the face geometry is known *a priori*, which avoids determining 252 facial shape parameters. Furthermore, if variations of local facial texture are also neglected, this will release around 100 parameters. The remaining parameters associated with the face model are: two translation parameters, a scaling parameter (related to the camera model also), three rotation parameters and four illumination parameters. In total, there are 10 parameters for the initial state of a face in a video. For these 10 parameters, the solution space has been dramatically reduced into 10^{20}. Unfortunately, the solution space is still huge. With a modern PC, performing 10^3 searches per second, it will take 10^{17} seconds to perform a blind search. Recall that there are only an estimated 10^{18} seconds in the history of the universe! Apparently, simply applying brute force to search all possible solutions is impossible. One strategy is to employ smart global optimization techniques to search for an optimal set of parameters [58, 59]. An alternative is to examine behaviors of all parameters and try to classify them. Figure 11.18 shows the cost surface with respect to two translation parameters. For those parameters, such as translation and scaling factor, that have multiple local minima, a global search is performed while for those parameters that have a global minimum that is significantly different from other minima, a local search is carried out.

In the following subsections we will examine how to employ 'smart' optimization techniques to solve the parameter estimation problem. The discussion focuses on how to build a cost function based on which optimization techniques can be applied.

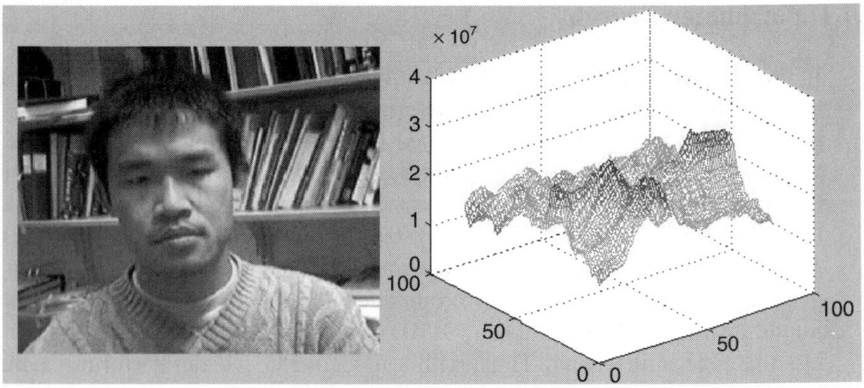

Figure 11.18 Cost surface with reference to two translation parameters

11.4.2 Forward or Backward Difference?

Since the cost function is based on the difference image, care has to be taken while designing such a difference operation. In the literature there are two approaches to perform the difference operation, sometimes also called alignment [27].

(a) Forward difference (Figure 11.19)

Here the reference image $\mathbf{I}_{\text{ref}}(\mathbf{x})$ is aligned to the input image $\mathbf{I}(\mathbf{x})$, where $\mathbf{x} = (x, y)^{\text{T}}$ are the image coordinates. If the warp defined by the model and its associated parameters $\mathbf{p}$ is denoted by $\mathbf{w}(\mathbf{x}; \mathbf{p})$, then the forward difference operation is defined as

$$\mathbf{E}(\mathbf{p}) = \sum_{\mathrm{x}} \{\mathbf{I}(\mathbf{x}) - \mathbf{I}_{\text{ref}}[\mathbf{w}(\mathbf{x}; \mathbf{p})]\}^2$$

The best parameter vector $\mathbf{p}$ can be obtained by minimizing $\mathbf{E}(\mathbf{p})$ with respect to $\mathbf{p}$. Such a difference operation has been widely used in model-based coding [24]. The major advantage is its good tracking performance and possibility to yield an analytic solution if the warp function $\mathbf{w}(\mathbf{x}; \mathbf{p})$ can be written as a linear function of $\mathbf{p}$. That is,

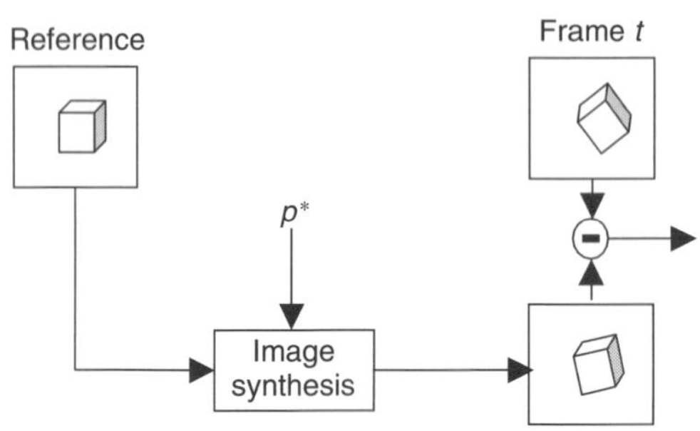

Figure 11.19 Forward difference operation

$\delta\mathbf{I}$ is a linear function of $\mathbf{p}$! This makes it possible to update the estimated parameters in real time.

(b) Backward difference (Figure 11.20)

The backward difference operation [48, 56] tries to align the input image $\mathbf{I}(\mathbf{x})$ to the reference image $\mathbf{I}_{ref}(\mathbf{x})$. If we still use $\mathbf{w}(\mathbf{x};\mathbf{p})$ to denote the warp transform specified by the model and its parameters $\mathbf{p}$, then the backward difference operation is defined as

$$\mathbf{E}(\mathbf{p}) = \sum_{\mathrm{x}} \{\mathbf{I}_{ref}(\mathbf{x}) - \mathbf{I}[\mathbf{w}(\mathbf{x};\mathbf{p})]\}^2$$

The benefits of this operation lie in the fact that all operations will take place within a normalized space, which makes it possible to avoid some real-time computations. Instead, a precomputing strategy can be used. It has been shown that the difference function $\mathbf{e} = \mathbf{I}(\mathbf{x}) - \mathbf{I}[\mathbf{w}(\mathbf{x};\mathbf{p})]$ can be written as a linear function of $\boldsymbol{\delta}\mathbf{p}$, $\mathbf{e} = \mathbf{R}\,\boldsymbol{\delta}\mathbf{p}$, where the $\mathbf{R}$ matrix can be learnt in advance from a training set.

11.4.3 How to Choose a Suitable Cost Function E(w)

No matter what kind of difference operations are used, the cost function is built on a comparison between the reference image and the reconstructed image:

$$\mathbf{e} = \mathbf{I}_{ref} - \mathbf{I}_{rec}$$

This can be explained as a simple vector difference as shown in Figure 11.21. It will be sufficient to use this operation if the illuminance effect has been already included in the model. Unfortunately, in most model-based coding algorithms, the illuminance factor

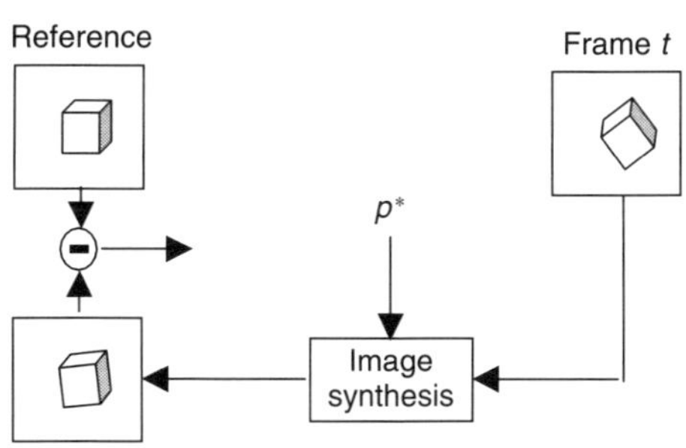

Figure 11.20 Backward difference operation

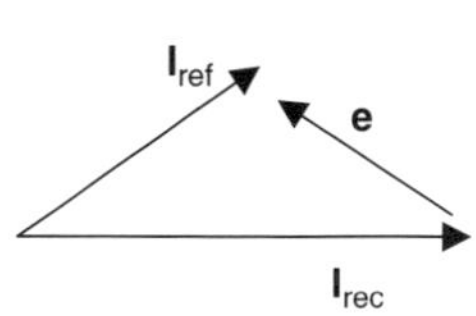

Figure 11.21 Vector difference operation

is not taken into account. It is sometimes argued that this criterion is too sensitive to illumination, especially when the input frame is far from the reference frame.

Therefore, in reality, we have to carefully determine two things (1) the function form and (2) the choice of the target image. The comparison is related to the means of measuring the vector distance.

Direct vector difference is a simple way to measure their distance. A better way is to employ vector projection to compute the distance [58]

$$\mathbf{e} = \mathbf{I}_{\text{ref}} - \alpha\, \mathbf{I}_{\text{rec}}$$

This projection operation is illustrated in Figure 11.22.

A direct benefit with this distance is that the global illuminance effect can be removed.

Here we only discuss the cases where the first image frame is taken as the reference frame, which is usually adopted in most model-based coding. As pointed out earlier, there is at least one drawback with the approach of using one single frame: there are some facial expressions that generate quite different facial skin deformations, such as eye closing or opening. Obviously, it is not sufficient to employ only one single facial texture to cover all texture variations. To use a reference face space instead of just a single face frame is a possible extension. For example, in References 48 and 57 a normal face space is used to track facial motion. Here a reference space could be used to check if the estimated parameters are correct when not even a single reference image is available or when more complex factors, like illuminance, need to be handled. Such an example is shown in Figure 11.23 in which the distance between the input image vector to the reference space is used to build the cost function [56].

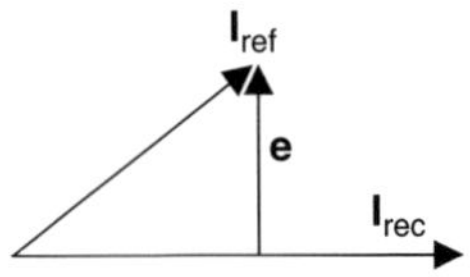

Figure 11.22 Vector projection operation

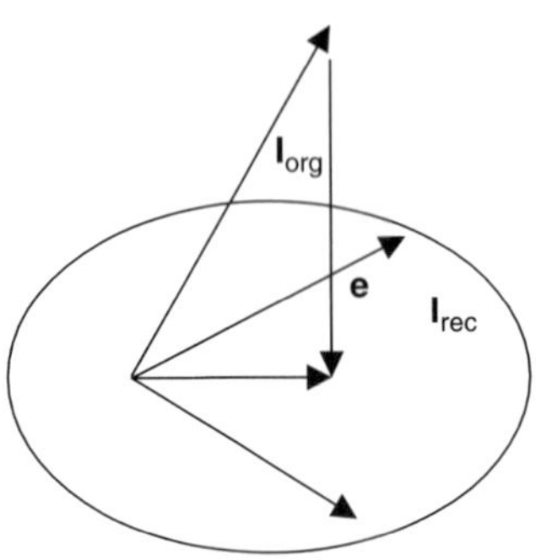

Figure 11.23 Projection onto a face space

11.4.4 Optimization Techniques

Once the cost function is determined, the following two types of optimization techniques can be chosen depending on the cases.

(1) Local optimization techniques

If the cost function is convex (Figure 11.24), which implies that only one minimum exists, the steepest descent technique is very efficient

$$\mathbf{P}_t = \mathbf{P}_{t-1} + \delta\mathbf{p}$$

where $\delta\mathbf{p} = -\dfrac{\partial E(\mathbf{p})}{\partial \mathbf{p}}$.

When a backward difference operation is used, the $\delta\mathbf{p}$ can be represented as [56]

$$\delta\mathbf{p} = \mathbf{Ue}$$

The update matrix $\mathbf{U}$ can be learned in advance and stored for computing $\delta\mathbf{p}$. Thus, a fast algorithm to update $\mathbf{p}$ can be obtained. This strategy has been successfully applied in facial feature extraction [56].

Local optimization techniques have been widely used in successive estimation of the set of parameters where a good initial set of parameters can be expected from previous computation. This topic is again dealt with in the next section.

(2) Global optimization techniques

Global optimization is useful when the cost function has many minima (Figure 11.25). As described in Reference 58 the various local minimization techniques go straight for the nearest minimum. They traverse downhill as fast as they can, but never uphill. To find the global minimum, it is necessary to be able to get out of local minima. A strategy is to allow occasional uphill jumps. Thus, there is a better chance of finding

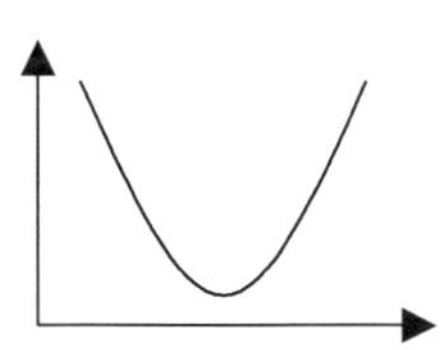

Figure 11.24 An ideal cost surface

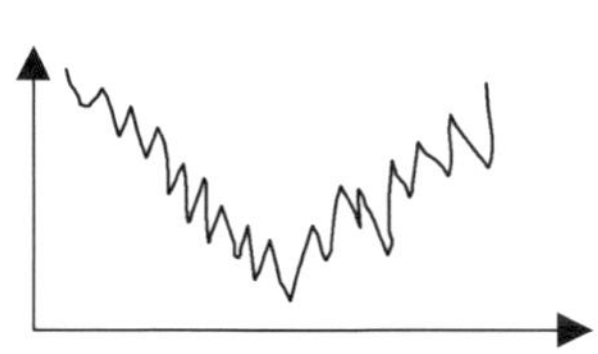

Figure 11.25 A cost surface with many minima

the global minimum. Optimization schemes that allow such uphill jumps are known as Metropolis algorithms. The Boltzmann probability distribution

$$\text{Prob}(\mathbf{E})\ \exp(-\mathbf{E}/kT)$$

can be used as an indicator to control uphill jumps. The algorithm is then called simulated annealing (SA) [60] when **T** is decreasing according to a 'cooling schedule'. The SA algorithm was inspired by thermodynamics and the way in which liquids freeze and crystallize [58, 60].

An alternative strategy to look for global optimization is to explore many options in parallel. This strategy is totally different from the one adopted by SA and can be described as 'repeatedly trying many changes around a single design'. The techniques adopting this strategy are called *genetic algorithms*, or GA [61, 62]. The state of a GA is given by a population, with each member of the population being a complete set of parameters for the function being searched [61].

Although both SA and GA are famous global optimization techniques, the difference in their optimization strategies determines their application fields. Reference 64 gives a good guidance on how to apply these two types of global optimization techniques for solving real-world problems: If the cost function has many equally good local minima. SA is a good choice since wherever the search starts out, a bit of randomness can direct the search out of small local minima and into one of the good ones. GA is more suitable to the case where one global minimum is much better than all other local ones as illustrated in Figure 11.26. In this case, it is better to spend less time improving any one set of parameters and more time working with an ensemble to examine a larger range.

Since one common feature of these two global optimization techniques is that they can robustly find good solutions in large search spaces using very few trials, it will be very interesting to know if they could solve the initialization problem. Although there have been a few trials of applying them on the problem of feature location and parameter estimation [64, 65], it is still not clear if they are capable of solving the problem. However, the example of using GA to solve the traveling salesman problem with 100 cities is very encouraging [66]: As is known the number of possible tours in any traveling salesman problem increases rapidly with the number of cities. So, for 100 cities, there are more than 10^{155} different possible paths through all the cities. The genetic algorithm has been used to search the space of possible tours and has discovered a very good one. Although it is probably not the perfect answer, it is of high quality. The important hint is that, in all, only 400 000, or one out of every 10^{150} possible solutions is examined, an infinitesimally small fraction of the whole! Therefore, it is reasonable to believe that the initialization problem can be solved with such global optimization techniques.

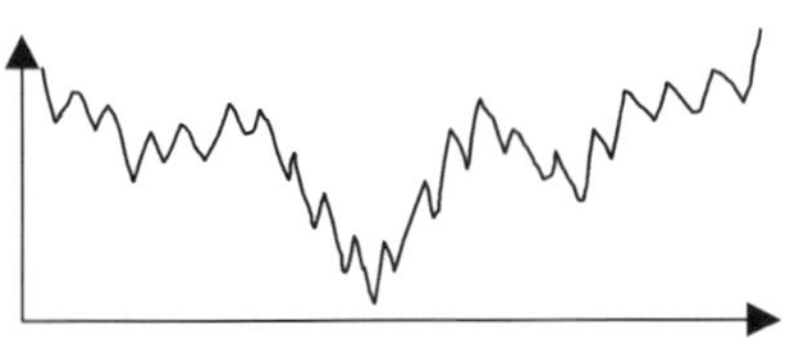

Figure 11.26 A cost surface suitable for GA

11.5 SUCCESSIVE ESTIMATION

For most initialization problems, the parameters associated with the image are assumed to be unchanging. Knowledge of the parameters estimated from previous frames is not explicitly exploited. In model-based coding, parameters are time-dependent and we have to dynamically provide the parameters after the initialization stage. This is the so-called *successive estimation* problem. In the literature, most successive estimation algorithms are related to the motion tracking problem, that is, how to track a moving head. Here, we will also take motion estimation as an example. We will describe how the successive estimation strategies developed for motion parameters can also be suitable for other parameters.

Let us first look at the famous *passive navigation* algorithm [18], which is very valuable for understanding the motion estimation problem. Under perspective projection, the optical flow field (u, v) is a nonlinear function of the 3-D motion and structure of the object and can be approximated by Reference 18

$$u = XY\omega_x - (1 + X^2)\ \omega_y + Y\omega_z - XT_z/z + T_x/z$$
$$v = (1 + Y^2)\ \omega_x - XY\omega_x - X\omega_x - YT_x/z + T_x/z$$

where (X,Y) are the coordinates of the image plane.

Using only this relationship it is very difficult to recover 3-D parameters from the optical flow field. In the model-based coding community, a clever way to handle this problem has been to use the 3-D wireframe model data to replace the depth. Once the depth z is known, the optical flow field becomes a linear function of the 3-D motion parameters. Least-Square (LS) methods or other robust methods can then be used to recover the motion parameters. The trick of using a geometric model to replace the depth z has also later been based on simpler models such as the planar model [39], ellipsoid model [67] and so on. In some of these papers the name 'regularization' is used to describe this principle.

11.5.1 Recursive Motion Estimation

3-D motion estimation from optical flow fields is also called *interframe motion estimation*, to emphasize that it provides only the relative motion between two frames. To obtain the absolute motion, or global motion, a *recursive motion estimation* system shown in Figure 11.27 was first used [28].

In this system the global motion is given in the following manner

$$\mathrm{P}_t = \mathrm{P}_{t-1} + \delta\mathrm{P}_{t-1}$$

where P_{t-1}, P_t are the absolute motion at time instant $t - 1$ and t, respectively. $\delta\mathrm{P}_{t-1}$ is the two-view motion. Although a 'regularization' term is applied in motion estimation, the motion parameters obtained from the two-view case are inherently noisy. Obviously, with the increase in the number of frames, the error of the estimated motion parameters will be accumulated, which will result in divergence in the motion estimation! This is

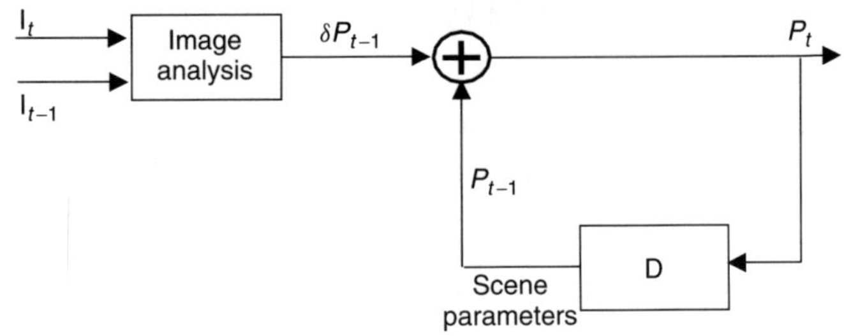

Figure 11.27 A recursive motion estimation scheme

the so-called *error accumulation* problem associated with the long sequence motion tracking [27].

11.5.2 Tracking System Based on the ABS Principle

In model-based coding a powerful system based on the ABS principle was invented to extract dynamic parameters [24, 28]

In the ABS system, there are three modules, *image analysis*, *image synthesis* and *parameter prediction* respectively. From the block diagram in Figure 11.28, one can see that this is a closed-loop system. The feedback mechanism is a powerful tool for handling error accumulation. The greatest advantage of this system is that it can provide satisfactory quality of the synthesized image owing to the following three reasons: (1) error accumulation is suppressed because of the feedback loop; (2) the error criterion is directly set up between the synthesized image and the observed one; and (3) parameter prediction makes the error due to the linearization smaller.

Since the whole image is taken as a reference, the ABS scheme is classified as an image-based tracking system. The main drawback with an image-based tracking scheme is its sensitivity to the light condition. In addition, underlying physical motion behind the 3-D facial motion cannot be explicitly utilized in the ABS scheme.

11.5.3 Tracking System Based on Kalman Filtering

An alternative approach [29, 45, 55] is to use Kalman filtering to trace facial motion by recursively following a set of feature points, which are supposed to be more robust to

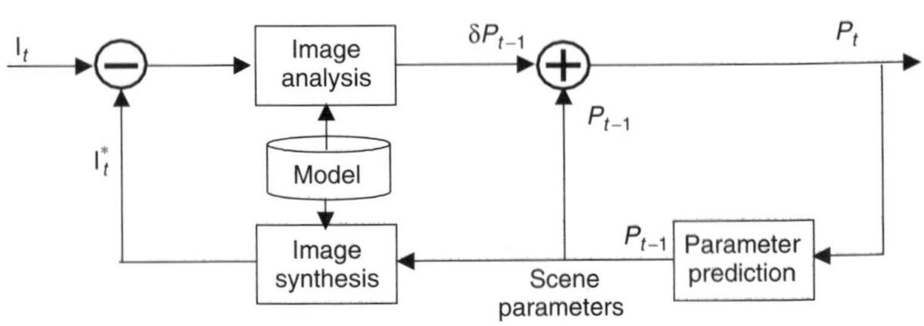

Figure 11.28 A tracking system based on the ABS principle

light variation. The Kalman filter allows the use of an explicit dynamic motion model

$$P_{t+1} = \Phi(t)P_t + n_t$$

and sample-based measurements

$$Y_t = H(P_t, t) + m_t$$

The Kalman filter solution is

$$P_t^+ = P_t^- + k(t)[Z_t - H(P_t, t)]$$

where $k(t)$ is the Kalman gain. Kalman filtering achieves an optimal combination of information from both motion model and sample-based measures. A motion-tracking system based on Kalman filtering is illustrated in Figure 11.29. It is seen that our ABS scheme is almost identical to the Kalman filter scheme.

In the employment of Kalman filtering for facial motion tracking, the key lies in the selection of feature points. There are some distinct feature points on our face, such as the corner of the mouth, tip of the nose and so on. A set of such feature points is defined in MPEG-4. A significant advantage of using standard feature points for tracking is that FAPs can be directly obtained. Unfortunately, most of these points are very vaguely defined. It is thus very difficult to track these points. Therefore, in the head tracking literature, the tracking points are typically the peaks, saddle points and pits of the intensity surface (online tracking points), which can be reliably tracked without aperture problems [45]. In an image frame, tracking points are selected on the basis of, for example, the Hessian matrix. In Reference 45, a clever way to identify and extract tracking points using sorting techniques is described. 3-D facial motion tracking systems based on this approach can be found in References 45 and 55. The drawback of this approach is that almost all feature points are selected from the first frame, that is, all the tracking points are located in the frontal face area. Thus, with large out-of-plane rotation, few or no feature points will be visible in the image. Thus, the tracking system will inevitably lose track. One way to solve this is to add new tracking points [68]. Another drawback of using such tracking points is that there is no way to associate a prior knowledge with these points.

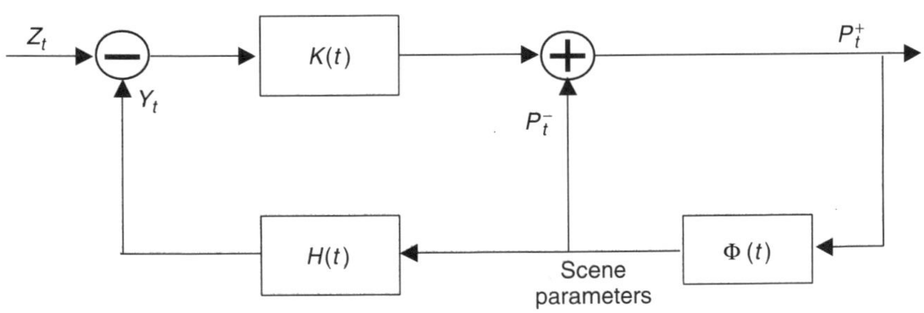

Figure 11.29 The Kalman filtering scheme

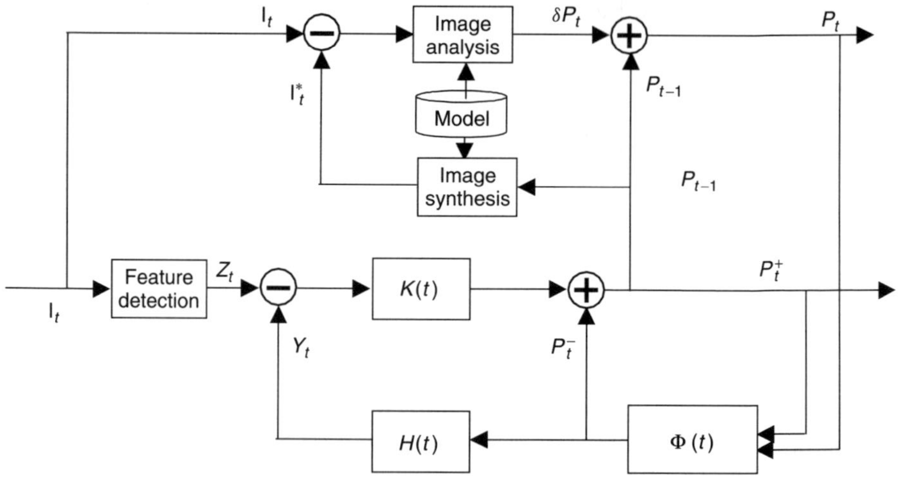

Figure 11.30 A tracking system based on a combination of ABS and Kalman filtering

11.5.4 Tracking System Based on a Combination of ABS and Kalman Filtering

A typical facial motion contains both global head motion and local facial expression variations. Considering that the system based on the ABS principle is good at handling local deformation like facial expression, while the system based on Kalman filtering is good at taking care of large motion, it seems to be a good idea to combine these two systems to handle general facial motion. A combined system is shown in Figure 11.30, from which one can find that Kalman filtering has been integrated into the ABS loop and that the major function of the Kalman filter is to provide global motion parameters. In fact, such an approach has been used to track 3-D human body movements [69]. Although not yet fully investigated there is reason to believe that this system should work well for solving the 3-D facial motion–tracking task.

11.6 HYBRID CODING

Model-based coding is of lossy nature. If lossless coding is needed, the coder can be enhanced by a second layer, that is, first we encode the parameters, and then encode the discrepancy between the real image and the synthesized image. The real image is thus seen as the sum of a synthetic and a residual image, where the residual image can be coded with waveform coding schemes. Such a coding strategy is also called hybrid coding. An example of combining model-based coding with transform coding is shown in Figure 11.31.

Besides improving the quality of the compressed video, additional benefits are [70]

- reduced image analysis complexity
- reduced image synthesis complexity
- reduced modeling complexity.

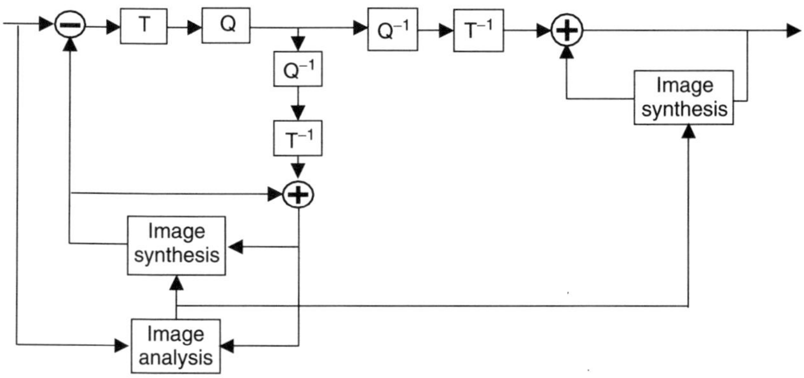

Figure 11.31 A hybrid coding scheme

Describing an image in a higher-level sense is, in general, a formidable task [22]. When applied to image coding, this task can, however, be simplified. By combining scene analysis with traditional coding techniques, possible errors occurring in the analysis and synthesis parts can be corrected using waveform coding. In addition, model-based coding is only targeted to restricted objects, for example, human faces while a talking head scene may contain other natural objects, such as background. To generalize the applicability of the coding scheme, the incorporated waveform coder can be used to encode the unmodeled objects [36, 71].

11.7 CONCLUSION

The face animation standard within MPEG-4 can be used as the decoder part of a model-based image-coding system. However, the encoding side is not specified by the standard. In this chapter we have focused on the main image analysis problems that need to be handled to implement a functional system. Although solutions exist for most of the described problems it is evident that the area is still under research. The existence of a well-defined decoder such as MPEG-4 face animation will most certainly encourage further work on the face analysis parts leading to faster and more robust encoders in the future.

REFERENCES

[1] D. Pearson, "Developments in model-based video coding", *Signal Processing: Image Communication*, **2**(4), 892–906 (1990).
[2] J. Ahlberg, WinCandide 1.3 user's manual, Report No. LiTH-ISY-R-2344, April, 2001.
[3] J. Ahlberg, Extraction and Coding of Face Model Parameters, Licentiate Thesis No. 747, Linköping University, Sweden, 1999.
[4] K. Aizawa, T. Huang, "Model-based image-coding: Advanced video coding techniques for very-low bit-rate applications", *Proceedings of IEEE*, **83**(2), 259–271 (1995).
[5] M. L. Cascia, S. Sclaroff, V. Athitsos, "Fast, reliable head tracking under varying illumination: an approach based on registration of texture-mapped 3D models", *IEEE PAMI*, **21**(6), (1999).

[6] F. Davoine, H. Li, R. Forchheimer, "Video compression and person authentication", in J. Bigun, G. Chollet, G. Borgefors, eds., *Audio- and Video-based Biometric Person Authentication*, Springer, Berlin, 1997, pp. 353–360.
[7] F. Davoine, C. Kervrann, H. Li, P. P'erez, R. Forchheimer, C. Labit, "On automatic face and facial features detection in video sequences", *Proceedings of the International Workshop on Synthetic-Natural Hybrid Coding and 3D Imaging (iwsnhc3di'97)*, Rhodes, Greece, September, 1997, pp. 196–199.
[8] D. Goldgof, H. Lee, T. Huang, Motion analysis of nonrigid surfaces, *Proceedings of IEEE conference on Computer Vision and Pattern Recognition*, 1988.
[9] H. Li, R. Forchheimer, Location of face using color cues, *Proceedings of International Picture Coding Symposium (PCS'93)*, March, 1993.
[10] N. Oliver, F. Berard, A. Pentland. Lafter: lips and face tracking, *IEEE International Conference on Computer Vision and Pattern Recognition (CVPR97)*, S. Juan, Puerto Rico, June, 1997.
[11] J. Ström, F. Davoine, J. Ahlberg, H. Li, R. Forchheimer, Very low bit rate facial texture coding, *International Workshop on Synthetic – Natural Hybrid Coding and Three-Dimensional Imaging*, Rhodes, Greece, September, 1997.
[12] W. J. Welsh, S. Scarby, E. Brigant. "Model-based coding of videophone images", *IEEE Colloquium on Realistic 3D Image Synthesis*, London, 1990, pp. 13–18.
[13] R. Forchheimer, O. Fahlander, T. Kronander, "Low bit-rate coding through animation", *Proceedings of International Picture Coding Symposium PCS'83*, March, 1983, pp. 113,114.
[14] R. Forchheimer, O. Fahlander, T. Kronander, A semantic approach to the transmission of face images, *Proceedings of International Picture Coding Symposium PCS'84*, number 10. 5, Cesson-Sevigne, France, July, 1984.
[15] F. I. Parke, "Parameterized models for facial animation", *IEEE Computer Graphics*, **2**(9), 61–68 (1982).
[16] B. T. Phong, "Illumination for computer generated pictures", *Communications of the ACM*, **18**(6), 311–317 (1975).
[17] P. Ekman, W. Friesen, *Facial Action Coding System*, Consulting Psychologists Press, Palo Alto, Calif., 1977.
[18] B. K. P. Horn, *Robot Vision*, MIT Press, McGraw-Hill, Cambridge, Mass., 1986.
[19] D. Marr, *Vision: A Computational Investigation into the Human Representation and Processing of Visual Information*, W.H. Freeman, San Francisco, Calif., 1982.
[20] K. Aizawa, H. Harashima, T. Saito, "Model-based analysis synthesis coding system for a person's face", *Signal Processing: Image Communication*, **1**(2), 139–152 (1989).
[21] W. J. Welsh, Model-based coding of moving images at very low bit rates, *Proceedings of International Picture Coding Symposium*, Stockholm, Sweden, 1987.
[22] R. Forchheimer, T. Kronander, "Image coding – from waveforms to animation", *IEEE Transactions on Acoustics, Speech, and Signal Processing*, **37**, 2008–2023 (1989).
[23] T. S. Huang, "Modeling, analysis, and visualization of nonrigid object motion", *Proceedings of 10th International Conference on Pattern Recognition*, Atlantic City, N.J., June, 1990, pp. 361–364.
[24] H. Li, P. Roivainen, R. Forchheimer. "3D motion estimation in model-based facial image coding", *IEEE Transactions on Pattern Analysis and Machine Intelligence*, **15**(6), 545–555 (1993).
[25] R. Nohre, Some Topics in Descriptive Complexity, Linköping Studies in Science and Technology, ISBN 91-7871-203-3, Linköping University, Sweden, Dissertation No. 330, 1994.
[26] H. G. Musmann, M. Hötter, J. Ostermann, "Object-oriented analysis–synthesis coding of moving images", *Signal Processing: Image Communication*, **1**(2), 117–138 (1989).
[27] H. Li, Low Bitrate Image Sequence Coding, Ph.D. thesis, Linkoping University, Sweden, 1993.

[28] P. Roivainen, Motion Estimation in Model-Based Coding of Human Faces, Licentiate Thesis LIU-TEK-LIC-1990:25, ISY, Linkoping University, Sweden, 1990.
[29] J. Aloimonos, "Purposive and qualitative active vision", *Proceedings of the DARPA Image Understanding Workshop*, Pittsburgh, Pa., September, 1990, pp. 816–828.
[30] J. Aloimonos, "Is visual reconstruction necessary? Obstacle avoidance without passive ranging", *Journal of Robotic Systems*, **9**(6), 843–858 (1992).
[31] M. Bertero, T. Poggio, V. Torre, "Ill-posed problems in early vision", *Proceedings of the IEEE*, **76**, 869–889 (1988).
[32] D. Ballard, C. Brown, "Principles of animate vision", in Y. Aloimonos, ed., *Active Perception*, Lawrence Erlbaum Associates, Hillsdale, N.J., 1993, pp. 245–282.
[33] F. I. Parke, K. Waters, *Computer Facial Animation*, A K Peters, Addision-Wesley, Mass., 1996.
[34] A. B. Carlson, *Communication Systems*, McGraw-Hill, New York, 1986.
[35] B. K. P. Horn, "Extended Gaussian images", *Proceedings of the IEEE*, **72**(12), 1671–1686 (1984).
[36] D. Pearson, "Texture mapping in model-based video coding", *Proceeding of the IEEE*, **83**(6), 829–906 (1995).
[37] M. Rydfalk, CANDIDE: a parameterised face, Technical Report, Department of Electrical Engineering, Linköping University, Sweden, October, 1987.
[38] J. N. Bassili, "Facial motion in the perception of faces and of emotional expression", *Journal of Experimental Psychology*, **4**, 373–379 (1978).
[39] M. J. Black, Y. Yacoob, "Tracking and recognizing rigid and non-rigid facial motions using local parametric models of image motion", *Proceedings of the 5th International Conference on Computer Vision*, IEEE, 1995, pp. 374–381.
[40] J. Ahlberg, H. Li, "Representing and compressing MPEG-4 facial animation parameters using facial action basis functions", *IEEE Transactions on Circuits and Systems for Video Technology*, **9**(3), 405–410 (1999).
[41] J. J. Lien, Automatic Recognition of Facial Expressions Using Hidden Markov Models and Estimation of Expression Intensity, Doctoral Dissertation, Technical Report CMU-RI-TR-98-31, Carnegie Mellon University, April, 1998.
[42] H. Li, A. Lundmark, R. Forchheimer, Video based human emotion estimation, *International Workshop on Synthetic-Natural Hybrid Coding and Three-Dimensional Imaging*, September, 1999.
[43] T. Jebara, A. Azarbayejani, A. P. Pentland. "3D structure from 2D motion", *IEEE Signal Processing Magazine*, **16**(3), 66–84 (1999).
[44] A. Azarbayejani, A. Pentland, "Recursive estimation of motion, structure, and focal length", *IEEE Transactions on Pattern Analysis and Machine Intelligence*, **17**(6), 562–575 (1995).
[45] J. Ström, Model-Based Head tracking and Coding, Linköping Studies in Science and Technology, Dissertation No. 733, Linköping University, Sweden, 2002.
[46] T. Broida, S. Chandrashekhar, R. Chellappa, "Recursive 3D *motion* estimation from a monocular image sequence", *IEEE Transactions on Aerospace and Electronic Systems*, **26**, 639–656 (1990).
[47] J. Ahlberg, Fast image warping for active models", Report No. LiTH-ISY-R-2355, May, 2001.
[48] T. F. Cootes, G. J. Edwards, C. J. Taylor, "Active appearance models", in H. Burkhardt, B. Neumann, eds., *Proceedings of the European Conference on Computer Vision 1998*, Vol. 2, Springer, 1998, pp. 484–498.
[49] M. Kirby, L. Sirovich. "Application of the Karhunen-Loeve procedure for the characterization of human faces", *IEEE Transactions on Pattern Analysis and Machine Intelligence*, **12**(1), 103–108 (1990).
[50] M. A. Turk, A. P. Pentland, "Face recognition using eigenfaces", *Proceedings of IEEE Conference on Computer Vision and Pattern Recognition*, June, 1991, pp. 586–591.

[51] M. Bichsel, A. Pentland, "Human face recognition and the face image set's topology", *CVGIP: Image Processing*, **59**(2), 254–261 (1994).

[52] J. Ström, Facial texture compression for model-based coding, Licentiate Thesis, LIU-TEK-LIC-1998-51, ISY, Linköping University, Sweden, August, 1998.

[53] P. Eisert, Very Low Bitrate Video Coding Using 3D Models, Ph.D. thesis, Friedrich-Alexander-Universitat, Erlangen-Nurnberg, 2000.

[54] P. Eisert, T. Wiegand, B. Girod, "Model-aided coding: a new approach to incorporate facial animation into motion-compensated video coding", *IEEE Transactions on Circuits and Systems for Video Technology*, **10**, 344–358 (2000).

[55] A. Azarbayejani, T. Starner, B. Horowitz, A. P. Pentland, "Visually controlled graphics", *IEEE Transactions on Pattern Analysis and Machine Intelligence*, **15**, 602–605 (1993).

[56] J. Ahlberg, "Facial feature tracking using an active appearance-driven model", *Euroimage ICAV3D, International Conference on Augmented, Virtual Environments and 3D Imaging*, Mykonos, Greece, May, 2001.

[57] N. A. Thacker, T. F. Cootes, Vision through optimization, BMVC Tutorial Notes, Edinburgh, September, 1996.

[58] W. Press, S. Teukolsky, W. Vetterling, B. Flannery, *Numerical Recipes C – The Art of Scientific Computing*, Cambridge University Press, Cambridge, 1995.

[59] A. Lanitis, A. Hill, T. F. Cootes, C. J. Taylor, "Locating facial features using genetic algorithms", *Proceedings of the International Conference on Digital Signal Processing*, Limassol, Cyprus, 1995, pp. 520–525.

[60] S. Kirkpatrick, C. D. Gelatt, M. P. Vecchi, "Optimization by simulated annealing", *Science*, **220**(4598), 671–680 (1983).

[61] S. Forrest, "Genetic algorithms: principles of adaptation applied to computation", *Science*, **261**, 872–878 (1993).

[62] M. Mitchell, J. H. Holland, S. Forrest, "When will a genetic algorithm outperform hill climbing?", in J. D. Cowan, G. Tesauro, J. Alspector, eds., *Advances in Neural Information Processing Systems*, Vol. 6, Morgan Kaufmann, San Mateo, Calif., 1994.

[63] N. Gershenfeld, *The Nature of Mathematical Modeling*, MIT Press, Cambridge, Mass., 2000.

[64] A. Hill, C. J. Taylor, "Model-based image interpretation using genetic algorithms", *Image and Vision Computing*, **10**(5), 295–300 (1992).

[65] A. Hill, C. J. Taylor, T. F. Cootes "Object recognition by flexible template matching using genetic algorithms", *2nd European Conference on Computer Vision*, Santa Margherita Ligure, Italy, May, 1992, pp. 852–856.

[66] D. Fogel, "Evolutionary computing", *IEEE Spectrum*, 2000.

[67] S. Basu, I. Essa, A. Pentland, Motion regularization for model-based head tracking, *International Conference on Pattern Recognition (ICPR '96)*, Vienna, Austria, 1996.

[68] J. Ström, T. Jebara, A. Pentland, Model-based real-time face tracking with adaptive texture update", Technical Report, LiTH-ISY-R-2342, Linköping University, Sweden, March, 2001.

[69] L. Klasén, H. Li, "Faceless identification: a model for person identification using the 3D-Shape and 3D-motion as cues", *12th Interpol Forensic Science Symposium*, The Forensic Sciences Foundation Press, Boulder, Colorado, October, 1998, pp. 400–402.

[70] H. Li, A. Lundmark, R. Forchheimer, "Image sequence coding at very low bitrates: a review", *IEEE Transactions on Image Processing*, **3**(5), 589–609 (1994).

[71] H. G. Musmann,. "A layered coding system for very low bit rate video coding", *Signal Processing: Image Communication*, **7**(4-6), 267–278 (1995).

12

A Facial Animation Case Study for HCI: The VHML-Based *Mentor System*

Andrew Marriott

Curtin University of Technology, Bentley, Western Australia

The outcomes of research on Talking Head (TH) Technology have recently made it possible to have TH human–computer interfaces (HCI) to applications and information. In fact, HCI are being replaced by human–human interfaces (H^2I) where users may, with natural language, interact with a lifelike, humane, believable computer-generated image that responds to them with computer-generated speech using textual information coming from a knowledge base.

Research on THs embraces the disciplines of computer graphics, dialogue management, text-to-speech synthesis systems and psychological modelling and uses the power of ubiquitous personal computers and their cheap high-performance graphics and audio systems. It is also firmly based on international standards for such things as displaying multi- and hypermedia information across the World Wide Web.

This technology is ideal for providing better interfaces in areas such as:

- *Universities to use Distance Education*: Custom-built THs modelled on the lecturer or bought in from other universities are used to help students understand the lesson, to provide one-on-one tutoring and to give accurate, consistent answers to queries. Each student has the lecturer as their tutor to guide them.
- *Web shopping*: The Virtual Shop assistant is a pleasant front end to the stock database, billing system, complaints department and so on.
- *Web hosting*: A Web site is represented by a TH rather than a URL (humans remember faces better than numbers). The TH has data-mined knowledge about the site, similar to a search engine.

- *Web guide*: On-line exhibitions such as a museum or art gallery where the user can have a knowledgeable Virtual Sister Wendy or any other art critic.
- *Travelling*: A Virtual travel agent provides information about accommodation and ticket arrangements. The travel agent could also book the tickets on your behalf.

As can be seen from other chapters in this book, if MPEG-4 technology is used, the interaction can take place across a very low bandwidth channel and hence becomes available to the home user. The quality and functionality of the TH can be very high even at this low bit rate. All these aspects make the TH believable, and hence a Virtual Lecturer or Virtual Distance Education Tutor is seen as being erudite and approachable, a Virtual SalesPerson in a Web page is seen as trustworthy and helpful and so forth.

This chapter details how to make a TH talk back to the user, how you can get users to listen to it and why you would want to do so. We present this through a series of experiments, user observations and successively refined implementations that will be shown throughout the chapter.

The chapter first analyses the background, general issues and requirements for TH interfaces and then, how our initial observations about how people relate to computer-generated THs helped us to create a more believable and realistic TH – a more humane TH.

The chapter then details how the evaluation of initial prototype systems fuelled the research being done at Curtin University in creating a Virtual Human Markup Language (VHML) that allows these interactive THs to be directed by text marked up in accordance with the Extensible Markup Language (XML) standard. This direction makes the interaction more effective. The language is designed to accommodate the various aspects of human–computer interaction with regard to Facial Animation, Body Animation, Dialogue manager (DM) interaction, Text-to-Speech production, Emotional Representation and Hyper- and Multimedia information. This chapter also points to audio and visual examples of the use of the language.

Importantly, the chapter details how a DM – the ***Mentor System*** – uses an MPEG-4-compliant Facial Animation Engine (FAE) along with a VHML knowledge base to produce a believable and responsive TH. Two significant design and implementation case studies are presented to highlight the functionality and versatility of the system. The case studies consist of two experiments:

1. *An adventure game storyteller*: This demonstrates the emotional capability of a VHML TH. The game is interactive in that the player can direct the story at certain branch points through a question–answer mechanism.
2. *The Mystery at West Bay Hospital*: An interactive detective story where the user asks questions to the THs to try to discover the murderer. The system demonstrates the DM functionality of the TH and its believability as a humane interface to information.

The chapter summarises the evaluation of the users' perception of the TH case studies to see how effective and believable they are.

Finally, we discuss planned future experiments using the VHML and the DM to produce further TH applications:

1. A Web-based information provider for enquiries about TH Technology.
2. An amusing TH front end to a weather forecast system called *Weather Woman*.

12.1 TALKING HEAD INTERFACES

Virtual or mythical THs have been around for a long time [1]. These THs act as mediators or communicators of a message and facilitators of outcomes. These virtual beings may be physical manifestations such as the robots described by Morris [2], detailed in literature such as *The Hero with a Thousand Faces* by Campbell [3] or in video games such as the people simulator *The Sims* by Maxis [4]. They can be as lifelike as in the Famous-3d Web site *www.sorryjohnny.com*, where you can make the 'Australian Prime Minister' talk, sing or cry but never say sorry, or as abstract but well-meaning as Microsoft's *Wizard Paperclip* (*www.iarchitect.com/metaphor.htm*). Their effectiveness in communicating does not normally correlate to their physical appearance.

Modern day mediators can be software programs that communicate with the user via textual, vocal or visual means. For example, desktop and interface agents emerged from the research in the Programming-by-Demonstration (PBD) discipline and resulted in PBD systems that learnt by watching what interaction took place between a user and a Graphical User Interface (GUI) [5–7]. The early Personal Assistants/Interfaces such as described by Etzioni [8] were soon applied to the World Wide Web to assist in browsing. The techniques were also applied to generalised browsing, calendar assistants, finding and communicating with people and Data Mining activities such as Usenet news filtering. Agent technology was beginning to be applied to many areas historically rich in information. See *www.cs.curtin.edu.au/~raytrace/papers/ajet/html/* for a bibliography of on-line papers on Software Agents and related technology.

One motivation for using THs as interfaces is that sound, graphics and knowledge can convey ideas faster than technical documents alone. Also, the use of an anthropomorphic agent may affect the manner in which a user interacts with the interface. Koda [7] in the article 'Agents with Faces: The Effects of Personification of Agents' investigated the most 'favourable' interface by the use of a poker game and four computer players, all with different personas. Two important results for THs were found:

- Faces made the game more engaging.
- The human face rated first in intelligence and comfort, and was second to the dog cartoon in likeability.

People liked faces in interfaces!

Automated responses to e-mail queries by consumers were also being generated by intelligent software agents, which processed the text of the query and replied using prewritten responses matching the enquiry [9]. The interfaces were starting to talk back to the users! But did the users want to listen?

An understanding of the effect of THs is clearly important in the field of information communication and its related areas, particularly in relation to cultural and behavioural issues, and especially in representations of gender, race, community and acquired behaviour [10]. Hence, it is important for the designers and implementers of THs to make informed decisions on how to make them fulfill their required roles more effectively in an interactive manner as facilitators of information.

Animated agents that are, for example, based on real video, cartoon-style drawings or model-based 3-D graphics often make presentations more lively and appealing and therefore make great improvements. André *et al.* [11] indicated that they

- made the human–computer interaction become more like the conversation styles known from human–human communication,
- made the interface more compelling and easier to use.

Similarly, Bickmore [12] indicated that sound, graphics and knowledge could convey ideas and feelings faster than technical documents. Reeves [13] suggests that the user interface is better if implemented with respect to what people would expect from the same kind of character in the real world, regarding, for example, politeness, personality and emotion. A *believable* TH was required.

However, it should be noted that with THs, the more believable the animated character appears, the greater the expectation that the user has. If the user gets the feeling that he or she is interacting with a human being, the user might get disappointed if the character is not as intelligent as expected or reacts in an inconsistent or unpredictable manner. Therefore, the consistent believability of the interaction is more important than the appearance of the TH. This is a TH Turing test in effect.

The virtual character also has to be well integrated with all other contents in the application or Web page to be able to act and react to the user's actions [14].

12.2 FIRST OBSERVATIONS

To determine how the TH could be improved as a user interface, a small experiment at a science fair was set up that would take a user's name and then tell a story to the user that incorporated their name. The users (both sexes) ranged from very young children (5 years old) to high school students (15 years old) and to parents/grandparents. The user was also able to type in any specific text for the TH to speak. Observations were made during the interaction and users were encouraged to fill up a brief questionnaire.

In summary, the evaluation of the five-point sliding-scale questionnaire revealed that the users thought that it was *moderately* friendly, *moderately* pleasant, *moderately* interesting, *moderately* well understood and *halfway between lifelike and robotic*! That is, it needed to be improved if it was to be anything other than novel.

Pelachaud [15] suggested that integrating such non-verbal behaviour as facial gestures, expressions and emotions, with expressive speech, would increase the realism of a TH. Therefore, research to improve the TH in the area of facial expression and realism, vocal expression and interactivity was started.

12.3 DESIGN OF A MORE BELIEVABLE TH, EXPERIMENTS AND EVALUATION

When we talk, we produce a complex acoustic signal that carries information in addition to the verbal content of the message:

- Emotion in speech has been described as 'the organism's interface to the world outside' [16].
- Vocal expression tells others about the emotional state of the speaker, as well as qualifying (or even disqualifying) the literal meaning of the words.
- Sudden pitch changes will make a syllable stand out, highlighting the associated word as an important component of that utterance [17]. A phrase's pitch will increase towards the end to denote a question [18].

A speaker will also pause at the end of key sentences in a discussion to allow listeners the chance to process what was said and for turn-taking.

Facial expressions change continuously in humans, and many of these changes are synchronized with the spoken discourse, for instance, scrunching one's nose when talking about something unpleasant. Facial expressions not only use lips to talk but raise eyebrows, move and blink eyes or nod and turn the head. Facial expressions are also inherently linked to emotion, personality and other behavioural variables [19]. Non-verbal cues may provide clarity, meaning or contradiction for a spoken utterance. They can also influence how we perceive others and how we are ourselves perceived. Non-verbal communication is an important means to convey meaning and information at the verbal, semantic and emotional level.

Because of these problems, it is necessary to be able to control the aspects of non-verbal communication that relate to content and structure of the spoken text, as well as the underlying behavioural aspects of human physiology present during communication. Similarly, listeners *expect* to hear vocal effects, paying attention not only to what is being said but how it is being said. Therefore, if our TH was to become more believable, it was necessary to design and implement some initial markup languages that could add simulated emotion effects and gestures to the Human–TH interaction.

Marriott [20] shows the results of the design, implementation and evaluation of this initial markup. In summary, participants' responses to a TH that had been marked up with emotional, facial and vocal tags as compared to an unmarked-up TH were

- 82.2% felt is was more understandable
- 86.7% felt it was more expressive
- 71.1% felt it was more natural
- 84.4% felt it was more interesting.

```
<sad>
        You <emph>said</emph> to me once <pause length="short"/>
        that pathos left you unmoved, but that beauty,
        <emph affect="b" level="moderate">mere</emph> beauty,
        could fill your eyes with tears.
</sad>
```

Figure 12.1 Text marked up with complex VHML tags to display emotion

This use of these initial markup languages to add value to an interaction was seen as very beneficial in HCI. The early markup languages controlled speech effects [Speech Markup Language (SML)], controlled facial animation and gestures [Facial Animation Markup Language (FAML)] and formed the basis for the development of the VHML.

12.3.1 Virtual Human Markup Language (VHML)

Although general in nature, the intent of **VHML** is to facilitate the natural interaction of a Virtual Human (VH) with a user via a Web page or a stand-alone application. The text that the VH is to speak is marked up with tags that direct the emotions, gestures, speech, face and body of the Virtual Human. VHML uses existing standards and describes new languages to accommodate functionality that is not catered for. An example of text that uses VHML is shown in Figure 12.1.

More information on VHML can be found at *http://www.vhml.org* and in Reference 21 and examples can be found at *http://www.interface.computing.edu.au/papers/jrpit-hci/*. The current specification level for VHML is 0.4.

All subsequent research on creating more believable THs used the VHML concepts in marking up the spoken text for the knowledge base and in interacting with the user.

12.4 SECOND OBSERVATIONS, EXPERIMENT ONE AND EVALUATION

Qualitative feedback from the first experiments also indicated that it was imperative that users be able to interact with the TH. Therefore a small demonstration (Figure 12.2) was created where users could interact with a basic ‘adventure game’.

The user was stranded on an island, had to find objects, fight monsters, make friends and so forth. At the end of the game, the user confronts a riddle-posing dragon. Get the riddle correct in three tries and the user is free. Get it wrong and the user is burnt to a cinder.

The knowledge base of the game was marked up with the pre-VHML tags to see what effect these tags would have on user interaction and acceptance of this application. The commands that the user could give to the application were very limited. The initial testing using three different THs (one with a very garish texture map) was done at a subsequent science fair with a similar audience as the first. Owing to the nature of this event, no formal evaluation was done, but the informal evaluation suggested that while

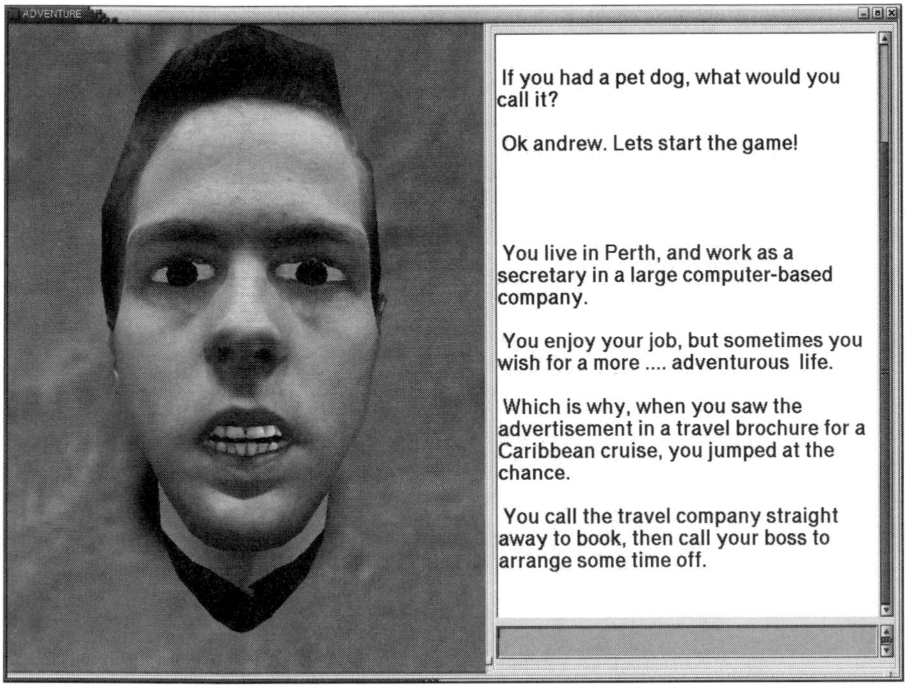

Figure 12.2 The adventure game interface

the 'emotion' displayed by the head was effective – even the garish TH fulfilled its task of engaging the audience – the demonstration was too long and the interactivity was too little. Most users could read the text faster than the TH could speak it and hence many were simply waiting for the TH to finish speaking to get to the next part of the game. Could we get away with no text but just the TH speaking?

Subsequent evaluations of the game tailored its complexity to the amount of time available to the user. It was also decided to investigate a more interactive TH experiment. A more sophisticated natural language DM was needed.

12.5 THE *MENTOR SYSTEM*

To further increase interactivity with the user, the TH was networked to the ***Mentor System*** – a suite of cooperating programs, databases and user interfaces running on a network of Unix workstations and PCs (Figure 12.3). It is a complex DM and has been designed to address the following elements of mentoring students:

- to determine if a software agent–based system can be developed using Java
- to determine if the Java-based system can be scaled and applied to a real-world, unconstrained environment
- to determine if the resulting system is beneficial to students using it.

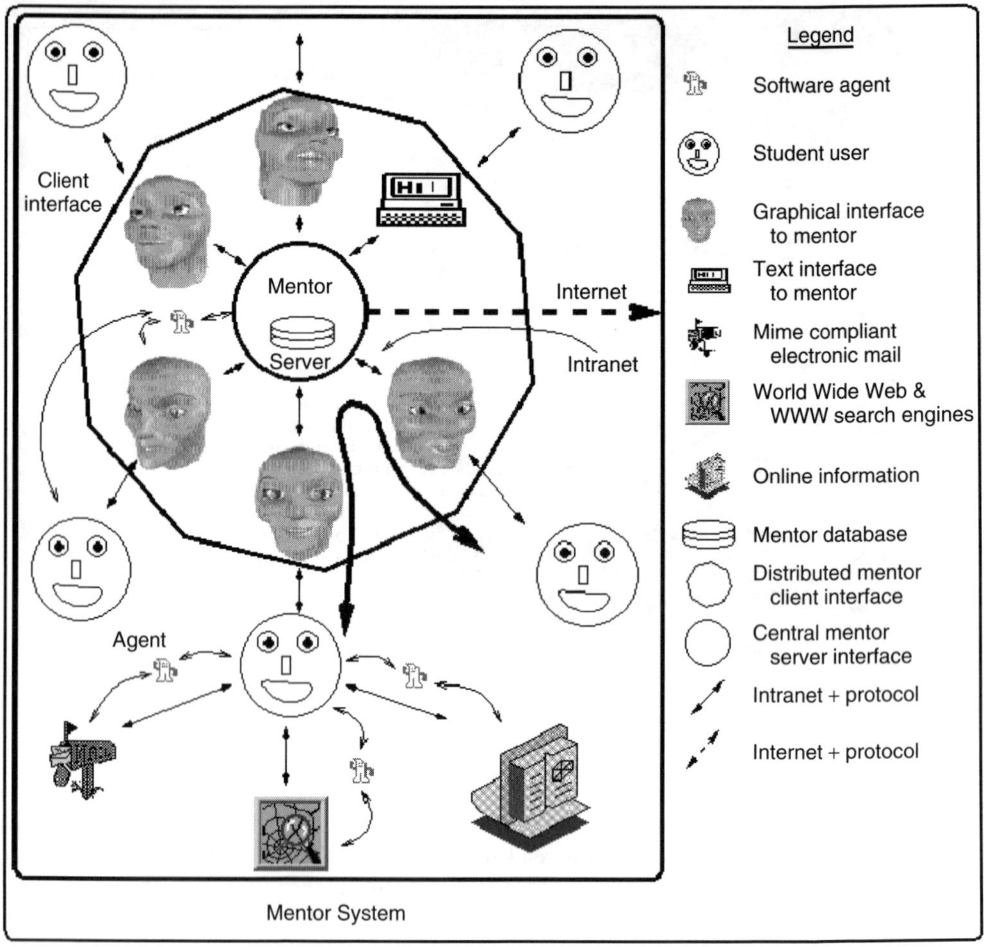

Figure 12.3 The ***Mentor System***

The system is about 70 000 lines of Java in 170 classes per 50 packages. It uses

- a client-server system with protocol between local agent and server;
- a database that maintains per-student, per-unit and global information;
- a suitable pedagogy that will help the students learn from the system;
- an intuitive text-based interface similar to the Unix talk program, a GUI and a TH Interface;
- reusable communicating agents that aid in information retrieval and management.

The entire system – a suite of cooperating network-based computer programs, databases and user interfaces – runs on a network of Unix workstations and PCs. Core to this is the central ***Mentor*** daemon or server – a Java-based mini operating system in its own right.

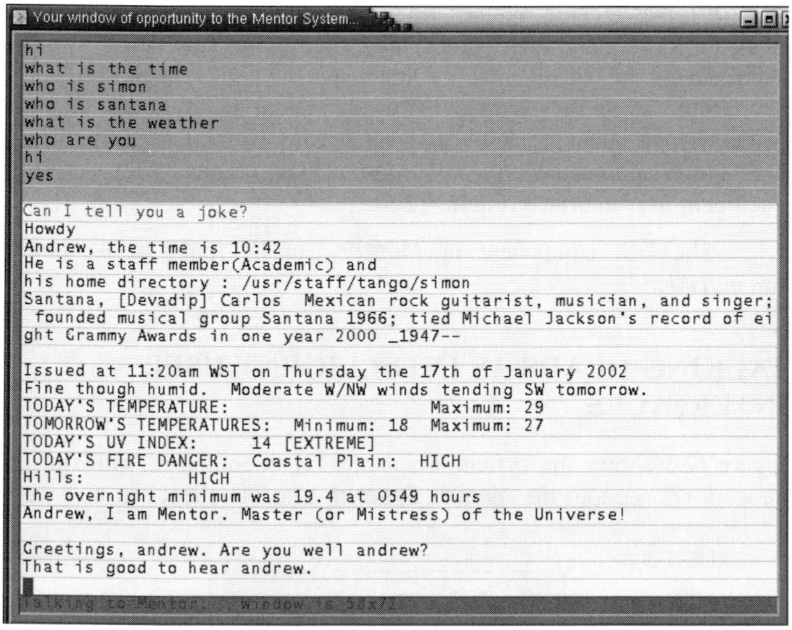

Figure 12.4 MentorClient – the normal user interface to the ***Mentor System***

An actual but contrived dialogue is shown in Figure 12.4 and shows some of the functionality of the system. The user input is in the top window and the ***Mentor System*** response is in the bottom window. Note that the ***Mentor System*** started the dialogue by asking the user if he/she wanted to hear a joke. The user ignored that and asked a few questions. Notice that a weather site was data-mined as was a 'famous people' site to find out who Carlos Santana is. Finally, in response to another greeting from the user, ***Mentor*** asks how the user is and appropriately acknowledges the reply from the user.

The ***Mentor System*** does not have Natural Language Parsing but has a limited ability to understand user requests via Perl-5 Regular Expression Pattern Matching and a state network. The user input moves the system from state to state to keep track of the conversation. It will greet the user in a non-deterministic fashion, occasionally using various parts of the user's full name in their own country-of-origin language. In a similar manner it will respond to various requests with varying phrases such as 'what is the weather', 'what is the time' and so on.

The system makes heuristic guesses about the subject of enquiries. Notice as well that the system knows the sex of the subject (and of any requested individual) and adjusts the output accordingly. A recent experiment indicated that the system correctly recognised more than 75% of the user requests. The system is not merely reactive but also offers timely, *unsolicited* advice on assignments, study deadlines and timetabling issues. It is proactive in the dialogue with the user.

The core of the above client interface is a loop that checks what the user has typed and what they have typed in the past against a list of ***Mentor*** topics. The base

Mentor topics include weather, time, multi-language greetings and good-byes, who is using the system and their names, how to be polite, how not to swear or be offensive and academic information. New domain-specific topics can be easily developed to cater for differing uses, for example, knowledge about an advanced computer graphics unit or knowledge about the work associated with a research group. This flexibility of a topic-based approach with domain knowledge, coupled with a TH, can produce an intelligent user interface (Figure 12.5).

The DM component of ***Mentor*** can be found at *www.mentor.computing.edu.au/DM* .

12.6 TALKING HEADS AS INTELLIGENT USER INTERFACES

Figure 12.5 shows the system when the TH is combined with (in most cases via a network connection) the ***Mentor System***. The TH client can now get the user input,

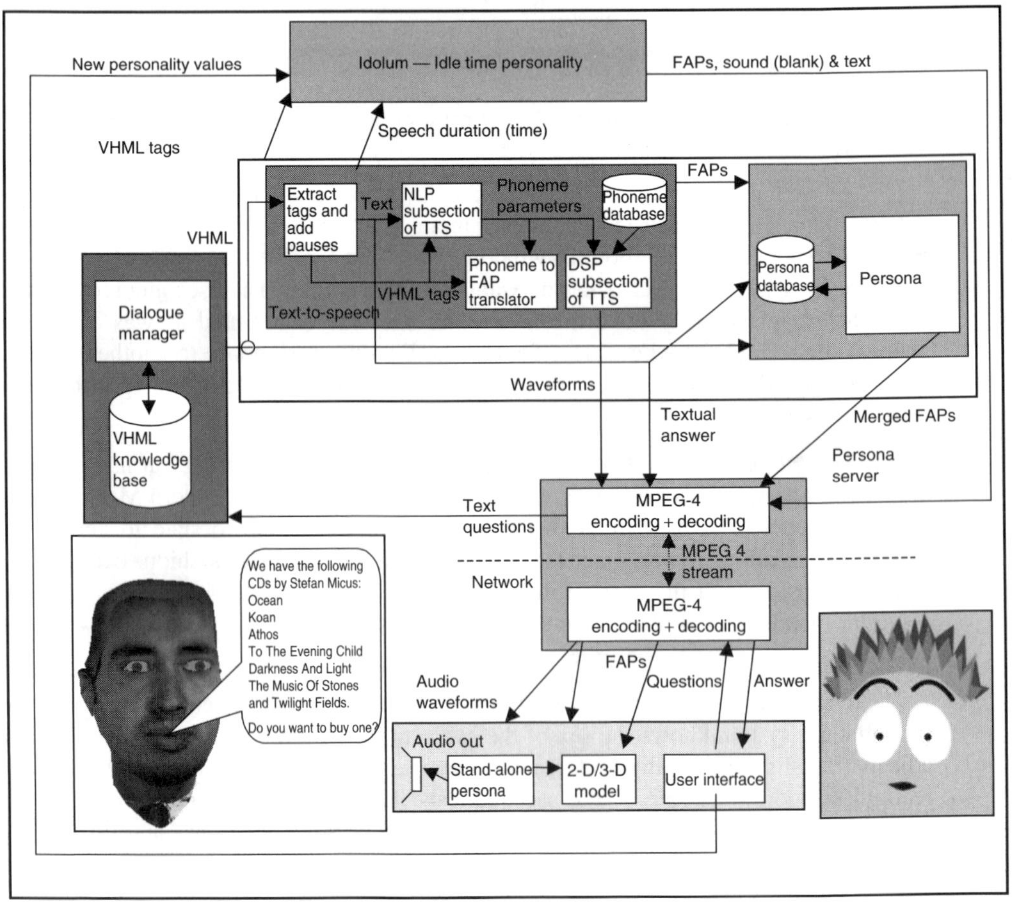

Figure 12.5 The Curtin interactive TH system using the ***Mentor System*** dialogue manager

send it to the ***Mentor System***, which processes it and returns the marked-up or tagged response. This is then sent by the TH client across to the TH server for conversion into Facial Animation Parameters (FAPs) and audio as before. The user may now interact with a TH by asking natural language questions and receiving a believable response.

In Figure 12.5, the information returned from the DM's knowledge base should be consistently marked up in a format such that it is easily parsed and categorized – an XML-based language – the VHML language in this case. It should be also noted that the final rendering and delivery of the marked-up text is context- /application-dependent:

- A straight textual interface for a very low bandwidth interface.
- An interactive Web-based multi- and hypermedia display.
- A voice-enabled system that lets the user hear the answer.
- A TH system with complex facial gestures and voice that shows the personality or emotion of the TH.
- An entire synthetic human figure with body language.

The information markup should stay the same but the way in which it is displayed should change depending upon the form of output – textual, vocal, TH, body language. The ***Mentor System*** allows for this final translation before delivery.

The text of Figure 12.6 may be the knowledge base response to the user enquiry 'What are you?' and has been marked up with a VHML sub-language – Dialogue Manager markup Language (DMML). The DMML text is never seen outside the DM itself but allows for a consistent naming of the relevant tags such as the user's name, sex and so forth.

In a normal interactive session, the TH client will (hopefully) know the name of the user. This name may be used to set a DM variable that can be used later in the conversation. In this example, <first_name/> is the name of the user who is making the enquiry, <welcome/> is a language-dependent greeting that has been set depending upon the user's home country or domain name, <mentorName/> is the name of the DM and so on. So the DMML response is converted by the DM to become the final plain text, for example,

Freda, Guten Tag. I am Mentor. I was developed by Andrew Marriott. . . .

The exact specification of DMML is still under development; it is being used by two DMs – ***Mentor*** and MetaFace. A mechanism has been developed that enables any DM static variables (such as the DM's name or purpose) to have any tag name, as long as the definitions get processed before they are used.

```
<first_name/>, <welcome/>. I am <mentorName/>.
I was developed by <mentorMaster/>.
<mentorDescription/>
<mentorPurpose/>
You can find out more about me from <mentorHomeURL/>.
```

Figure 12.6 A segment of XML marked-up text showing the use of atomic tags

The text in Figure 12.1 contains VHML tags that could be used to add vocal, facial and emotional effects to the response. These tags/effects would need to be suppressed for a text-only display but would add emotion to a voice and/or facial rendering.

VHML specifies emotions/actions to facilitate the direction of a VH interacting with a user. For example, a VH that has to give some bad news to the user – 'I'm sorry Dave, I can't find that file you want' – may speak in a sad manner, with a sorry face and with a bowed body stance. Similarly, a different message may be delivered with a happy voice, a smiley face and a lively body. VHML tags such as <smile>, <angry>, <surprised>, <look_down> have been specified to produce the required vocal, facial and emotional actions.

12.6.1 Rendering

Notice that the `<mentorHomeURL/>` in Figure 12.6 would probably become a URL and hence the pure text may contain '*http://www...*'. Therefore, in the 'rendering'

1. a Web-based display may turn that into a link,
2. a vocal display may have to change the text into something like 'You can find out more about me *from the link www.blah*',
3. a TH may say the above and also open up a new browser window,
4. a VH may point to the link and so forth.

XML Schema Language (XSL) is a style sheet language designed to be used with XML data and documents to help in the different 'renderings'. Unlike HTML, which defines the rendering or display behaviour for each of its elements, XML says absolutely nothing about how the data is to be displayed. XSL allows the author to apply formatting operations to XML elements. XSL is a language in which the DM author can indicate that the `<emph>` element should be ignored or should be rendered [by the Face and/or the text-to-speech (TTS)].

It must be noted that a TH is not normally viewed in isolation and hence there must be a way for the TH to interact not only with the user but also with its environment. The *target* attribute of the VHML <p> tag can be used to specify that the rendering of the VHML data is directed to a specified 'target'. It is up to the application to know the semantics and syntax of the specified target string. For example, the string could be used to indicate a different frame for a Web-based TH interface (see Point 3 in previous discussion) or it could be used to direct the data to different THs in a multi-headed application (see later discussion on the detective story).

In this, and other ways, VHML has been designed for the future – how will THs be used in future interactions, what is needed to make these interactions more engaging.

12.7 THIRD OBSERVATIONS, EXPERIMENT TWO AND EVALUATION

Given the effectiveness of the TH with marked-up text to engage the user, it was decided to develop an experiment that would combine the emotiveness of the TH with

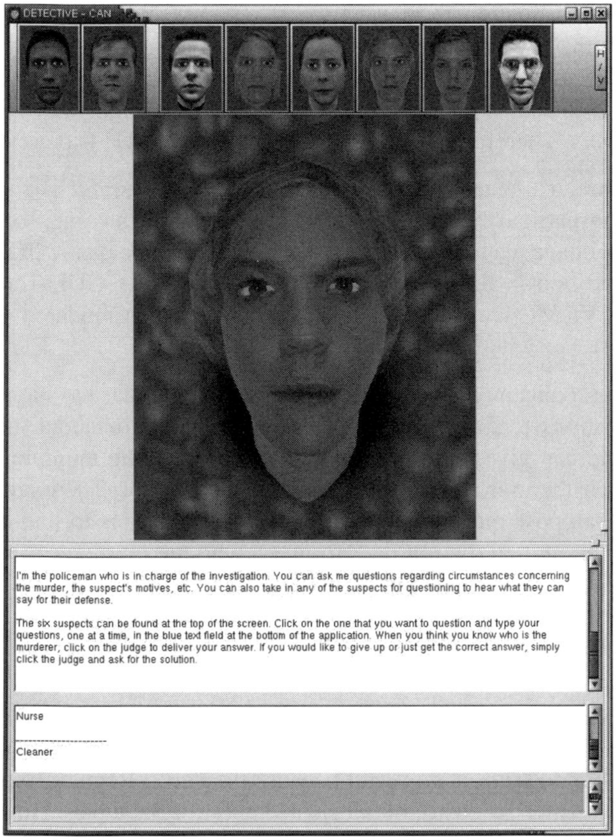

Figure 12.7 The Mystery at West Bay Hospital interface

the natural language interaction of the ***Mentor system***. The domain knowledge of the experiment would be marked up using VHML (or the pre-VHML tags). The experiment would be highly interactive and would test the user's ability to solve a detective mystery by questioning suspects. A synopsis of the mystery can be seen below and the interface can be seen in Figure 12.7.

The Mystery at West Bay Hospital.

A murder has been committed at the West Bay Hospital. John Smith was found dead in his bed, obviously choked.

All people involved – his roommate Paul Windsley, his colleague Amy Goldman, his girlfriend Patricia Stone, the doctor Kevin Goldman, the nurse Alice Duffy and the cleaner Susan Leonard – are seen as suspects. You are a well-known detective and you will receive help from Tom Cartier the policeman as well as the Judge.

The evaluation from experiment one had indicated that displaying the spoken text took the user's attention from the TH, and it is felt that further investigation is required regarding the best use of textual display with a TH. However, since the goal of the user is to actually solve the mystery, he or she might want to read earlier questions and answers. Therefore, the plain text spoken by the TH is presented to the user as well.

The idea behind *The Mystery at West Bay Hospital* is similar to existing applications, in particular *Cluedo*. To pose a question to anyone, the user clicks the corresponding image, causing the character to appear at the centre of the GUI and then types in the question in the text field at the bottom of the GUI. The character's response, from the VHML knowledge base of ***Mentor***, is then rendered (spoken, displayed and acted out).

The policeman has some knowledge about what has happened and the user can pose questions to him concerning the crime scene or about the suspects. There is a judge who can give the user hints on how to find the murderer and give information on whether the user accuses the correct person or not. There are six suspects to whom the user can pose questions and the goal for the user is to find out which one of these suspects committed the murder. To guess who the murderer is, the user tells the judge his/her guesses. The user gets three chances to guess who the murderer is and when those guesses are used up, the judge asks if the user wants the correct solution before the application finishes.

12.8 DIALOGUE MANAGEMENT TOOL (DMT)

The ***Mentor System*** is powerful because the Perl-5 Regular Expression Pattern matching is augmented by custom-built Java classes for the topics. However, the development and implementation of specialist Java classes is not seen as an ideal solution for the production of interactive THs by non-programmers. The DMT [21] was developed in order to create a useful tool for creating and maintaining dialogues. It also uses an XML-based markup language – the Dialogue Management Tool Language (DMTL) – to represent the dialogue and its states as a network. The structure of DMTL is shown in Figure 12.8 [21].

A new topic class for the ***Mentor System*** was developed to process DMTL-managed dialogues. The output from the DMT was transformed into a complex Java data structure class that *extended* this new DMTL-processing class. All the power of the ***Mentor System*** could now be used by the DMTL dialogues. Users could concentrate on the dialogue between user and VH rather than worry about the underlying details of the pattern-matching.

The DMT lets a user create a *dialogue* that contains multiple *topics* (with *subtopics*) composed of various *states* that can match a *stimulus* (such as a question) from the user and return a *response* (perhaps marked up in VHML). Depending upon the response, the dialogue between user and DM may move into a new state. The response may also *signal* the DM to end the dialogue, iconify the user interface and so forth.

These topics and their stimulus/response states are then used by the DM (in some DM-dependent way) to interact with the user. For example, a topic for each

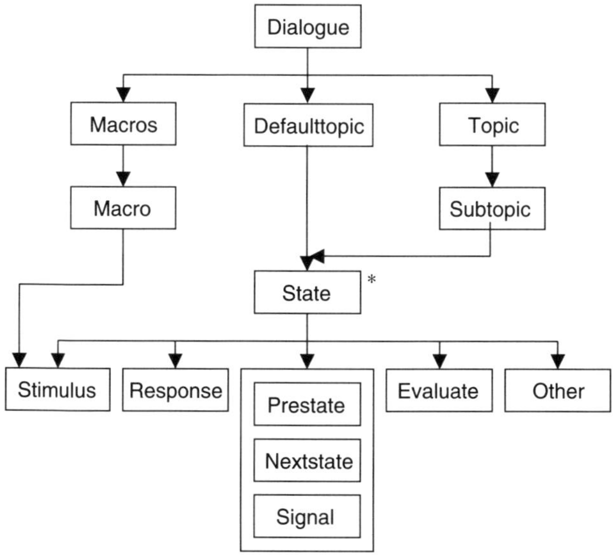

Figure 12.8 The structure of DMTL

virtual character in the mystery story was developed to cater for the user questions and the answers. In developing the mystery story and the supporting infrastructure, a number of problems arose. Dialogues tended to become large and complex, with many topics, subtopics and states. Since the DM has to match the user input against all the stimulus in the dialogue, this became an efficiency problem. To avoid this, the `keywords` attribute for the `subtopic` was introduced to limit the search to only those subtopics that match the user input. If no keywords are specified for a subtopic, it means that everything needs to be matched.

Similarly, the `evaluate` attribute for a `subtopic` element means that some conditions have to be fulfilled in order to process that `subtopic`. This could be as simple as 'has this node been visited before' or a complex Boolean expression involving several DM variables. For the subtopic to be considered, the expression has to be true.

Another way to cut down on the search space is to classify the states so that only those that are appropriate are searched:

- `active`: A state that invokes a question, without having to be triggered by a stimulus. For example, the question 'Do you want to know more about VHML?' *Not searched.*
- `entry`: A state that can be invoked anytime during the dialogue if the stimulus matches. This is the default state type. An example of this is 'What is VHML?' *Always searched.*
- `linked`: A state that is connected to other states by using `<nextstate>` or `<prestate>`. The state is linked because the stimulus depends on having some kind of context to be understood correctly. An example is the question 'What is that?' The 'that' corresponds to something introduced earlier in the conversation and

the DM should know what it is. A linked state can never directly match the initial user input; it has to be linked from another state. *Reduced search.*

- `visitswitch`: A state that points to several other states and works in a similar way as a case statement in C or Java. Which state the dialogue should move into can, for example, depend on whether the state has been visited before. *Searched on the basis of evaluation.*

It should be noted that a `stimulus` can be of several different types: `text`, `audio`, `visual` and `haptic` (*text* is the default). Instead of having 'Yes' as a text stimulus, there can be a visual stimulus when the user nods. Also, multiple `responses` with varying *weights* can be specified and the DM can randomly choose which one to use. This enables the speech of the TH to be non-deterministic, given the same input.

To keep track of which state is active, fully qualified names are used. For example, a *state* called *description* in a subtopic whatis in a topic VHML has the fully qualified name VHML.whatis.description. By using these names it is possible to exactly specify which state is being referenced.

Macros are used in order to make it easier when creating stimuli. Since natural language is complex, there are many different ways to express the same question. For example, the macro 'WHATIS' can be used in the stimulus 'WHATIS VHML?'. This is then defined to match 'What is VHML?', 'Can you please tell me about VHML?', 'What is the Virtual Human Markup Language?' and so on. The macro specifies the *intent* of the stimulus by providing an (exhaustive) list of possibilities. The DM may use these or may use the *intent* to generate a suitable matching mechanism (e.g. a Perl-5 regular expression in the ***Mentor System***).

The DMT and associated software can be found at *www.vhml.org/DMT*.

12.9 DISCUSSION AND EVALUATION

The *dialogue* in the mystery story has approximately 1000 states. With this size, the DMT was necessary to keep track of these stimulus/response/state transitions. However, to test the system better, it may be appropriate to include more states and more responses. It became very obvious that the perceived quality of the application went up as the number and range of stimuli increased.

Also, since the users of the *Adventure Game* became very interested as soon as the TH started to address them with their names, this could have become a way to further engage the user. This trivial addition of <first_name/> to the responses was not done though! Currently the dialogues have not been marked up in any way other than to get different voices for the players. This is something that will be done as soon as the VHML implementation is finished so as to make the application more engaging.

12.9.1 Results

Firstly, the evaluators were asked to read the front page of the questionnaire that gave a synopsis of the mystery. Secondly, the policeman told the initial story, and thereafter the participant could start posing questions to the characters. They were told to try to solve the mystery and that they could quit whenever they wanted.

For technical reasons beyond our control, the initial evaluation only used seven people (further evaluation is planned). Therefore, it is not possible to draw any strong conclusion. However, some initial observations from the experiment are as follows:

- The macros and stimulus did not seem to be general enough. The participants found that the mystery was too complicated, mostly owing to the lack of answers or owing to the DM's inability to understand the user's request.
- The dialogue was nowhere near complete – lack of answers to questions that people are bound to ask and no real 'hints' to questions that the characters can answer.
- Using THs in this kind of application is very suitable and was appreciated.
- Users are probably reading more than looking and listening to the THs.
- Having a story and a mystery that is to be solved seemed to engage the users: several people put much effort into trying to solve the mystery (10 to 45 minutes).
- The participants liked the application more than the research group did (possibly because of lack of familiarity).
- Different faces and voices were important for the enjoyment of the story.

A more detailed analysis of this small data set can be found in Reference 22 with further formal evaluation being planned. However, the results are reminiscent of the first science fair evaluation: moderately this, moderately that and so on. With each few steps forward, we go one step back. At least we now have an interactive TH that is more humane.

12.10 FUTURE EXPERIMENTS

Given the inconclusive but positive nature of the previous experiment, it was decided that further improvement and subsequent evaluation was needed for the mystery story.

This was to be done via improvements in the manner in which the stimulus macros were changed into Perl-5 regular expressions. The mystery story could also use a different DM – MetaFace. This could provide a comparison of the functionality/effectiveness of the two DMs.

It was also decided to invest significant time and resources into developing a new experiment that had a richer domain knowledge and a better stimulus range to see if this would better engage a user. The TH Technology project will produce a Web site that will interact with visitors via a VH that can answer multi-modal questions from a limited domain. Its development will be based upon a number of related areas concerned with video/audio analysis and MPEG-4 Facial (and Body) Animation.

It will use the disciplines of

- MPEG-4 human animation
- multi-modal analysis and input
- VHML to produce emotional, believable, humane VHs

- text-to-speech synthesis
- dialogue management.

The experiment will be concluded with a case study evaluation of the HCI issues such as effectiveness and affectiveness arising from a TH being used to deliver information in response to user enquiries.

The FAE from the University of Genova will be used as the displayer of the photo-realistic 3-D TH, while Spiky Boy from Curtin will be used as the displayer of the abstract 2-D VH. The emotional moods and personality of the VH will be controlled by the Idolum Persona server at Curtin. Two possible TTS systems will be used to evaluate the TH Technology Project: the Festival NLP with the MBROLA DSP and the Elan SpeechCube.

The stimuli–responses will be developed in VHML and maintained using the DMT. The domain knowledge will be marked up in VHML so that the output can be rendered via diverse media such as text-only, sound-only, TH and VH. Two DMs from Curtin will be used: the ***Mentor System*** (*www.mentor.computing.edu.au*) and the MetaFace (*www.metaface.computing.edu.au*).

The domain knowledge will be concerned with

- The InterFace project (*www.ist-interface.org*)
- TH Technologies such as MPEG-4 Facial and Body Animation, Famous-3D and so on
- VHML.

In a similar way, we intend to develop a TH application called *WeatherWoman* that will data-mine a weather site and comment on the weather, which will be marked up in a suitable way – it will sound sad when it is raining in winter but happy when it is raining in summer. This will test the flexibility and functionality of the DM and also test the effectiveness of a specific type of TH representation – abstract or photo-realistic.

12.11 FUTURE WORK

The most important step next is the implementation of a working subset of VHML. Currently the tags are still SML- and FAML-based. Parallel to this work will be the implementation of more tags for both vocal and facial emotions. This low-level development of the TH tags must be accompanied by research on high-level tag classes that embody an overall feel for a body of text. Currently, the marking up of text is manually intensive, tedious and prone to error. Current research is investigating the automatic marking up of text for a given subject–context type of presenter. A long-term goal is for THs that can remember how certain text/words were marked up in the past and use this knowledge to transparently process minimally tagged text in a similar manner, that is, the THs learn from experience.

Also, since VHML aims to become an international standard, it should be possible to be able to write the elements in any language or use synonyms for the words. For example, use a Swedish word `<arg>` instead of the English VHML tag name `<angry>` or a synonym `<joyful>` instead of `<happy>`. Figure 12.9 shows how by

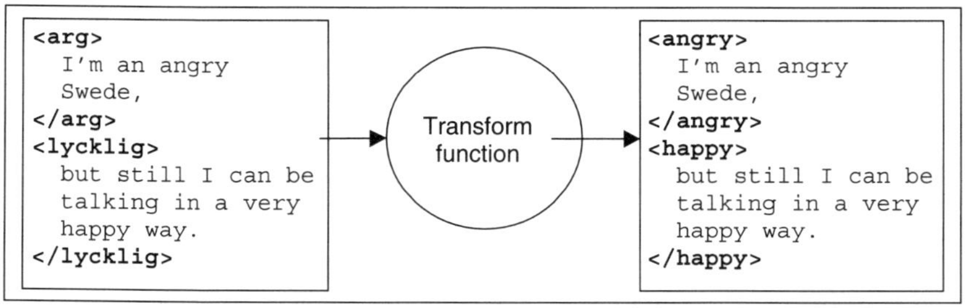

Figure 12.9 An example of transforming Swedish tags into VHML

using an XSL style sheet the original language-specific text is transformed to valid VHML. A specific style sheet has to be constructed for each language as well as for synonyms.

Similarly, if interactive THs are to be produced by non-programmers, the DMT will need to be improved to provide better flexibility and functionality to cater for more complex dialogues. The work on VHML and the DMT will extend the research done by Gustavsson *et al.* [21].

Finally, more work needs to be done on creating various personalities and models for the TH. An important aspect of believability of a computer character is to show consistent behaviour [22]. It needs a permanent personality. What non-psychologists tend to think of as personality is really a combination of the static, long-term controls of a personality with the shorter-term emotions/moods.

Currently the TH's idle time personality models the Big Five Personality Theory – the five factors being extraversion, neuroticism, conscientiousness, agreeableness and openness. Each trait represents a range of values, for example, 'agreeableness' represents a range of values from agreeable to disagreeable. The idle time personality – Idolum – currently modifies the TH's behaviour on the basis of the emotional content of the marked-up input text. Sad text makes the TH sadder and this sadness lasts for a time depending upon the TH's personality. Further development needs to be done on this personality system.

12.12 CONCLUSION

If we have a TH or VH, users expect to interact with it. Hence we must have a DM that is capable of understanding their natural language dialogues. The DM must be proactive and not just reactive, since this is what happens in normal dialogues. The DM must also be capable of directing the VH to interact with its environment, pointing out other information such as diagrams, text, Web sites and so on.

The DM must have the relevant domain knowledge to be able to answer the user's queries. It must be easy to build and maintain this domain knowledge, and hence we need tools for non-programmers to do this. We have seen that the results of using the ***Mentor System*** as a DM are positive.

Since humans are familiar with human faces and the way in which they react, the VH must also act and react in this commonplace, believable fashion. The VH need not be photo-realistic, but it must be believable. Therefore the VH must be capable of consistent, believable actions [23, 24]:

- *Situated liveness*: the agent should continue to show believable behaviour, suitable to its environment, at all times.
- *Controlled visual impact*: a higher visual impact involves more active and obvious behaviours. These are more interesting, but their use must be controlled as they can also be distracting.
- *Complex behaviour patterns*: the agent's behaviour patterns must be complex so that behaviours do not appear obvious and mechanical.
- *Natural unobtrusive behaviour*: users can be distracted by blatantly unnatural behaviours, including such simple aspects as a lack of breathing or blinking.

This means that a consistent temporal personality must be implemented. Until these personalities are capable of 'real' insightful thought, we believe that the emotional/humane component of their actions can be produced through the marking up of the domain knowledge base in a form that directs the VH. The VHML is a step in that direction.

Finally, if you want users to listen to your VH, we believe that you need to assess what kind of VH your application needs, given the intended audience, the information content and the delivery mechanism, design your interface along with the domain knowledge so that the responses can interact with the environment and finally to put it all together in an engaging package.

ACKNOWLEDGEMENT

VHML owes its existence to the work done primarily by John Stallo. Quoc Huynh and Simon Beard also contributed to its development. The current specification of VHML at *www.vhml.org* also owes a huge debt to Emma Wiknertz, Linda Strindlund and Camilla Gustavsson, whose efforts and research made the language more solid, homogenous and complete. Thank you.

REFERENCES

[1] M. Senior, *The Illustrated Who's Who in Mythology*, Guild Publishing, London, 1985.
[2] B. Morris, *The World of Robots*, Gallery Books, New York, 1985.
[3] J. Campbell, *The Hero with a Thousand Faces*, Fontana Press, London, 1993.
[4] M. Stepnik, Reviews: the sims, *PC PowerPlay*, **47** (April), 58–63 (2000).
[5] P. Maes, R. Kozierok, "Learning interface agents", *Proceedings of AAAI'93*, MIT Press, Cambridge, Mass., 1993, pp. 459–465.
[6] H. Lieberman, *Attaching Interface Agent Software to Applications*, AAAI Press, 1994.
[7] T. Koda, P. Maes, Agents with faces: the effects of personification of agents, *HCI'96*, The British HCI Group, London, UK, September, 1996.
[8] O. Etzioni, D. Weld, "A softbot-based interface to the internet", *Communications of the ACM*, **37**(7), 72–76 (1994).
[9] J. Hibbard, "Web service: ready or not", *Information Week*, 18–22 (1998).
[10] J. Murray, *Hamlet on the Holodeck: The Future of Narrative in Cyberspace*, Free Press (Simon and Schuster), New York, 1997.
[11] E. André, T. Rist, J. Müller, WebPersona: a life-like presentation agent for educational applications on the world-wide web, *Proceedings of the Workshop Intelligent Educational Systems on the World Wide Web, 8th World Conference of the AIED Society*, Kobe, Japan, August, 1997, *http://www.cs.curtin.edu.au/~raytrace/phdpapers/aied97/Andre/Andre.html*.
[12] T. W. Bickmore, L. K. Cook, E. F. Churchill, J. W. Sullivan, Animated autonomous personal representatives, *The Second International Conference on Autonomous Agents (Agents '98)*, Minneapolis, St. Paul, May, 1998.
[13] B. Reeves, C. Nass, *The Media Equation*, Cambridge University Press, Cambridge, 1996.
[14] I. S. Pandzic, "Life on the web", *Journal of Software Focus*, **2**(2), 52–59 (2001).
[15] C. Pelachaud, N. Badler, M. Steedman, "Generating facial expressions for speech", *Cognitive Science*, **20**(1), 1–46 (1996).
[16] K. L. Scherer, "Speech and emotional states", in J. K. Darby, eds., *Speech Evaluation in Psychiatry*, Grune and Stratton, New York, 1981, pp. 189–220.
[17] T. Dutoit, *An Introduction to Text-to-Speech Synthesis*, Kluwer Academic Publishers, Dordrecht, 1997.
[18] L. Malandro, L. Barker, D. Barker, *Nonverbal Communication*, 2nd ed., Random House, New York, 1989, pp. 32–50.
[19] P. Ekman, "About brows: emotional and conversational signals", *Human Ethology: Claims of a New Discipline: Contributions to the Colloquium*, Cambridge University Press, Cambridge, 1979.
[20] A. Marriott, S. Beard, H. Haddad, R. Pockaj, J. Stallo, Q. Huynh, B. Tschirren, "The face of the future", *Journal of Research and Practice in Information Technology*, **32**(3/4), 231–245 (2001b).
[21] C. Gustavsson, L. Strindlund, E. Wiknertz Dialogue management tool, *The Talking Head Technology Workshop of OZCHI2001, The Annual Conference for the Computer-Human Interaction Special Interest Group (CHISIG) of the Ergonomics Society of Australia*, Fremantle, Australia, November, 2001.
[22] K. Isbister, C. Nass, "Consistency of personality in interactive characters: verbal cues, non-verbal cues, and user characteristics", *International Journal of Human-Computer Studies*, **53**, 251–267 (2000).
[23] W. L. Johnson, J. W. Rickel, J. C. Lester "Animated pedagogical agents: face-to-face interaction in interactive learning environments", *International Journal of Artificial Intelligence in Education*, **11**, 47–78 (2000).
[24] J. C. Lester, B. A. Stone, Increasing believability in animated pedagogical agents, *First International Conference on Autonomous Agents*, Marina del Rey, Calif., February, 1997.

13
PlayMail – Put Words into Other People's Mouth

Jörn Ostermann

AT&T Labs-Research, Middletown, New Jersey, USA

ABSTRACT

PlayMail is an E-mail service that translates the text and emoticons of an e-mail into a video with an animated talking face. Using a web-based interface, the e-mail sender may select the face that presents the e-mail. Furthermore, the user may submit an image to the server and create his own face model. MPEG-4 face animation technology is used at the server to enable this service.

13.1 INTRODUCTION

MPEG-4 provides exciting technology for face animation. However, the lack of MPEG-4 players installed on the desktop of client terminals limits its use in commercial applications at this point in time. On the other hand, applications that use MPEG-4 technology at the server and ubiquitous technology at the client can easily be made available to the public. Using MPEG-4 technology at the server side only requires that the server renders the MPEG-4 contents and converts them into a commonly available format for audio and video.

There are several companies that offer technology for delivering video including audio from a server to a client. Microsoft's streaming format Advanced Systems Format (ASF) allows delivering to a client that has the windows media player installed [1]. Real Networks uses its Real Server and Real Player to deliver the same contents [2]. Apple promotes its QuickTime server and players [3]. These three streaming solutions perform similarly in terms of coding efficiency and latency. Their latency of more than three seconds makes these solutions suitable for broadcasting and noninteractive applications. They are not suitable for interactive services. For interactive services, one could rely on video coding applications based on H.323 [4] or on Microsoft NetMeeting [1]. However, these solutions usually do not penetrate firewalls.

For face animation, any of these solutions require the rendering of a synthetic face model and a synthetic voice into a regular video sequence at the server. Whereas a standard PC is able to render 30 or more voices simultaneously and in real time [5], PCs are not yet able to render several graphics models in parallel and code them into video at the same time. As a consequence, server-based face animation requires a dedicated server for every simultaneous one or two sessions. This makes real-time interactive services expensive. However, e-mail applications do not have this real-time requirement. E-mails need to be delivered in a reasonable time, but we can tolerate a delay of several minutes between sending and receiving an e-mail.

PlayMail is a multimedia enhanced e-mail service that translates text messages into animated videos with face models reading the message [6]. In order to communicate emotions, you may use several predefined *Emoticons* such as :-) for smile or :-(for frown in the text. The face models are created, customized and selected by the user. Using a web-based interface, the user creates a face model from a frontal picture of a friend, teacher or celebrity that he submits into his private or public database of face models. The PlayMail recipient will receive a small e-mail with a link to a web page that has the message video embedded.

In Section 13.2, we discuss the architecture of the PlayMail service. Creating and delivering PlayMail messages is presented in Section 13.3. In Section 13.4, we present how a PlayMail user may create his private face models to be used in the PlayMail service. We conclude with Section 13.5.

13.2 SYSTEM ARCHITECTURE

PlayMail is an E-mail service that users can use without special requirements for their computer. Therefore, the PlayMail service relies as much as possible on commonly installed software on the client side; in order to avoid that, users have to download and install new software. As such, the sender of a PlayMail is required to use a standard web browser with support for JavaScript and Java. The receiver of a PlayMail needs to have available an e-mail client to read his e-mail, a web browser and a streaming client such as the Real Player [2] or the Windows Media Player [1] to view the PlayMail. The server technology is more complex and customized for the task.

Figure 13.1 shows the system architecture of the PlayMail service. The client sends a http-request with the information for sending a PlayMail to the server. Since the server may not be able to compute all PlayMails immediately, it places all requests in a first in first out (FIFO) queue. For each request, the server starts its animation engine with the user-selected face model. Our face models are described using indexed facesets, transform nodes and face animation tables (FAT). Furthermore, the server sends the text of the PlayMail message to a text-to-speech (TTS) synthesizer. The TTS translates the emoticons in the text into MPEG-4 bookmarks. Then it creates the audio as well as the markup information including phonemes, related timing information, bookmarks and stress marks as described in MPEG-4. With this markup information, the animation engine renders the video of the face animation. The streaming audio and video server stores the PlayMail message for retrieval. In order to accommodate

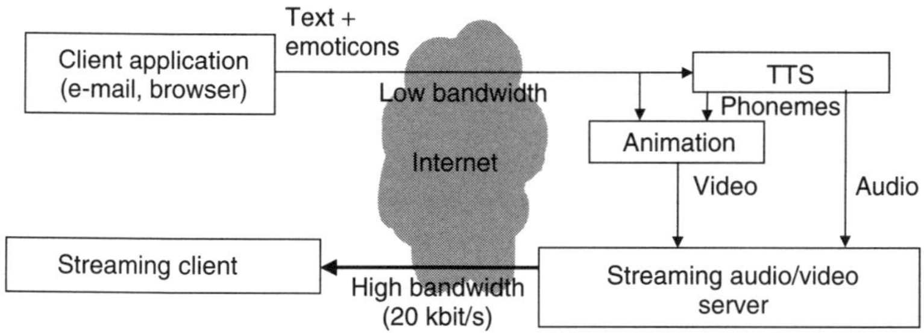

Figure 13.1 Overview of the system architecture with the client and the server connected via the Internet

recipients with different network access speeds, the streaming server compresses each audio/video stream with five different bandwidths between 20 kbit s^{-1} to 80 kbit s^{-1}. Finally, the server sends a regular e-mail to the recipient providing the URL of the PlayMail message.

When retrieving the PlayMail message, the server probes the connection to the client and starts streaming at the highest available bandwidth that can be sent to the receiver. The bandwidth gets adapted in case of network congestion during playback.

One of the exiting features of the PlayMail is the capability for a user to create his own face models with the option of making these face models available to the public (Section 13.4). To this extent, the server acts as a hosting service for specialized contents. It provides for password-protected user access.

13.3 PLAYMAIL MESSAGES

At this point, sending PlayMails requires going to the website of the PlayMail [6]. Figure 13.2 shows the web page for sending the message. It shows the desired fields for sender's name and the e-mail addresses of the sender and the recipient. The character for presenting the message is selected by clicking onto one of the images in the left frame of the page. This replaces the PlayMail logo in the center frame with the image of the face model, which is Mona Lisa in the example of Figure 13.2. Each face model comes with a default gender that determines the voice of the speech synthesizer. However, the user may choose to override this default. The large window takes the actual message. As mentioned in Section 13.2, the server uses MPEG-4 bookmarks to synchronize facial expressions with the spoken text. Since these bookmarks are not very convenient for the user, we use simple emoticons on the web page. The user may choose to use one or more of the emoticons shown on the right side. After the user is done with composing the message, he may submit the request to the server by pressing the 'Send the PlayMail' button.

Our emoticons are defined as a prefix code. The server translates them into MPEG-4 bookmarks. These bookmarks define the amplitude and the duration of an

Figure 13.2 User interface for sending PlayMails

expression. We enable the user to control the amplitude of the expression by repeating the last character of the emoticon as in :-) and :-)). For emoticons like smile and frown, the actual expression may start a couple of words prior to the location of the emoticon and end a couple of words later. Other emoticons such as a head nod are executed at the time when the word after them is spoken.

Since the speech is synthesized, it is difficult to provide the intended intonation. Therefore, simple icons for controlling the speech synthesizer are required, with '–' for a pause being the simplest.

Figure 13.3 shows the window that presents the PlayMail message to the user. Compared to a regular e-mail, the text window is replaced by the video of the talking head. Users that access their messages without media playing capability may access the text of the message by clicking on the appropriate button. A reply button brings the recipient to a page as shown in Figure 13.2 with the e-mail addresses and subject line already filled in.

Figure 13.3 A face model looking concerned while speaking a PlayMail message

13.4 PLAYMAIL FACE MODEL CREATION

PlayMail provides more entertainment if users are enabled to put words into the mouth of people they relate to. Therefore, it is essential to enable each user to create face models of his liking. And these models need to be kept private to the creator or made publicly available. We first describe the process as the user sees it and then we describe the algorithm behind the face model creation.

13.4.1 User Interface

We provide a simple user interface in which the user marks several feature points (FP) on an image that he submitted to the server (Figure 13.4). This process is guided by red dots that are placed on a prototype face model seen on the right side of Figure 13.4. As soon as a sufficient number of points are marked, the user can create the face model. The new model is created on the server. A preview of the model with different facial expressions gives feedback such that the user may decide to better align some FP (Figure 13.5). Finally, the user may choose the background for the face model. Currently, we offer a choice of black background, image around the face and the entire image as background (Figure 13.6).

It appears that animating a face within the image that it was created from is the preferred background of users. It enables to have the face animation in context, helping to explain holiday pictures or showing off in a new dress.

Figure 13.4 The user marks feature points in the face manually in order to create a face model. Dots marked in the right generic face model (see, right upper eyelid) guide this process

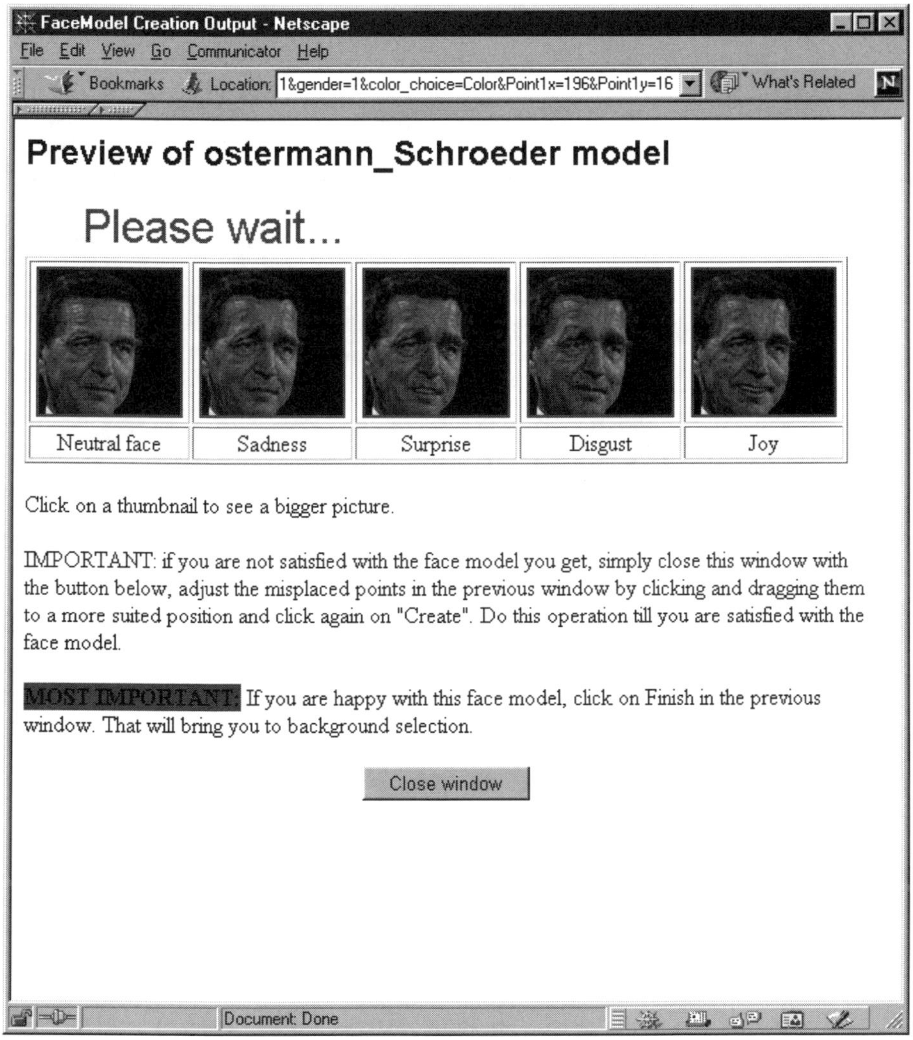

Figure 13.5 Preview of model created in Figure 13.4

13.4.2 Interpolation Function

For several FP, we know their new location as marked in the image. An interpolation function is defined, which computes the new positions for all those vertices of the indexed faceset of the generic face model that were not identified as FP. We use interpolation of surfaces from scattered data using radial basis functions (RBFs) [7]. This method has provided good results to a similar problem [8]. We built the interpolating function

$$s(x) = \sum_{i=1}^{N} b_i \Phi(x - x_i) + \sum_{l=1}^{4} c_l p_l(x)$$

Figure 13.6 Background selection for face model

from the following system:

$$\begin{bmatrix} A & P \\ P^{\mathrm{T}} & 0 \end{bmatrix}\begin{bmatrix} B \\ C \end{bmatrix} = \begin{bmatrix} Y \\ 0 \end{bmatrix} A = \begin{bmatrix} \Phi(x_1 - x_1) & \Phi(x_1 - x_2) & \Lambda & \Phi(x_1 - x_N) \\ \Phi(x_2 - x_1) & \Phi(x_2 - x_2) & \Lambda & \Phi(x_2 - x_N) \\ \mathrm{M} & \mathrm{M} & \mathrm{O} & \mathrm{M} \\ \Phi(x_N - x_1) & \Phi(x_N - x_2) & \Lambda & \Phi(x_N - x_N) \end{bmatrix}$$

$$P = \begin{bmatrix} 1 & x_{x1} & y_{x1} & z_{x1} \\ 1 & x_{x2} & y_{x2} & z_{x2} \\ \mathrm{M} & \mathrm{M} & \mathrm{M} & \mathrm{M} \\ 1 & x_{xN} & y_{xN} & z_{xN} \end{bmatrix} \quad B = \begin{bmatrix} b_1 \\ b_2 \\ \mathrm{M} \\ b_N \end{bmatrix} \quad C = \begin{bmatrix} c_1 \\ c_2 \\ \mathrm{M} \\ c_4 \end{bmatrix} \quad Y = \begin{bmatrix} y_1 \\ y_2 \\ \mathrm{M} \\ y_N \end{bmatrix}$$

N is the number of FP or vertices x_i we use to build the function; we need to know their old and their new location. x is any of the head mesh vertices whose interpolated position is unknown and $s(x)$ is its interpolated location. 0 represents a matrix or a vector with null values, b and c are the coefficients of the interpolation function, x_{xi}, y_{xi}, z_{xi} are the coordinates of vertex x_i, and $y_i = s(x_i)$ is the new position of x_i. $\phi(x_i - x_j)$ represents a RBF centered in x_i. $p(x)$ is a polynomial of 4th degree $(1 + x + y + z)$ added to the general RBF interpolation to better adjust global translations. This addition forces the constraint $\sum_{j=1}^{N} b_j p_j(x_i) = 0$ to make the complete system solvable. $x_i - x_j$ is the distance between two vertices of the predefined head model computed with the Euclidean norm in R^3. Calculating the distances among vertices for the entire head model is a compute-intensive task; it is only performed once and the distance values are stored in a file.

From all RBFs, we are interested in those that have the property of being monotone decreasing to zero for increasing values of the radius. This limits the influence of each of the FP (on which the RBFs are centered) to a region around its location on the face model. It is physically understandable that the FP located in the chin should not interact with points located on the upper part of the head. For our process, we have been testing and applying a family of Compactly Supported RBFs (RBFCS) [9]:

$$
\begin{aligned}
\phi_{0,0} &= (1-r)_+ \in C^0 \cap \mathrm{PD}_1 \\
\phi_{1,0} &= (1-r)_+^3(1+3r+r^2) \in C^2 \cap \mathrm{PD}_1 \\
\phi_{1,1} &= \mathrm{D}\phi_{1,0} = (1-r)_+^2(2+r) \in C^0 \cap \mathrm{PD}_3 \\
\phi_{2,0} &= (1-r)_+^5(1+5r+9r^2+5r^3+r^4) \in C^4 \cap \mathrm{PD}_1 \\
\phi_{2,1} &= \mathrm{D}\phi_{2,0} = (1-r)_+^4(4+16r+12r^2+3r^3) \in C^2 \cap \mathrm{PD}_3 \\
\phi_{2,2} &= \mathrm{D}^2\phi_{2,0} = (1-r)_+^3(8+9r+3r^2) \in C^0 \cap \mathrm{PD}_5 \\
\phi_{3,0} &= (1-r)_+^7(5+35r+101r^2+147r^3+101r^4+35r^5+5r^6) \in C^6 \cap \mathrm{PD}_1 \\
\phi_{3,1} &= \mathrm{D}\phi_{3,0} = (1-r)_+^6(6+36r+82r^2+72r^3+30r^4+5r^5) \in C^4 \cap \mathrm{PD}_3 \\
\phi_{3,2} &= \mathrm{D}^2\phi_{3,0} = (1-r)_+^5(8+40r+48r^2+25r^3+5r^4) \in C^2 \cap \mathrm{PD}_5 \\
\phi_{3,3} &= \mathrm{D}^3\phi_{3,0} = (1-r)_+^4(16+29r+20r^2+5r^3) \in C^0 \cap \mathrm{PD}_7
\end{aligned}
$$

13.4.3 Algorithm

Our system uses a predefined generic 3-D head model and a front picture of a face. The conditions in which the picture was taken, focal length of the camera distance to the individual and so on are unknown. The adaptation process follows these four main steps (Figure 13.7) [10]:

1. *Calibration*: During this step the camera parameters are established. We adjust the focal length of a pinhole camera model based on the head width determined by feature points located on the ears of the person. Furthermore, we translate and scale the 3-D model such that eye corners and nose tip of the 3-D model get projected onto the corresponding feature points marked on the image.
2. *Feature Point Marking*: For a good adaptation of the generic face model to the image, we require at least 25 feature points positioned. They define the outline of the face, the chin, nose, mouth and eyes. We provide the option to position more feature points in order to accommodate faces that deviate to a larger extent from a normal human face.
3. *Interpolation*: We assume that vertices of the generic face model need only to be modified in their x- and y-coordinates. In other words, we do not change the depth of generic face model. On the basis of the location of the feature points in the

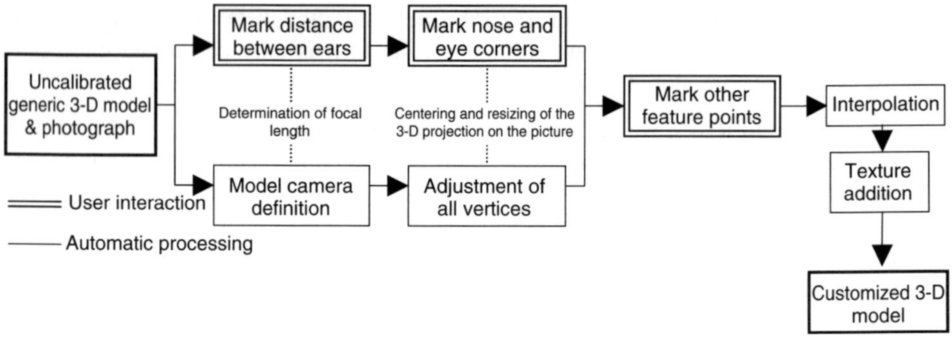

Figure 13.7 Block diagram of the face model creation process

image and the z-coordinate of the corresponding feature point, we can determine the new 3-D location of each feature point. Knowing this new location and the original location of all feature points, we compute an interpolation function that allows computing the 3-D position of all vertices of the face models.

4. Finally, we add texture coordinates to the model that maps the image correctly onto the face model.

The adapted face model is placed into the gallery of face models of the user. It is ready to be used for sending PlayMails without further processing. A user typically takes about three to four minutes to create a face model.

Our face models use the FAT for defining the animation of the model. At this point, we do not adapt these tables to the new models. However, these tables can also be adapted using RBFs [11]. Other approaches are presented in Reference 12 and in Chapter 4 of this book.

13.5 CONCLUSION

PlayMail relies on conventional software on the user terminal and several applications on the server for creating face models as well as converting the text messages into video with synthesized audio. Because of this design, no software installation at the user's terminal is necessary. However, this comes at the price of an increased server load. Our experience lets us believe that a standard 600 MHz PC can synthesize, compress and convert into two streaming formats – one PlayMail message every two minutes.

PlayMail is an exciting sticky service targeted toward young Internet users. It enables users to create cool face animations. PlayMail is easy to use and fun. The service was rated very high in consumer studies. Several newsletters recommended PlayMail as pick of the week. The service is available at *http://playmail.research.att.com*.

REFERENCES

[1] Microsoft, *http://www.microsoft.com*.
[2] Real Networks, *http://www.real.com*.

[3] Apple, *http://www.apple.com*.
[4] ITU-T Recommendation H.323, Packet-Based Multimedia Communications Systems, 1998.
[5] AT&T Labs Natural Voices™, *http://www.naturalvoices.att.com*.
[6] PlayMail, *http://playmail.research.att.com*.
[7] R. Schaback, "Creating surfaces from scattered data using radial basis functions", in M. Daehlen, T. Lyche, L. L. Schumaker, eds., *Mathematical Methods for Curves and Surfaces*, Vanderbilt University Press, Lillehammer, 1995, pp. 477–496.
[8] F. Lavagetto, R. Pockaj, "The facial animation engine: toward a high-level interface for design of MPEG-4 compliant animated faces", *IEEE Transactions on circuits and systems for video technology*, **9**(2), 277–289 (1999).
[9] Z. Wu, "Multivariate compactly supported positive definite radial functions", *Advances in Computational Mathematics*, **4**, 283–292 (1995).
[10] A. C. Andres del Valle, J. Ostermann, "3D talking head customization by adapting a generic model to one uncalibrated picture", *ISCAS 2001*, Vol. II, IEEE, Sydney, Mai, 2001, pp. 325–328.
[11] M. V. Mani, J. Ostermann, Cloning of MPEG-4 face models, *International Workshop on Very Low Bitrate Video Coding (VLBV 01)*, Athens, October, 2001.
[12] I. S. Pandzic, Facial animation framework for the web and mobile platforms, *Web3D Symposium 2002*, Tempe, Arizona, February, 2002.

14

E-Cogent: An Electronic Convincing aGENT

Jörn Ostermann

AT&T Labs-Research, Middletown, New Jersey, USA

ABSTRACT

Subjective tests indicate that animated faces increase the trust that users have in the information presented on the computer. We predict that the use of animated faces will allow E-commerce Web sites to enhance their sales. E-cogent is an Electronic Convincing aGENT that assists when doing online shopping. E-cogents talk to the customer, give advice and suggest further actions in order to help the customer to buy. Architectures for supporting efficiently the use of face animation (FA) in real-time interactive services are presented.

14.1 INTRODUCTION

Computer simulation of human faces has been an active research area for some time, resulting in the development of a variety of facial models and several animation systems [1–7]. The advances in animation systems, such as those mentioned above, have prompted interest in the use of animation to enrich the human–computer interface. This prompted ISO to support animation of talking faces and bodies in the MPEG-4 standard [8–12]. One important application of animated characters has been to make the human–computer interface more compelling and easier to use. For example, animated characters have been used in presentation systems to help attract the user's focus of attention, to guide the user through several steps in a presentation and to add expressive power by presenting nonverbal conversational and emotional signals [13, 14]. Animated guides or assistants have also been used with some success in user help systems [15–17] and for user assistance in web navigation [18].

Character animation has also been used in the interface design of communication or collaboration systems. There are several multiuser systems that currently use avatars that are animated representations of individual users [6, 7]. In many cases, the avatar authoring tools and online controls remain cumbersome. The social cues that are needed to mediate social interaction in these new virtual worlds have been slow to develop and have resulted in frequent communication misunderstandings [19]. Nevertheless, the enormous popularity of Internet chat applications suggest considerable future use of avatars in social communication applications.

In this chapter, we focus on using avatars in E-commerce. An interactive E-commerce application requires a user interface and a dialog management system [20, 21] that evaluates the user input and determines the output to the user. The avatar is part of the user interface. The avatar needs to interact in real time with the customer. Several architecture choices for integrating an avatar into the user interface are presented. The choice of a particular architecture depends on the available server-client bandwidth, client software and server resources. The cost of adding avatars to an electronic commerce Web site requires benefits to justify the expense. While we cannot yet prove that an avatar helps to increase sales, we conducted subjective tests that measure how the trust between a user and the computer is influenced by an avatar.

In Section 14.2, we present the 'Social dilemma' experiment that shows how FA and text-to-speech (TTS) can influence the trust and collaboration of users when interacting with a computer. In Section 14.3, we present the client and server architecture for supporting Web-based applications such as electronic commerce with TTS synthesis and avatars. There are several rules that should be considered when designing a dialog that involves an avatar. We highlight some of them in Section 14.4. We conclude with some remarks in Section 14.5.

14.2 'SOCIAL DILEMMA' GAME EXPERIMENT

The 'social dilemma' game is used to measure trust and cooperation between two individuals or in our case between an individual and a computer. A variation of this game is also known as the 'prisoner's dilemma' game. Using the 'social dilemma' game, we investigated whether user trust or cooperation with a computer is influenced by the presence of text, speech or face animation. Users played the 'social dilemma' game with the computer as partner (Figure 14.1). The rules of the game are as follows: Partners have to choose project Green or project Blue in a secret ballot. If both partners choose Green, both receive \$6. If both choose Blue, both receive \$3. If one partner chooses Green and one chooses Blue, the first receives \$0, whereas the latter gets \$9. According to these rules, we say that a partner who chooses Green is cooperating because he favors both partners to earn money. The goal of the game is 'Earn as much money as possible'. We do not define whether the partners should maximize their own money or the joint money.

The game starts with the computer and the partner introducing each other. For each round of the game, the user and partner discuss what they will do in that round of the game. After they have communicated their intentions, they secretly select a project using a ballot window. After both partners have selected a project, the scores appear on the scoreboard (Figure 14.1).

Participants in the study played the game using one of three interfaces: The computer partner, named Jason, was either represented using text only, TTS only, or an avatar consisting of TTS and FA.

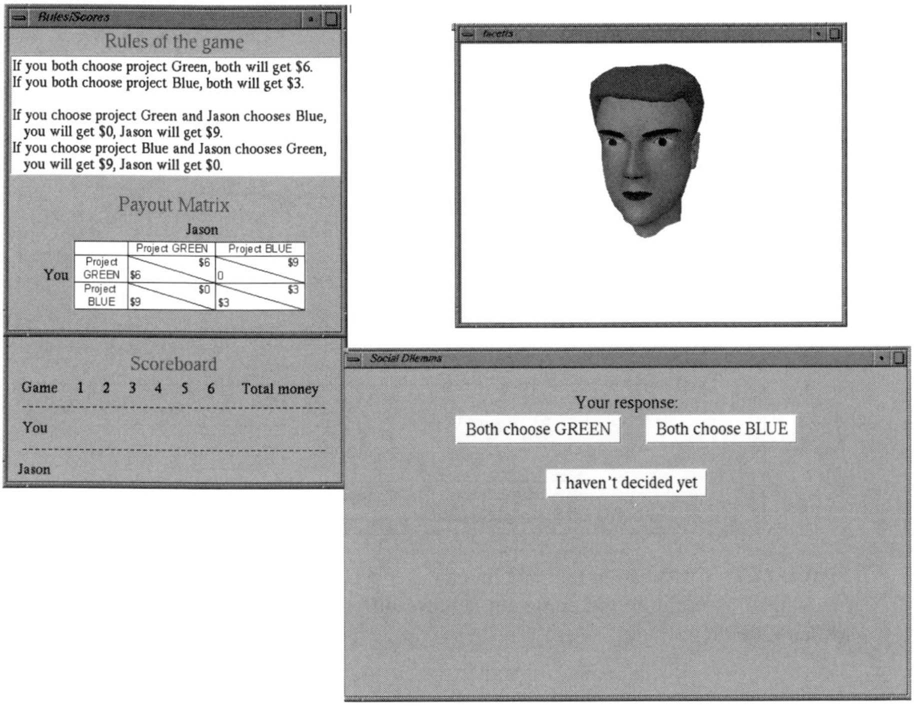

Figure 14.1 Screen shot of the game. The left window permanently displays the rules of the game and the current score. The top right window shows the avatar Jason. The bottom right window displays an instant of the dialog box that is used as part of the discussion phase [From Ostermann and Millen (2000) [22]]

14.2.1 Experimental Setup

Fifty students and employees of Princeton University participated in the experiment. They were paid for their participation. The median age was 25 and the majority of subjects were self-described heavy users of computers at work and at home.

Each participant played six rounds of the game. The computer always initiates the dialog prior to a ballot. Using different words, the computer suggests cooperating. The human partner indicates his intentions by typing free text and confirming his intention by pressing one of three buttons (similar to Project Green, Project Blue, I don't know, Figure 14.1). As far as the ballot is concerned, the computer always cooperates (Project Green), unless the human partner suggests that both choose Blue. However, the human partner does not know the behavior of the computer.

Upon completion of the game, users were asked to fill out a short computer-administered questionnaire.

14.2.2 Experimental Results

We define the cooperation rate of a partner as the percentage of choosing Green in his actual decision. Figure 14.2 shows the average cooperation rate of the human partner

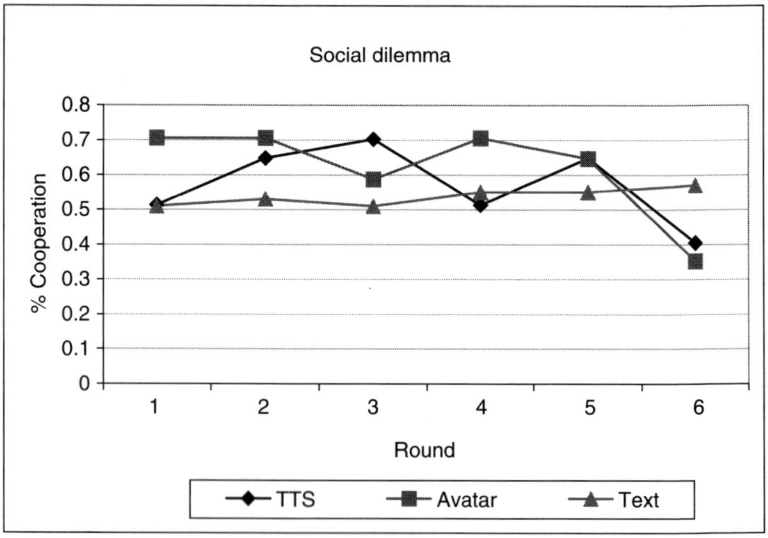

Figure 14.2 Cooperation rate of a human user as a function of the round of the game. Different modalities representing the computer achieve different cooperation rates [From Ostermann and Millen (2000) [22]]

as a function of the round of the game. As can be seen, the cooperation rate is highest when the computer is represented using TTS and FA. A TTS representation yields a higher cooperation rate than text only. The decrease of cooperation for the last round (here round 6) is expected as described in the literature [23]. Users know that a lack of cooperation in the last round of the game will not result in future consequences.

The evaluation of the questionnaires showed a weak trend in which users prefer the interface with FA and TTS. While the results of this questionnaire do not justify the use of avatars, the change in cooperation rate from 52% for the text interface to 67% for the talking avatar interface justifies the use of FA in e-commerce. Apparently, users are not aware of how they are influenced by FA or avatars.

Independent of the user interface, the computer usually played in a cooperative spirit. This can make the computer appear stupid if the human partner continuously plays in a noncooperative manner by choosing blue. We asked users about the intelligence of their game partner. Compared to a representation of an avatar, the computer partner was perceived as being smarter when represented with text only. We have two possible explanations for this: (1) The avatar did not show any emotion in response to the new scores, that is, to the choice of the human partner. (2) People expect more intelligence as soon as they see a face. This has obvious consequences for the choice of the avatar and the underlying control system or dialog management system. We expect that cartoonlike avatars are more suitable for simple interactive systems while photo-realistic avatars [1] would require a more intelligent dialog management system.

In a separate social dilemma experiment, the avatar provided visual feedback to the human partner when the scores of a round were made available on the scoreboard to the human partner. The avatar showed a joyful expression in case of collaboration, a sad

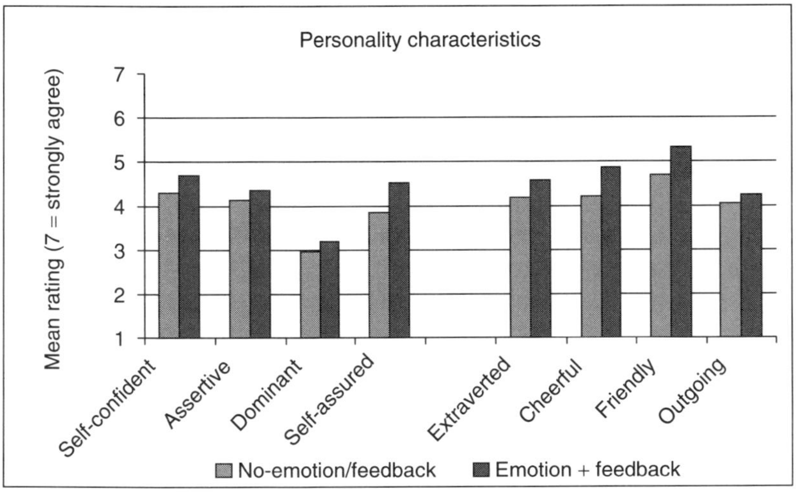

Figure 14.3 Personality ratings of an avatar in the social dilemma game

expression in case of no collaboration and a surprised expression in case of the human partner choosing green and the computer choosing blue, which could only happen if the human partner suggests that both of them choose blue. This feedback increases the personality ratings of the avatar in all characteristics such as self-confidence and friendliness (Figure 14.3). It appears that an avatar should not only talk but also provide appropriate emotional responses.

14.3 ARCHITECTURES FOR WEB-BASED APPLICATIONS USING TTS AND FACIAL ANIMATION

We can imagine several architectures that enable Web-based FA for interactive services on a client. They differ in terms of software and hardware requirements for the client as well as bandwidth requirements for the server-client connection. Furthermore, the architecture determines the load each client puts on the server. We first present the architecture with the lowest bandwidth and server requirements. An alternative solution requires more server resources but gives the content provider more freedom in choosing the voice of an avatar. The third architecture requires minimal client resources but a lot of server resources and bandwidth.

14.3.1 Client with TTS and Face Animation Renderer

The client requires a Web browser, a TTS and an FA renderer that displays the animated avatar on the screen (Figure 14.4). In order to enable synchronized speech and facial animation, the TTS provides phonemes and related timing information to the renderer. Using a coarticulation model [2], the renderer can then animate a model downloaded from a server and let it move its mouth synchronously to the speech of the TTS. The

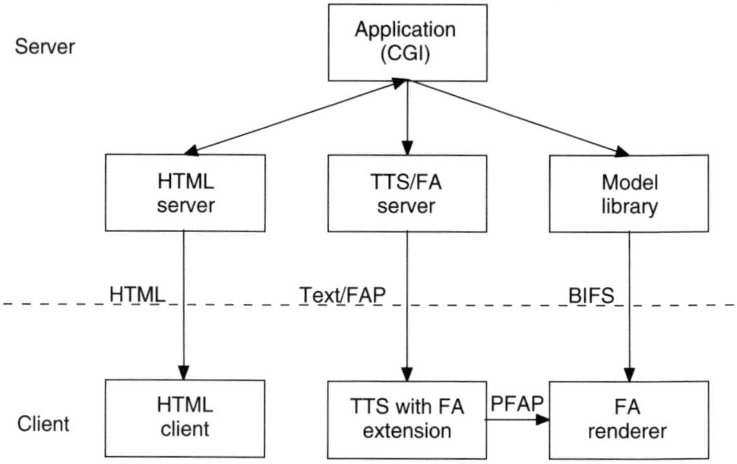

Figure 14.4 Architecture for using TTS and face animation (FA) in an Internet environment. The functionality of the client could be implemented as an MPEG-4 player

TTS handles bookmarks that contain facial animation parameters (FAPs). The TTS receives the text and bookmarks from the server (see Chapter 2).

The server for this client comprises a web server, a TTS/FA server and a database of face models (Figure 14.4). They are controlled by the application that would include a dialog management system. When the application is started due to a client's request, it downloads a face model including face animation tables (FAT) [8] (Chapter 1) from the model library to the client. Furthermore, the application initializes the TTS of the client with the desired voice. Then the application provides the HTML pages for the server and the text with embedded FAPs to the TTS/FA server. This TTS/FA server can be implemented by using MPEG-4 [8–10] (Chapter 1).

This architecture has the following advantages and disadvantages with respect to the avatar:

- The look of the avatar is controlled by the content provider
- The voice of the avatar is limited to choices of TTS at the client
- Very low server load
- Architecture scalable for a large number of users
- Very low additional bandwidth requirements during interaction (text with FAPs only)
- High software and hardware requirements on the client.

At this point in time, it is not yet feasible to download high-quality voices in excess of 100 MB into the TTS at a client because of size limitations. This limit of voices to the ones preinstalled at the client will cause content providers prefer a solution in which the voice is synthesized on the server.

14.3.2 Client with Face Animation Renderer

Modern TTS synthesizers with high voice quality use concatenative speech synthesis. They can synthesize about 50 sentences in parallel on a high-end Pentium 3 PC [24].

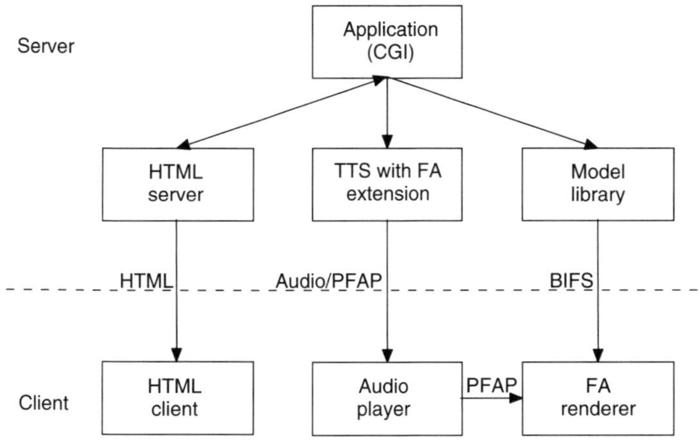

Figure 14.5 Architecture for using TTS and face animation (FA) in an Internet environment with server-based TTS synthesizer

Considering that many parts of a dialog such as greetings, acknowledgements and so on may be precomputed, the use of speech synthesis becomes affordable for electronic stores and customer service. Figure 14.5 shows the overview of the architecture. Compared to Figure 14.4, the TTS is now located at the server or in the network. The TTS streams the synthesized audio as well as related phoneme and markup information as defined in the MPEG-4 TtsFAPInterface (Chapter 1) [10] to the client. Internet Engineering Task Force (IETF) is currently considering a payload format for PFAP streams (phoneme facial animation parameter streams) that carry the information of the TtsFAPInterface as a Real-time Transport Protocol (RTP) payload. A PFAP stream requires a bandwidth of about 500 bit s^{-1} in addition to the bandwidth of the audio stream. This architecture has the following advantages and disadvantages with respect to the avatar:

- The look of the avatar is controlled the by content provider
- The voice of the avatar is controlled by the content provider
- High server load due to server-based TTS
- Architecture scalable for a large number of users
- Additional bandwidth requirements during interaction for audio and PFAP streams
- Limited software and hardware requirements on the client.

This architecture eases the load of the client and gives full control over the voice selection to the content provider. This will be the preferred solution for avatars in electronic commerce applications as soon as FA renderer is ubiquitous.

In case there is no FA renderer available at the client, it needs to be installed or the server decides to render the avatar itself and streams audio and video to the client. This is the most costly architecture in terms of server resources and bandwidth requirements (Figure 14.6).

Server/client resources versus bandwidth

Server resources	Server media	Bandwidth	Client
TTS Animation graphics	Audio Video	80 kbit/s	A/V decoder
TTS Animation	Audio FAP	10 kbit/s	Audio decoder, graphics
-	Text	100 bit/s	TTS Animation graphics

Server ↑ Client ↓

Figure 14.6 Shifting resource requirements from the server to the client decreases the bandwidth requirements for animating the avatar

14.4 VISUAL DIALOG

At AT&T, we implemented an E-cogent application that helps customers to choose a mobile phone. The customer is asked a couple of questions regarding phone price, weight and talk time. Then the E-cogent presents available choices (Figure 14.7). The user may choose to see the details of the phones, proceed to buy a phone or go back to start over. In case the user starts over, he/she is presented with a screen as shown in Figure 14.8. Compared to the first screen that the user saw, we have now a new button called *Most popular phones*. The avatar will verbally highlight this new choice to the user and recommend selecting that option. However, the user still has the choice to specify his preferences.

Initial attempts to integrate avatars into existing dialog systems failed to create a convincing experience because no effort was made to adapt the dialog to the new modality of a talking avatar. A successful dialog management system for avatars needs to carefully consider what information is presented verbally and what information needs to be printed on the screen. In the following, we list a couple of guidelines for a dialog that involves an avatar. These guidelines are not meant to be exhaustive but represent a snapshot of our current knowledge:

- The avatar must appear to be alive, that is, it has to move while not speaking. When waiting for input, the avatar should have a listening posture.
- The avatar may under no circumstances, slowdown or interfere with the interaction, with the user. To that extent, it appears useful to design the Web site such that it is functional even without the avatar.
- The avatar has to speak whenever a new Web page is presented. However, it should not read what is written on the screen. It should emphasize highlights and avoid long sentences. For example, if the user requests technical details for a phone on

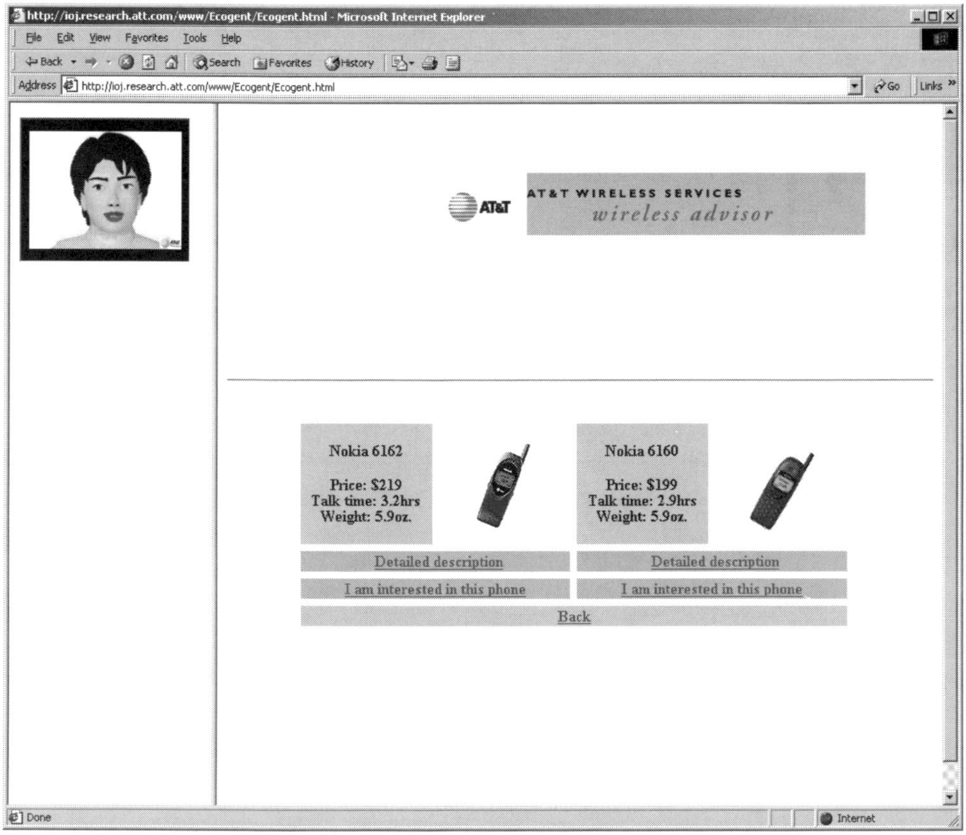

Figure 14.7 E-cogent presents several phones that fit the needs of the customers

Figure 14.7, the avatar will merely say something like 'Here are the technical details'. For Figure 14.8, we use a comment like 'You are back here. Perhaps you would like to see our most popular phones'.

- The avatar may not repeat the same sentence. Therefore the server stores state information in order to keep track of the dialog. This enables the server to choose different sentences with identical meanings when presenting the same Web page again.
- Monitoring the click trail enables the server to easily detect customer frustration and issue sentences and facial expressions that soothe the customer.
- The server may use the avatar like a regular sales person. As such, the avatar can also engage in sales promotional dialog.

These guidelines indicate that an avatar might be most efficiently used to guide a customer to get the information and service he is looking for. Finally, the avatars may regain the attention of the customer even when the customer is not looking at the screen anymore.

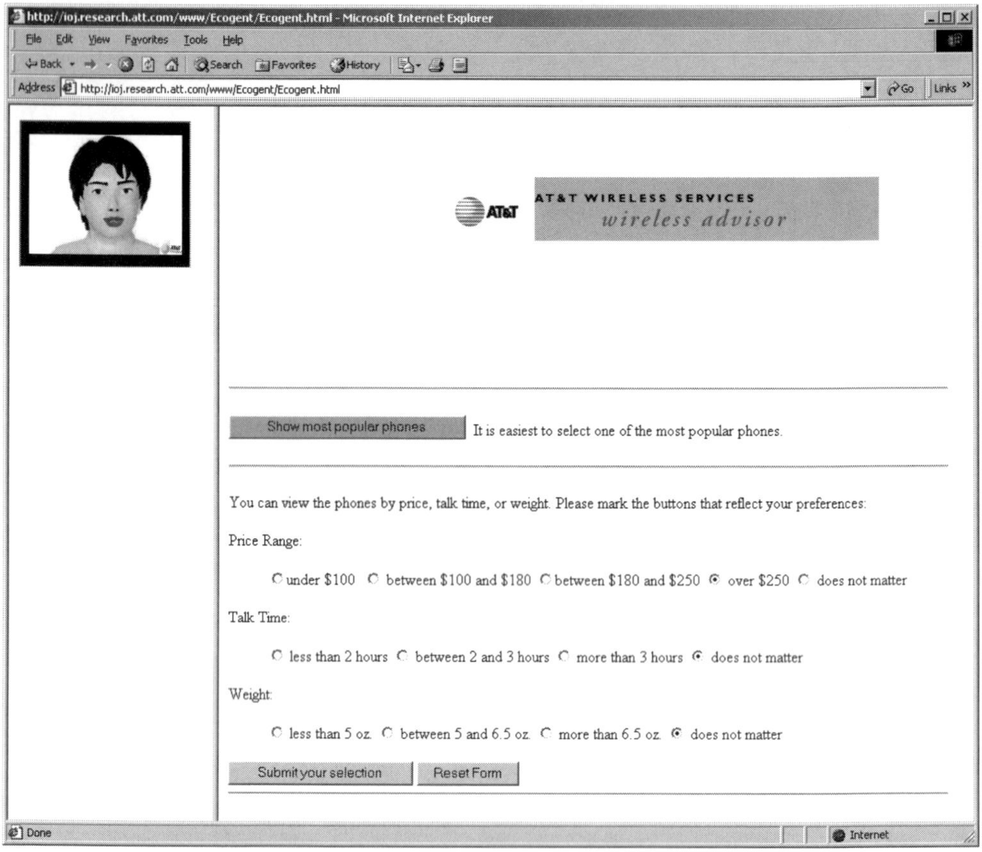

Figure 14.8 E-cogent verbally highlights the new button on the page that leads to the most popular phones

14.5 CONCLUSION

In this paper, we presented two architectures that integrate TTS and FA into Internet applications. A server sends a face model to a client. The client animates this face model and synthesizes speech according to the text and embedded bookmarks that it receives from the server. The bookmarks enable the animation of non-speech-related actions such as head nodding and smiles. This architecture puts the least load on the server and bandwidth requirements for animating the avatar are about 100 bits s^{-1}. We show that most content providers would actually prefer to use a speech synthesizer that is located in the network and not on the client in order to have full control over the synthesized voice. In this case, adding an avatar to an existing Web site requires the additional bandwidth for streaming the audio and a PFAP stream with Phonemes and markup information (500 bits s^{-1}) to the client.

In subjective tests, we evaluated this technology simulating a game geared toward measuring cooperation comparing the performance of interfaces using facial animation, TTS and text. According to the results, the use of facial animation in the design of interactive services was favorably rated for most of the attributes in these experiments. An important result for E-commerce is that users cooperate more with a computer if it is represented with a talking avatar instead of TTS only or text only. We measured an increase of the cooperation rate from 50% for text to 70% for a talking avatar in a first human–computer interaction. The average cooperation rate over 5 consecutive interactions increased from 52% to 67%. On the basis of these results we expect that talking heads will increase the performance of Web stores and Web-based customer service.

ACKNOWLEDGMENTS

Dr. David Millen helped in organizing and evaluating the subjective tests. Reha Civanlar, Eric Cosatto and Hans-Peter Graf helped in conceiving the E-cogent store for mobile phones.

REFERENCES

[1] E. Cosatto, H. P. Graf, "Photo-realistic talking-heads from image samples", *IEEE Transactions on Multimedia*, **2**(3), 152–163 (2000).

[2] M. M. Cohen, D. W. Massaro, "Modeling coarticulation in synthetic visual speech", in M. Thalmann, D. Thalmann, eds., *Computer Animation '93*, Springer-Verlag, Tokyo, 1993.

[3] P. Eisert, S. Chaudhuri, B. Girod, "Speech driven synthesis of talking head sequences", *3D Image Analysis and Synthesis*, Erlangen, November, 1997, pp. 51–56.

[4] P. Kalra, A. Mangili, N. Magnenat-Thalmann, D. Thalmann, "Simulation of facial muscle actions based on rational free form deformation", *Proceedings Eurographics 92*, pp. 65–69, 1992.

[5] M. Kampmann, B. Nagel, Synthesis of facial expressions for semantic coding of videophone sequences, *Computer Graphics International (CGI98)*, Hannover, Germany, June, 1998.

[6] I. S. Pandzic, T. K. Capin, E. Lee, N. Magnenat-Thalmann, D. Thalmann, A flexible architecture for virtual humans in networked collaborative virtual environments, *Proceedings Eurographics 97*, Budapest, Hungary, 1997.

[7] J. R. Suler, From ASCII to Holodecks: psychology of an online multimedia community, *Presentation at the Convention of the American Psychological Association*, Chicago, Ill., 1997.

[8] ISO/IEC IS 14496-1: MPEG-4 Systems, 1999.

[9] ISO/IEC IS 14496-2: MPEG-4 Visual, 1999.

[10] ISO/IEC IS 14496-3: MPEG-4 Audio, 1999.

[11] A. M. Tekalp, J. Ostermann, "Face and 2-D mesh animation in MPEG-4", *Signal Processing: Image Communication*, **15**(4-5), 387–421 (2000).

[12] P. van Beek, E. Petajan, J. Ostermann, "MPEG-4 synthetic video", in A. Puri, T. Chen, eds., *Multimedia Systems, Standards and Networks*, Marcel Dekker, New York, 2000.

[13] E. Andre, T. Rist, J. Muller, "Guiding the user through dynamically generated hypermedia presentations with a life-like character", *Intelligent User Interfaces '98*, San Francisco, Calif., 1998, pp. 21–28.

[14] T. Rist, E. Andre, J. Muller, "Adding animated presentation agents to the interface", *Intelligent User Interfaces '97*, Orlando, Florida, ACM Press, New York, 1997, pp. 79–86.

[15] Y. Arafa, P. Charlton, P. Fehin, A. Mamdani, "Personal service assistants with personality", *Proceedings of HCI International '99 – Volume 2*, Lawrence Erlbaum Associates, Munich, Germany, August, 1999, pp. 147–151.

[16] A. Don, T. Oren, B. Laurel, "Guides 3.0", *In CHI-93, Video Proceedings*, ACM Press, pp. 447, 448.

[17] S. Gibbs, C. Breiteneder, V. deMey, M. Papathomas, "Video widgets and video actors", *Proceedings of the UIST '93 (ACM SIGGRAPH Symposium on User Interface Software and Technology)*, Atlanta, Ga., ACM Press, New York, 1993, pp. 179–184.

[18] A. E. Milewski, G. E. Blonder, System and Method for Providing Structured Tours of Hypertext Files, US Patent # 5760771, June, 1998.

[19] B. Damer, C. Kekenes, T. Hoffman, "Inhabited digital spaces", *Proceedings of CHI '96, Vancouver*, ACM Press, New York, pp. 9–10.

[20] R. Pieraccini, E. Levin, W. Eckert, AMICA: the AT&T mixed initiative conversational architecture, *Proceedings of EUROSPEECH 97*, Rhodes, Greece, September, 1997.

[21] P. Kalra, An Interactive Multimodal Facial Animation System, Ph.D. Thesis nr. 1183, EPFL, 1993.

[22] J. Ostermann, D. Millen, "Talking heads and synthetic speech: an architecture for supporting electronic commerce", *ICME 2000*, New York, 2000, p. MA2.3.

[23] S. Van Mulken, E. Andre, J. Muller, "An empirical study on the trustworthiness of life-like interface agents", *Proceedings of HCI International '99 – Volume 2*, Lawrence Erlbaum Associates, Munich, Germany, August, 1999, pp. 153–156.

[24] AT&T Labs Natural Voices™, *http://www.naturalvoices.att.com*

15

alterEGO: Video Analysis for Facial Animation

Eric Petajan

face2face animation, Inc., Summit, New Jersey, USA

face2face (f2f) animation, inc. provides software applications for the analysis of talking head video and animation of synthetic faces from MPEG-4 facial animation parameters (FAPs). The video analysis system (alterEGO) extracts head, lip and eye movements from ordinary video without facial markers and outputs an FAP file. A variety of 3-D animation platform plug-ins are also provided for high-quality character animation driven by FAPs. Real-time face animation is also supported. f2f applications can be used to animate any character over any network, or for films/TV.

15.1 SYSTEM OVERVIEW

The automatic analysis of an unconstrained video of a talking face has not been achieved with sufficient quality for commercial applications. The human face is a rapidly moving, deformable, multilayered structure with a wide variety of color, texture, surface properties and geometric structure. Furthermore, the analysis of facial movements requires a high frame rate in order to capture rapid lip and eye movements. This chapter describes a face analysis system that produces commercially valuable FAP data by imposing simple constraints on the view of the face.

The primary goal of the face2face alterEGO system is to produce FAPs that reflect the lip, eye region and head movements with enough accuracy to drive animated faces that are entertaining, and induce a feeling of human connection in the viewer. These quality criteria are achieved when animating any MPEG-4-compliant face model regardless of its simplicity.

15.2 FACE TRACKING INITIALIZATION

The first step toward robust analysis of facial movements is to locate the face in the scene. The alterEGO requires the user to manually locate the pupil centers and mouth corners in the first frame of the video by simply ‘clicking’ on face in the frame. The initial eye and mouth positions are then used to form mouth and eye windows. In addition, a nostril-tracking window is centered at the midpoint between the nostrils based on the expected nostril position relative to the eyes and mouth. The initial nostril-tracking window size is large enough to capture the nostrils regardless of head pose and facial type (assuming that both nostrils are visible).

While automatic face location algorithms were developed, they were not deployed in the current version of alterEGO to avoid any problems with multiple faces in the scene. Given the simplicity of the manual face location interface, this manual step does not impede the processing of video clips by the animator.

The generation (and interpretation) of FAPs requires the 3-D neutral face feature point positions to be established. alterEGO requires that each video clip start with at least two seconds of neutral face (eyes open, mouth closed, neutral facial expression etc). A set of 3-D feature point positions are estimated during the two-second training period by finding a closed mouth line, pupil centers, eye corners, upper and lower eyelid positions and eyebrows. The nostril centers, pupil centers and mouth line are then used to estimate head pose and scale relative to the video frame. Absolute head scale cannot be estimated without additional information (such as camera parameters) but the calculation of the FAP values does not depend on the absolute scale.

15.3 NOSTRIL TRACKING

The pair of nostrils are the only face feature that are not affected by facial articulation, skin color, facial accessories (glasses) or facial hair. The nostrils are also the least effected by lighting variations because holes do not reflect light. The pair of nostrils provides relative facial scale (separation distance) and facial orientation. Pitch, yaw and roll of the head can be estimated from the vertical nostril position, horizontal nostril position and nostril angle in the plane, respectively. During the two-second training period, the head pitch and yaw are assumed to be zero and the body is assumed to remain still during the clip. The head roll is computed directly from the angle of the line connecting the nostril centers.

The nostril-tracking algorithm [1] finds the best nostril candidate while sweeping through a multidimensional search space. During the first stage, a set of red, green and blue intensity thresholds are adjusted up or down until the nostrils are found in succeeding steps. The pixels in the nostril-tracking window are tested for nostril area and skin color area. Nostril area is simply the number of pixels that are below color nostril threshold and skin color area is the number of pixels that are labeled as skin color in a predetermined skin color lookup table using the most significant six bits of the RGB values as table indices. If either test fails then the nostril thresholds are adjusted and the tests are applied again. If the nostril area and skin color area are within acceptable ranges then nostril size, separation and contiguity tests are applied. This is done using horizontal and vertical projections or histograms of nostril pixels as shown in Figure 15.1. A pair of nostril size values is determined for each nostril by thresholding the projections for each nostril and computing the run length of projection

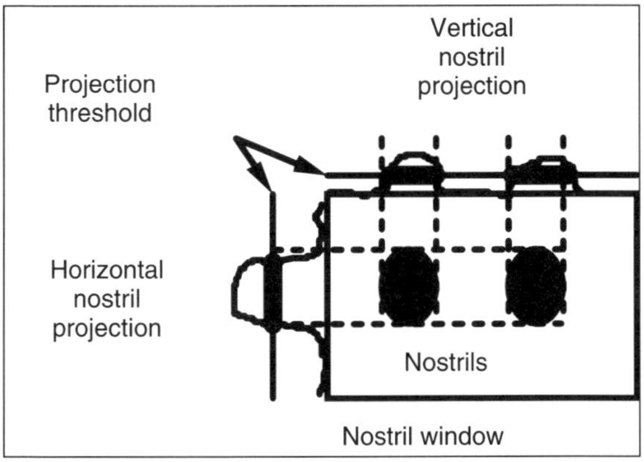

Figure 15.1 Nostril analysis and tracking

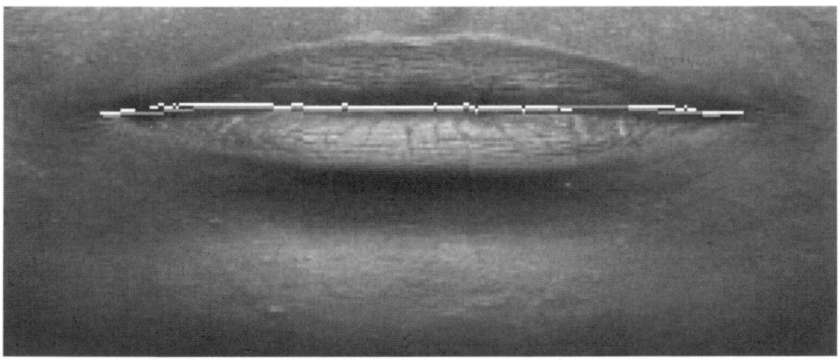

Figure 15.2 One field of a mouth image taken from a nostril aligned mouth window

values that exceed the threshold. All threshold values are tested and the value that produces the best match from a template is used.

15.4 THE MOUTH WINDOW

A set of mouth-analysis algorithms are applied to mouth images as shown in Figure 15.2. Each mouth image is taken from a video image using a mouth window with orientation and position that is aligned with the nostrils. Assuming standard definition interlaced video, each field of the mouth images is processed independently. The inner lip contour results from each field are then merged in preparation for frame-based FAP calculation.

The nostril visibility constraint conveniently limits the mouth image asymmetry caused by head yaw to acceptable limits. In addition, nostril visibility constraints on head pitch enforce a clearer view of the upper lip when the subject has a mustache. In other words, if the nostrils are visible, a bilaterally symmetric mouth image can be formed without resorting to 3-D model-based approaches to mouth image analysis.

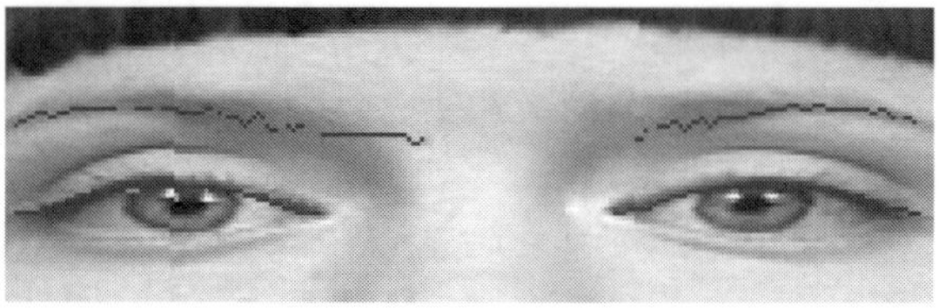

Figure 15.3 Symmetric eye image from eye window

15.5 THE EYE WINDOW

As shown in Figure 15.3, a symmetric eye image is taken from a nostril-aligned eye window. As with the mouth window, the nostril visibility constraint limits asymmetry in the eye image. The pupil centers, eyelids and eyebrows are identified in the eye image by applying anatomical constraints to a collection of color- and gradient-based features. Figures 15.1 and 15.2 show graphic overlay of the mouth line, and upper eyelid and eyebrow lines, respectively. These face features are formed from local gradient valleys that are discovered after applying slope, color, position and continuity constraints.

15.6 LIP AND TEETH COLOR ESTIMATION

During the two-second training period at the beginning of each clip, the lip color is measured in the vicinity of the mouth line. When the mouth is open, shadows inside the mouth are identified if their colors are not lip colors and below the luminance intensity of the lip colors. Teeth colors are computed by first synthesizing a teeth color table from color ranges and then removing lip colors from the teeth color table.

15.7 THE INNER LIP CONTOUR

The mouth line, mouth shadow regions and teeth regions are combined to estimate the inner lip contour for each field of the mouth image. When the mouth is closed, only the inner lip contour consists only of the mouth line.

15.8 THE FAP ESTIMATION

During the two-second training period, the neutral MPEG-4 standard face feature positions are established. These include eight points of the inner lip, eye corners, upper and lower eyelid centers, pupil centers and three eyebrow points per eyebrow. After the training period, the FAP values for a given frame are computed from the normalized difference between the neutral feature point positions and those from the given frame.

15.9 FAP SMOOTHING

The FAP estimates contain noise that appears in the FAP sequences. For example, Figure 15.4 shows a plot of the vertical deviation of the top and bottom mid-lip points

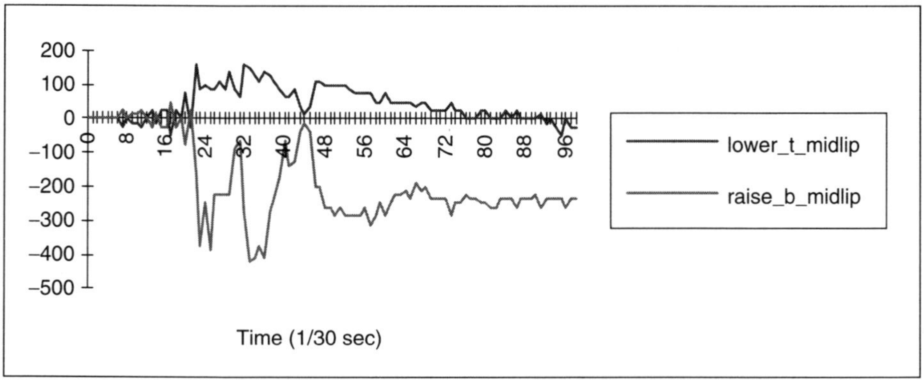

Figure 15.4 First 100 frames of midlip FAPs from 'Opossum 2' with no smoothing

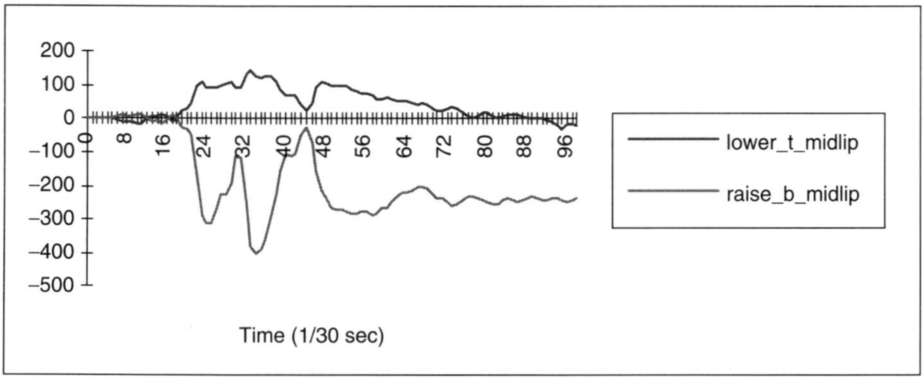

Figure 15.5 First 100 frames of midlip FAPs from 'Opossum 2' with smoothing

as a function of time (frame number at 30 Hz). Some noise is visible in Figure 15.4, which has been smoothed in Figure 15.5. Note that when the values of the FAPs at a given time sum to zero, the mouth is closed. AlterEGO adapts the amount of temporal smoothing to the scale of the face relative to the video frame. As the pixel resolution of the face is reduced with smaller scale, the amount of smoothing is increased.

15.10 ANIMATING FACES WITH FAPs

MPEG-4-compliant animated faces are being developed worldwide. One approach to animating face models from FAPs is to use 'bones' in a commercial 3-D animation-authoring tool. The bones are organized in a 'bones rig' that is controlled by a minimum of position and orientation parameters. f2f provides such a bones rig with the Animate plug-ins for the most popular animation platforms. Standard FAP files are imported via the Animate plug-in and the denormalized FAP values drive the rig, which in turn deforms the surface of the face. FAPs are denormalized by computing the face animation

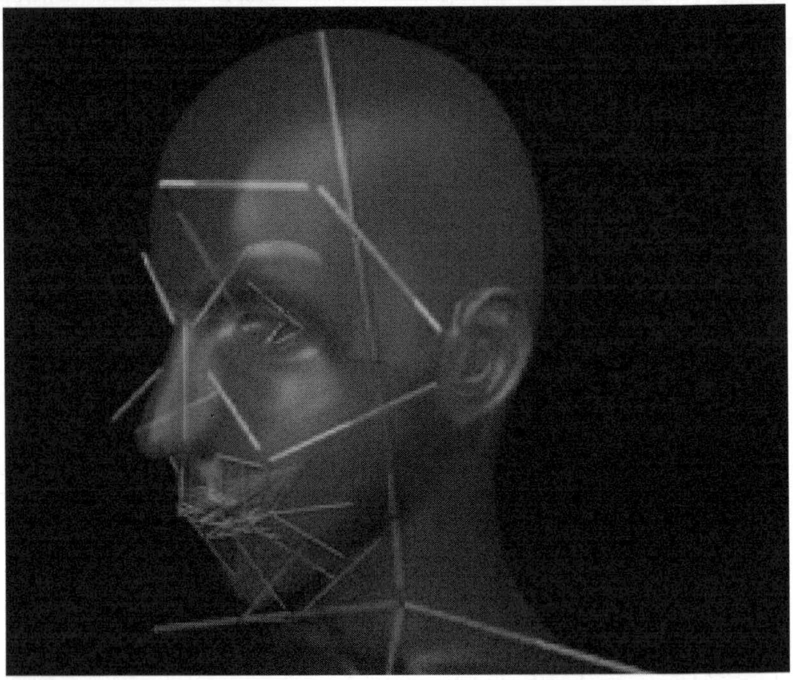

Figure 15.6 face2face bones rig shown under skin for surface deformation

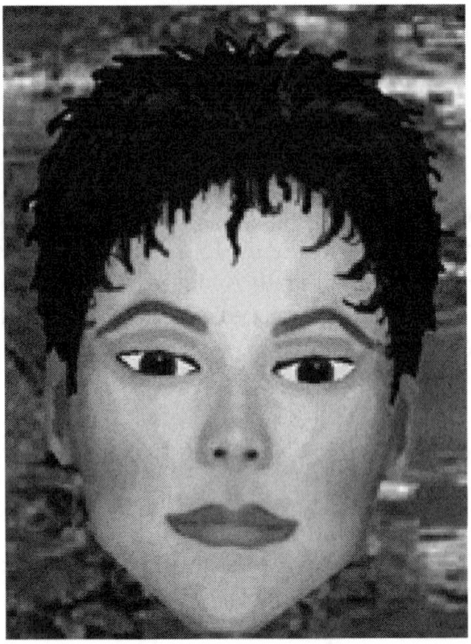

Figure 15.7 face2face texture mapped fixed mesh real-time face model

parameter units (FAPUs) from the user place facial definition parameter (FDP) feature points on the face model. Figure 15.6 shows the f2f bones rig applied to a head with translucent skin. This approach to face animation builds on the tools currently used by 3-D character animators.

f2f has also developed a face modeling and animation system that does not require a 3-D-authoring package and renders in real time on PCs. This face model includes a fixed topology mesh with geometry that adapts to the feature points located on the facial texture. One of the many face models developed with this system is shown in Figure 15.7.

15.11 SUMMARY

The f2f animation alterEGO facial motion capture system was described and its application to both high-end animation production and real-time PC-based animation were shown. A simple set of constraints on the subject and view of the subject were explained and shown to work together to maximize the quality of the results. The FAP files generated by alterEGO can be encoded for streaming using the face and body animation (FBA) visual stream coding tools included in Part 5 of the MPEG-4 standard. The same FAP file can be imported into the f2f Animate plug-in or used to animate any other MPEG-4-compliant face model. The FAP-based approach to face animation provides high quality for all networked and nonnetworked application domains including film/television, games, internet and wireless. FAP-based animation is being used today for television programming and games. When the MPEG-4 FBA-enabled players are widely available, the same FAP data will be streamed to real-time versions of the animated faces.

Finally, when a real-time implementation of alterEGO is available, live chat from call centers to consumers and between consumers (online games) and visually enhanced speech recognition [2,3] will be enabled.

REFERENCES

[1] H. P. Graf, E. Cosatto, D. Gibbon, M. Kocheisen, E. Petajan, "Multi-modal system for locating heads and faces", *Proceedings of 2nd International Conference on Automatic Face and Gesture Recognition*, IEEE Computer Society Press, 1996, pp. 88–93.

[2] E. D. Petajan, "Automatic lipreading to enhance speech recognition", *Proceedings of the IEEE Conference on Computer Vision and Pattern Recognition*, IEEE, 1985, pp. 40–47.

[3] A. Goldschen, O. Garcia, E. Petajan, "Continuous optical automatic speech recognition", *Proceedings of the 28th Asilomar Conference on Signals, Systems, and Computers*, IEEE, 1994, pp. 572–577.

16

EPTAMEDIA: Virtual Guides and Other Applications

Fabio Lavagetto[1] and Roberto Pockaj[2]

[1]*Università degli Studi di Genova, Genova, Italia,* [2]*EPTAMEDIA srl, Genova, Italia*

There are a number of applications in which the facial animation engine (FAE) technology can be used. In this chapter, we report some examples of applications currently developed by EPTAMEDIA srl, exploiter of the FAE technology, all of them based on the EptaPlayer and EptaPublisher, the two major products of EPTAMEDIA. The major focus will be on two specific applications concerning the use of a talking animated interface within web applications for the presentation of tourism contents and for sending animated postcards. The other examples that are mentioned concern the realization of information kiosks easy to upgrade with new contents, web call centers where answers to FAQs are provided by the virtual face, distance learning applications, advanced natural man–machine interfaces, cultural/historical presentations within museums or exhibitions, advertising and news casting, virtual speakers for interactive CD-ROMs and so on. Before entering into a description of the applications, we provide a short description of EPTAMEDIA srl followed by an introduction to some basic technical details related to the integration of the FAE within web applications and to the modality for content creation.

16.1 EPTAMEDIA Srl

EPTAMEDIA is a spin-off of DIST, the Department of Telecommunications, Computers and System Science of the University of Genoa, Italy, founded in February 2001 by a group of engineers, Ph.D.s and Professors. Through the participation to the MPEG meetings from 1996 to 1999, EPTAMEDIA founders have been involved in the definition of the MPEG-4 specification in the area of facial animation (FA), bringing a significant contribution to the standardization activities. The main activity of EPTAMEDIA is software development for advanced Man–Machine Interfaces and Multimedia. The expertise of EPTAMEDIA in facial animation has been consolidated through a long and successful collaboration with DIST and through an agreement for the commercial exploitation of products in the field of Facial Animation. EPTAMEDIA is mainly a provider of the technology of the FAE to third parties that want to integrate it in proprietary applications or products. EPTAMEDIA has developed two programs that already integrate the FAE technology: EptaPlayer, a plug-in to display FA sequences within browsers, and EptaPublisher, an authoring tool for the creation of animation sequences. In addition to that, EPTAMEDIA provides additional services, from the creation of animation sequences to the design of 3-D models, to customers who want to use the FAE technology within their products.

EPTAMEDIA has a wide expertise in creating Web sites based on the most advanced multimedia technologies that have been recently enriched due to the FAE technology that allows the implementation of natural user interfaces based on talking heads. A talking face, in fact, is able to transfer more information in less time and in a more effective way than through plain text. EPTAMEDIA has developed software for the synthesis of facial movements from phonetic information. In fact, by interfacing this software with a speech synthesizer, EPTAMEDIA is now capable of creating animation of 3-D faces starting from plain text. New content for Web sites can be created by simply providing EptaPublisher with the plain text. In addition, now EPTAMEDIA has the possibility to use special hardware for capturing the movements of a real actor and therefore creating hi-quality, attractive animation sequences, in which the whole facial mimics can be captured, encoded and reproduced by the EptaPlayer on the Web site. Encoding tools for the two approaches can be possibly provided to users that need to create a significant amount of animation data.

EPTAMEDIA can also provide 3-D face models based on the customer's request: the desired face can be embedded into a customized version of EptaPlayer. Though most of the applications mentioned throughout these pages refer to the Internet, EptaPlayer can be successfully used in many other areas, such as kiosks, interactive CD-ROMs and so on; EptaPlayer could also be embedded in several electronic devices as user interface or tutor. Being the core code of the EptaPlayer written in ANSI C, and using OpenGL as graphic interface, it can be easily ported on different platforms. EPTAMEDIA can be contacted at the address *www.eptamedia.com.*

16.2 EPTAPLAYER: HOW CONTENT IS RENDERED

As explained in Chapter 5, the FAE developed by DIST represents a stand-alone software capable of calibrating and animating 3-D facial models based on the MPEG-4 specifications. Through the agreement signed with DIST, EPTAMEDIA has developed further this technology finalizing EptaPlayer, an application based on the FAE that allows the visualization of FA contents within Web pages. It is a software for real-time animation of 3-D virtual characters, either human or cartoonlike. Animation is driven by a set of facial animation parameters (FAPs), either the 68 parameters recently standardized by MPEG-4 (ISO-IEC 14496-2, Simple FA Visual Profile) or a subset of them. EptaPlayer is available as plug-in for Netscape Communicator or as ActiveX control for MS Internet Explorer on Windows (TM) platforms.

Default technical characteristics:

- Supported browsers: Netscape Communicator v. 4.7.x and Microsoft Internet Explorer v. 4.x, 5.x and 6.x;

- Supported operating systems: Windows (TM) 98, 2K, NT 4, Me;
- No special hardware required; Pentium III with audio board or equivalent recommended;
- Audio/video sync guaranteed on low performance PC by frame-skipping mechanism;
- Controlled by a simple scripting interface from HTML pages, for selection of different face models, backgrounds and animation sequences;
- Display of textured models and backgrounds;
- Reproduction of the same content on any selected model;
- Proprietary decoding scheme for animation parameters at 2 kbps @ 25 fps, independent from window size;
- Audio decoding based on DSP Group TrueSpeech 8.5 (TM) voice codec.

Optional features:

- MPEG-4-compatible decoder (Version 1, Part 2 Visual, Simple FA Profile) on request;
- Possibility to create a customized 3-D face; Possibility to integrate different audio decoders.

16.3 EPTAPUBLISHER: HOW CONTENT IS AUTHORED

There are several ways to create animation sequences. Figure 16.1 shows different approaches for the creation of animation sequences.

In the first case (Figure 16.1a), a text-to-speech (TTS) synthesizer is used to create synthetic audio from plain text. Together with the audio samples, the TTS provides the sequence of pronounced phonemes and their duration. This information is

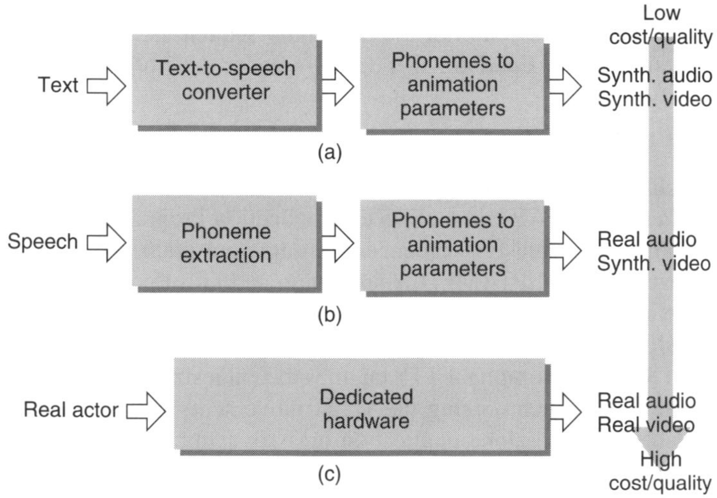

Figure 16.1 Creation of animation sequences

Figure 16.2 EptaPublisher-Text allows the user to insert a text, select a language, a TTS and a facial model. Before recording the animation file, the user can preview the generated sequence

used by a proprietary phoneme-to-FAP converter to infer the mouth movements corresponding to the pronounced text. In the second case (Figure 16.1b), natural audio is used as input. By processing the audio with a phoneme recognizer, the sequence of pronounced phonemes and, from that, the mouth movements can be obtained.

The third case (Figure 16.1c) makes use of dedicated tracking hardware, capable of capturing audio and facial movements of a real actor. In this case, the captured facial movements are encoded into an animation parameter and then used to drive the virtual face. The quality of this last approach is obviously higher if compared with the other solutions; in addition, facial expression are also captured and synthesized, unlike the former cases, in which only mouth movements can be synthesized.

EptaPublisher (Figure 16.2) is the authoring tool developed by EPTAMEDIA and is used for the creation of the animation sequences. It is available in its Text and Live version, while EptaPublisher-voice is under development.

16.3.1 EptaPublisher-Text

EptaPublisher-Text (Figure 16.3), assessable through a simple and friendly user interface, relies on the FAE and on Speech Application Program Interface (SAPI) compliant TTS engines to create FA sequences starting from plain text only. Owing to the software developed by EPTAMEDIA and integrated into EptaPublisher-Text, it is possible to create animation sequences in English and Italian, simply by typing text on the keyboard. Most common European languages will be supported soon. EptaPublisher-Text relies on SAPI 4.0-compliant TTS engines to synthesize the text and to get the necessary information for synchronizing the facial movements with speech. EptaPublisher-Text is available as stand-alone application or as dynamic link libraries (DLL) that can be integrated into other applications for Windows (TM) platforms. In the stand-alone version, EptaPublisher-Text integrates EptaPlayer for immediate preview of the authored content.

Default technical characteristics:

- Supported operating systems: Windows 2K;
- No special hardware required; Pentium III or equivalent recommended;
- Supported languages: Italian and English;
- Proprietary encoding of animation parameters at 2 kbps @ 25 fps;
- Audio encoding based on DSP Group TrueSpeech 8.5 (TM) voice codec.

Optional features:

- MPEG-4 FAP encoding (Version 1, Part 2 Visual, Simple FA Profile) on request;
- Possibility to integrate different audio encoders

The most recent version of EptaPublisher-Text includes an additional functionality that allows the emotional animation of the model, both in terms of emotional facial expressions and emotional speech synthesis. The text supplied to the EptaPublisher-Text can be interleaved with suitable tags that specify the emotional content of the message that must be reproduced by the animation. The emotional tags are interpreted by the system and translated in consistent visual and acoustic emotional cues.

The objective is reached by matching the facial expression of the model to the desired emotion and by modifying accordingly some parameters of the TTS. The set of emotions supported by EptaPublisher-Text is currently limited to the six emotions defined in MPEG-4: joy, sadness, surprise, anger, fear and disgust. Future upgrades of EptaPublisher-Text will improve the rendering of the emotions as well as the user interface for selecting the emotional contexts.

16.3.2 EptaPublisher-Live

EptaPublisher-Live is able to generate animation contents based on the real speaker's voice recording and on real-time capturing of his/her facial movements. The functional description of EptaPublisher-Live provided in Figure 16.4 shows the acquisition of facial movements by means of a motion tracking system integrated on a light helmet that the speaker wears on his/her head, and the consequent translation into MPEG-4 FAPs capable of generating an animation sequence through the EptaPlayer. The face and head tracking system, composed of the helmet, a video/IR camera and a gyroscope together with the associated software driver, is commercialized by a German company, X-IST Real-time Technologies GMbH (*http://www.x-ist.de*).

Rotations of the head are measured directly by the gyroscope while two-dimensional facial movements are estimated from the displacement of a set of reflective markers that are suitably distributed on the speaker's face. As shown in Figure 16.5, we use 10 markers whose displacements, together with the gyroscope measurements, are then translated into 19 FAPs. The reference marker positioned on the speaker's nose is used to detect undesired movements of the helmet during the acquisition.

The translation of the marker displacements into FAPs and the subsequent animation are done in real time.

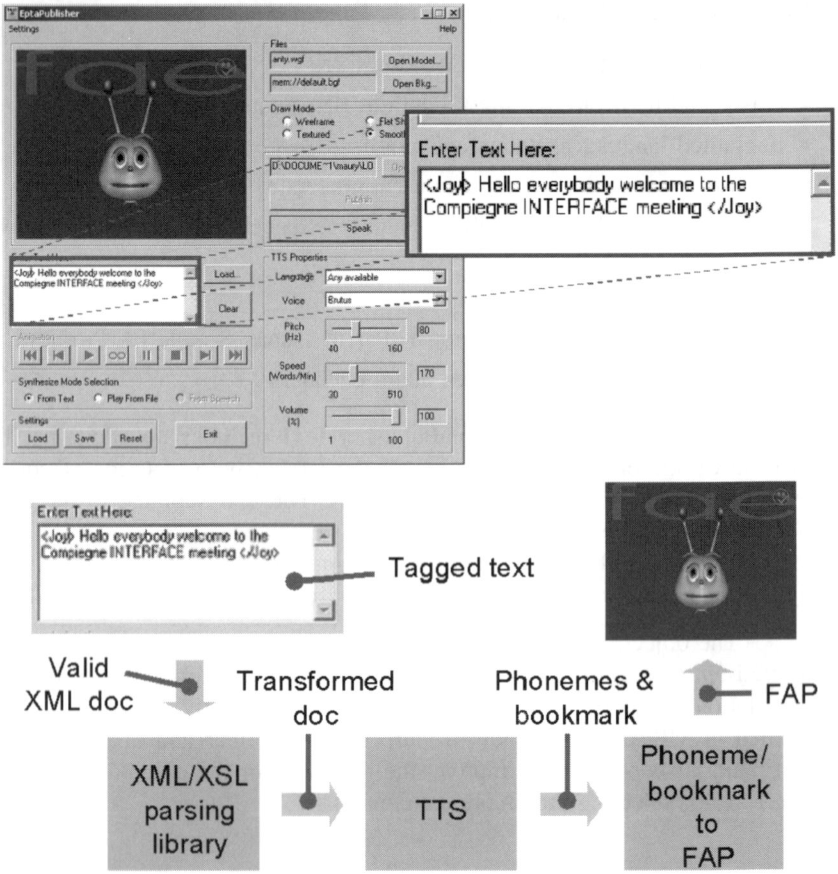

Figure 16.3 EptaPublisher-Text allows the user interleave the text with suitable tags that specify an emotional context. Tags are parsed and translated into audio (TTS parameters) and visual (FAPs) emotional cues

Figure 16.4 Functional description of EptaPublisher-Live

16.3.3 EptaPublisher-Voice

EptaPublisher-Voice, currently still under development, will be soon offered as a further modality to create FA contents. It will allow for the creation of animation sequences from natural audio: recording one's voice will be sufficient to generate a synthetic sequence in which a virtual face repeats the recorded sentence.

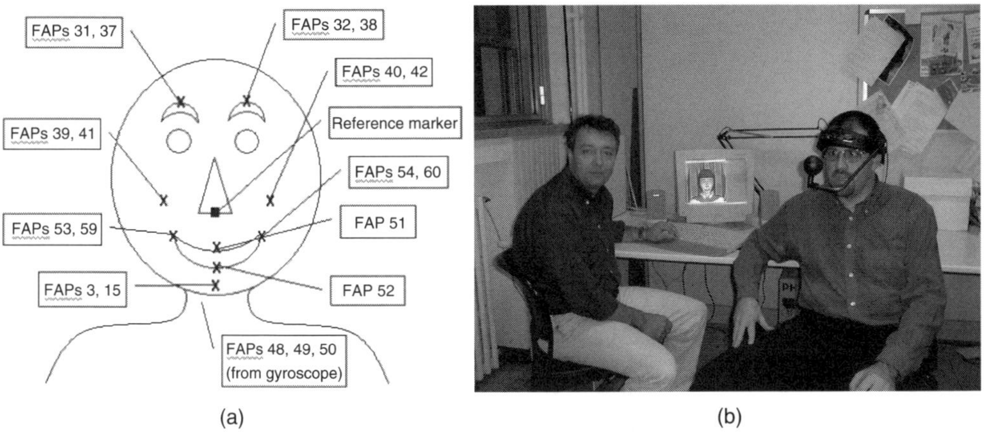

Figure 16.5 (a) Sketch describing the distribution of the reflective markers on the speaker's face, with indication of the FAPs that are estimated from each of them. (b) Photo of the capturing set-up, with the speaker wearing the acquisition helmet and his facial movements reproduced by the synthetic model displayed on the monitor. The photo has been shot with the light on, while in real acquisitions, the room is dark to facilitate IR data processing

16.4 POSSIBLE APPLICATIONS

There are a number of applications in which EptaPlayer and EptaPublisher can be used. We report just some examples of possible applications in this page. EPTAMEDIA can support companies that want to integrate the EptaPlayer/EptaPublisher technology within their products.

- *Web sites*: useless to spend more words on that. EptaPlayer gives the possibility of generating animations on the fly on the client PC at a download cost of 10 kbps or less.
- *Kiosks*: a novel way to provide information to people; easy to upgrade with new contents.
- *Web call centers*: a lot of answers to FAQs can be easily provided by the virtual face.
- *Distance learning*: EptaPlayer/EptaPublisher can be integrated in distance learning applications.
- *Advanced man–machine interfaces*: where there is the need of a more efficient and friendly interface between man and computer.
- *Cultural/historical presentations*: within museums, exhibitions and so on, EptaPlayer can provide up-to-date information to the visitors.
- *Advertising*: a novel approach for advertising through the Web, within mall centers and so on.
- *Interactive CD-ROMs*: the same multimedia content can be interactively reproduced by different faces, maybe chosen by the user. Minutes of animations stored in very few kilobytes.

Figure 16.6 EptaPlayer/EptaPublisher have been used here to implement a demonstration of their possible use within an e-commerce application. The talking head guides the customer through the commercial site and provides all the needed support to the purchase

The most relevant application scenarios for the EPTAMEDIA technology can be grouped as follows.

16.4.1 E-commerce Applications

Many existing web commercial sites could benefit immediately from this technology (Figure 16.6). Instead of using conventional media for announcing, describing and offering products, the use of an animated face could provide a significant added value either in terms of naturalness and interactivity. The animation quality requested for these applications is moderate, therefore reachable immediately. In most cases, a very simple animated character is enough, very likely a fantasy character rather than a humanlike one. The requirements for media synchronization and rendering performances are also moderate. Any entry-level platform is enough for the scope.

The market is enormous and very articulated. Possible customers vary from simple individuals who want to improve their personal Web page to companies interested in integrating their Web site functionality up to the Internet provider that designs, sells and installs e-commerce portals.

The creation of content can be done reasonably off-line and this activity represents a self-standing business on its own. For such reasons, it can be considered a follow-up activity with a separate market. The creation of content can be done, as an example, using suitable hardware/software (hw/sw) studio operating on-line and could involve professional actors/speakers.

16.4.2 Multimedia Contents Production

The industry of multimedia publishing is increasing dramatically. All that was on paper up to now is migrating to the electronic format assessable on-line or on a CD-ROM/DVD. The face/body animation technology can play a very important role for any product requiring an interacting virtual reader or teacher. Main examples of application are related to using the 'virtual actor' to teach foreign languages, dancing or singing, to train people to use specific tools, to help the user in different operations or consulting and to tell stories to children.

The investments for integrating such technology into this kind of application are variable, ranging from negligible figures for CD-ROM or DVD-based applications to medium figures when real-time data streaming over IP is requested.

Also, in this case, the creation of contents can be seen as a separate business. The level of quality requested is rather high, as is the level of naturalness. The hw/sw profile requested for the rendering platform is therefore a bit higher than entry-level. In case of IP transmission, the bandwidth requested for real-time performances is anyway very low due to the very compact bit stream produced by this technology.

16.4.3 Web Virtual Guides

The Web site described in Figure 16.7 has been recently created by EPTAMEDIA under a contract with the Province of Genova for describing and promoting the cultural/natural heritage of the surroundings of Genova. The animation has been generated using the natural speech of a speaker whose facial movements have been captured using EptaPublisher-Live. The Web site is at present only in Italian but will be soon extended to English.

The Web site reported in Figure 16.8 refers to a similar initiative currently submitted to the municipality of Genova and oriented to provide a virtual guide for visiting Genova scheduled to be Capital City of European Culture in 2004.

16.4.4 Newscasting

Another application concerns the dissemination of information about daily news, sport events, cultural heritage or medical care (Figure 16.9). Cost-effective systems might be implemented, based on multilingual TTS that would allow almost real-time news updating by simple text input. More sophisticated systems would require the use of EptaPublisher-Live with true speakers for realistic voice and facial animation. The market is very big especially as far as commercial spots are concerned. Main customers for this technology are seen in e-newspapers, e-TV programs, news agencies and e-publishing houses.

16.4.5 Tele-Learning

A promising field of application for the FA technology is that of tele-learning. In this framework, a virtual lecturer can present one or more topics to the student as a valid

Figure 16.7 The Web site shown in this picture has been created by EPTAMEDIA under a contract from the province of Genova

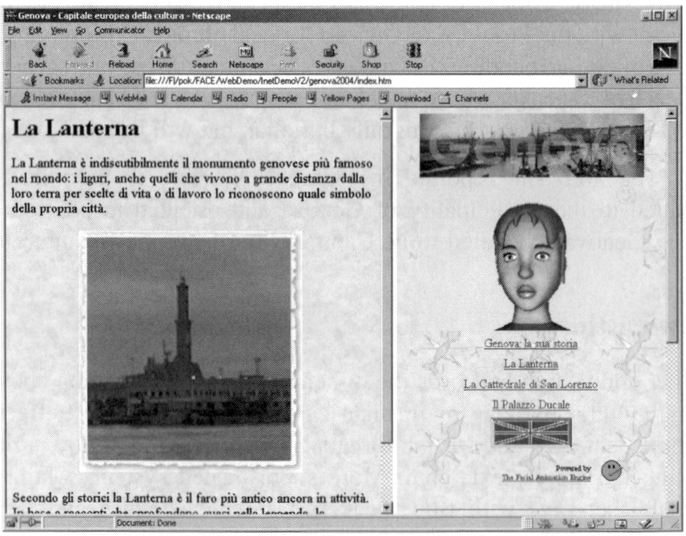

Figure 16.8 In this Web site, the talking head provides geographical, historical and cultural information to the tourist, guiding him/her through a virtual tour of the city of Genova

alternative to the transmission of video lessons or, in alternative, a virtual tutor can answer students' questions and/or supervise them during exercises. In this case, the animated head acts as a front-end to a knowledge base that can be extended over time (Figure 16.10). A hybrid system might use a dialog manager to answer the most common questions, and then occasionally redirect them to a human operator when

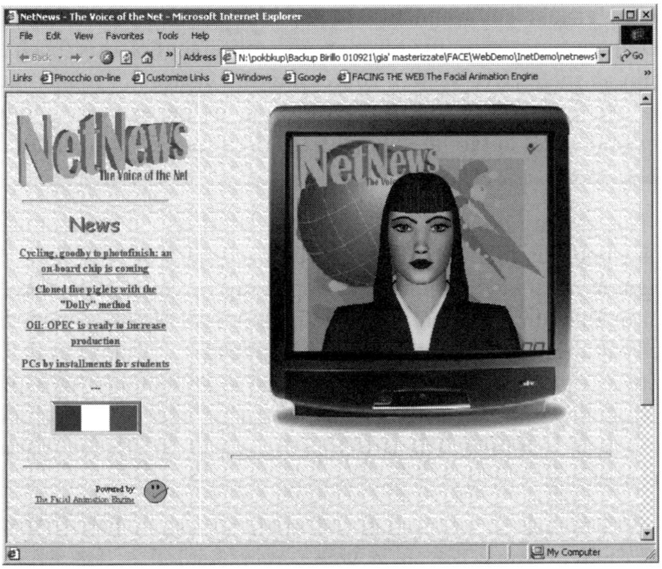

Figure 16.9 EptaPlayer/EptaPublisher have been used here to implement a demonstration of their possible use for web news casting. The virtual speaker announces in different languages news supplied by the press agency

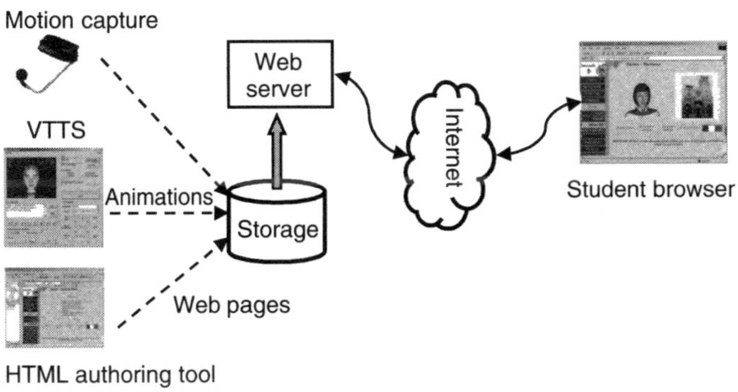

Figure 16.10 Authoring and deployment of web-based courses with the EPTAMEDIA FA technology. A dedicated server is not required, as the animation and audio streams can be stored on any web server

no suitable answer can be found inside the knowledge base. Implementing a tutoring system requires also the capability to monitor user behavior and react immediately to his/her inputs. The potential market for this kind of applications is very large.

16.4.6 Entertainment

Applications in the entertainment field have a huge potential, both for self-contained products such as gadgets or toys and for Internet-based products operating within net-shared virtual environments. Other interesting fields of application concerns the entertainment industry in the business of amusement resorts (such as Disneyland parks) or within an interactive system addressing the visitors of malls, museums or exhibition centers. An interesting area of application, as an example, addresses entertainment systems or devices apt to interact with children for educational, training or rehabilitation purposes.

Net-gaming applications represent definitely an interesting area of application. Face and body animation (FBA) can be used here to reproduce the virtual appearing of net-gamers (close to his/her true appearing or completely fake) and let their clones interact in shared mixed realities. The quality issue is of primary importance here. The level of naturalness requested for these applications is close to that of computer games. Interactions must rely on very sophisticated analysis tools for capturing human parameters. Speech, facial and body gestures must be tracked and reproduced at distance, in most cases, in real time. The stress is also posed on efficient parameter streaming and synchronization.

With reference to this kind of applications, EPTAMEDIA has recently implemented EptaPublisher-Text in a web-server version under a contract with the Carlo Collodi Italian Foundation, owners of the rights over Pinocchio. The realized system provides the functionality for sending multimedia cards via e-mail, which are pronounced by Pinocchio. The text of the message, together with information about the sender and e-mail address of the destination is supplied to the Pinocchio server (*http://www.pinocchioland.com*) through the user interface shown in Figure 16.11. The sender has also the possibility to preview the card before sending it. Once the message is sent, an e-mail is delivered to the destination address to inform about the presence of a message from Pinocchio and to provide the link to the dynamic page created by the Pinocchio server and containing the message. If the browser of the destination client

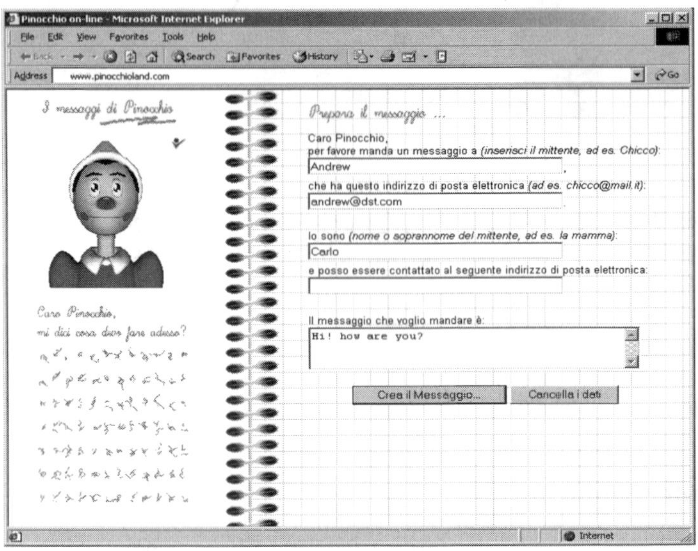

Figure 16.11 Main page of the Web site realized for the Collodi Foundation

has the EptaPlayer plug-in already installed, the message can be viewed immediately by clicking on the link. Otherwise, a dialog window is opened for the automatic download and installation of the suited plug-in.

The system is presently working in Italian only. Next releases will be extended to other languages and will also include the functionality that allows the user to interleave the message text with emotional tags.

16.4.7 Web Call Centers

Web call centers are representing increasingly a valid alternative to conventional telephone call centers. The possibility to transmit efficiently voice over IP and of browsing voice messages with video and graphics can make this solution preferable to the telephone-based one. A major technology in this kind of applications is that of speech synthesis. The economic convenience of this type of solution depends on the principle of maximizing the efficiency of the system while minimizing the number of the operators. A residual presence of real operators is left for managing exceptional situations. Most of the answers to users are predefined and prerecorded in suitable databases. They are extracted from the base of knowledge through intelligent queries and algorithms of artificial reasoning. The opportunity for integrating the FA technology in this kind of application is very promising and the market is potentially large. The quality of the animation must be rather high, together with the naturalness of expressions and of the synthesized speech.

16.4.8 Synthetic Video Over Mobile

FBA technologies are definitely suitable for a variety of mobile applications. The very compact parameter bit stream required for the animation makes its use feasible also for 2G mobile phones. 3G mobile technology, addressing large bandwidth and oriented to multimedia transmission, can exploit synthetic video for many applications. The computational power of the terminal equipment is expected to increase along with the technology of their displays. The hw/sw profile of the terminal should be able to allow efficient animation in a short time. As an example, FBA on the mobile terminal could be used for reading short messages or announcing weather forecasts. Major competitors in mobile telephony are constantly looking for innovative technologies to integrate into their systems. A variety of innovative network or terminal-based applications could be built on top of FBA.

This market is huge. Mobile telephony, through 3G technology, is expected to overcome wired telephony in terms of the number of subscribers, in a few years from now. The access to Internet through a 3G mobile terminal is also expected to get closer, in terms of number of accesses and duration of connections, to that conventionally performed through a wired terminal.

Appendix 1

Evaluating MPEG-4 Facial Animation Players

Jörgen Ahlberg, Igor S. Pandzic and Liwen You

Linköping University, Linköping, Sweden

1 ABSTRACT

In this appendix, we describe a method to investigate how well animated face models can express emotions when controlled by low-level MPEG-4 Facial Animation Parameters (FAPs) reproducing the facial motion captured from real persons acting out the emotions. We propose a benchmark procedure for evaluating the expressiveness of a face model and compare it to other face models as well as to real video. The benchmark is reproducible, and the software and data used are available to the public.

2 INTRODUCTION

Despite the fact that the MPEG-4 standard has existed for a few years now, as have several MPEG-4-compliant face animation systems, there has been little effort made to evaluate the standard (or the face models used) in terms of subjective quality of reproduced facial motion. Even if it is clear that the actual feature point motion can be reproduced, it is not evident that the synthetic faces have the expressiveness of real, human faces. On the contrary, synthetic faces can rarely convey emotions as well as a real face. Yet, this is of great importance for the deployment of synthetic faces if they are going to be widely used in any of the applications mentioned above. For example, experiments show that animated human faces add value to computer interfaces, provided that the face animation is good [1]. An unnatural-looking face animation might instead worsen the impression given by a computer interface.

We propose in the paper, a benchmark for measuring the subjective quality of a face model in terms of how well it can convey emotions (via facial expressions) to human observers. The procedure is reproducible, and all our test data is available to the public. In the following sections the goal of the experiment is defined, the acquisition of the test data and the execution of the test are described and the relative and absolute quality measures for face models is defined. Since the author encourages others to repeat the test with their own face models, one section explains how to acquire our test data and reproduce the experiment.

We specifically want to measure how well a given face model can express emotions when controlled by low-level MPEG-4 FAPs captured from real persons acting out the emotions. The expressiveness should be judged by the accuracy rate of human observers recognizing the facial expression being shown. We want to relate the result to the *ideal case*, the *random case* and the *real case*. The ideal case is when all emotions are correctly recognized and the random case is when the recognition results are completely random, that is, drawn from a uniform distribution. The real case corresponds to the recognition rate for the real faces recorded on video, that is, the faces whose motion the synthetic faces try to reproduce. The recognition rate for each face model is supposedly better than the random case but worse than the real case.

3 ACQUIRING TEST DATA

In order to perform the test, it was necessary to record video sequences of persons acting to show different emotions. Additionally, the corresponding FAP sequences were required so that synthetic sequences reproducing the facial motion from the real video sequences could be created. Thus, the 3-D motion of the head and a subset of the MPEG-4 facial feature points had to be tracked.

For this tracking, we used a head tracking equipment in which IR-reflecting markers are attached to the faces (at the feature points to be tracked) of the actors. The system uses four IR-sensitive cameras to recover the full 3-D motion of each feature point, operating at about 50 frames per second when tracking 30 markers. In our case, 22 feature points, including feature points around the mouth, eyebrows and the nose were tracked. Additionally, special glasses that had five markers on them were worn by the actor to recover the 3-D rigid head motion.

The actor then showed facial expressions corresponding to the six basic emotions available as MPEG-4 high-level FAPs (fear, anger, surprise, sadness, boredom, happiness). The actor expressed each emotion for a few seconds, interspaced with neutral facial expression. Also, the actor showed these expressions while reading a sentence. The sentence chosen was 'The small room was completely empty'. This sentence was carefully selected as being easy to say in each of the aforementioned emotional states – it is easy to imagine being surprised as well as angry and so on due to the small room being empty.

Thus, animation sequences of head- and feature-point motion were recorded for each actor. Simultaneously, the actors were also filmed using an ordinary video camera, thus creating the real sequences for the test. After removing sequences judged as useless as a result of tracking failure or too low quality acting, 21 sequences were left. From those sequences, MPEG-4 FAP files were created.

3.1 Synthetic and Natural Video Sequences

As mentioned in the preceding section, two types of digital video sequences were created – real and synthetic ones. The real sequences were created by digitizing the video sequences recorded by the ordinary video camera while doing the face and facial feature tracking. The synthetic sequences were created by two face-animation systems to test their two face models. For this experiment, we have chosen the web-based facial animation player (using the face model Jorgen) described in Chapter 4 and the Facial Animation Engine (using the face model Oscar) described in Chapter 5. The FAP sequences created from the face and facial feature tracking were input into these two systems, producing 42 synthetic sequences.

4 SUBJECTIVE EVALUATION BY HUMAN OBSERVERS

To evaluate the sequences, a number of subjects watched the real and the synthetic sequences and tried to recognize the facial expression that the actors/face models were showing. The experiment was performed so that a group of subjects entered a room with a video projector and a canvas. Each test subject was given a (paper) form, in which each video showing one of the facial expressions should be judged. When ready (equipped with a form and a pencil), the subjects were shown the instructions and a few training video sequences. Then, for each video sequence, the following was shown:

1. A screen indicating that a particular video sequence number n was to be shown.
2. The video sequence was shown one or more times. The number of times was determined so that the total playing time for each sequence was approximately 10 seconds.
3. A screen indicating that video sequence number n was shown and that it was time to fill in the form at row n.

To make the experiment practical, it was automated as much as possible. Using a SMIL (Synchronized Multimedia Integration Language) script [2], playable by RealPlayer [3], all the instructions and video sequences to be shown to the subjects were ordered and timed in advance. The SMIL scripts were generated automatically by a Matlab-program, taking as input the number of models to be tested, the number of subject groups and so on.

Each video was watched by more than 100 subjects in this way.

5 THE BENCHMARK

The results of the face models' ability to convey emotions should be measured both in an absolute way as well as in relation to the performance of the real videos. We have thus chosen the error measure computed as described in the following text.

First, compute the dispersion matrix for each model and for the real videos. Those three test matrices (the two synthetic ones and the real one) should be compared with the ideal dispersion matrix and the random matrix, corresponding to perfect recognition and totally random recognition.

Table 1 Absolute and relative expressive performances of the models and the real video

Animation	Oscar model (FAE)	Jorgen model (MpegWeb)	Real video
AEP	9.1	9.4	58.1
REP	15.6	16.2	n/a

Note: REP: Relative Expressive Performance

The distance from a test matrix to the ideal matrix is computed as the L1-norm of the difference, and is then scaled so that the random matrix gets the performance value zero and the ideal matrix gets the performance value 100. We call this measure the Absolute Expressive Performance (AEP), expressed as

$$\mathrm{AEP}(X) = 100 \cdot \frac{\|R - I\|_1 - \|X - I\|_1}{\|R - I\|_1},$$

where X is the (synthetic or real) test matrix, R is the random matrix, and I is the ideal matrix. The L1-norm of a matrix is defined as

$$\|X\|_1 = \sum_{i,j} |X_{ij}|$$

where X_{ij} is the element at row i and column j in the matrix X.

The relative measure of a face model's expressive performance is defined as

$$\mathrm{REP}(X) = 100 \cdot \frac{\mathrm{AEP}(X)}{\mathrm{AEP}(X_{\mathrm{real}})}.$$

The results for the experiment is shown in Table 1.

To evaluate the validity of the results, a dispersion matrix has also been computed for each subject watching the sequences. Thus, the standard deviation can be estimated and a t-test for statistical significance [4] is done. It is found that the difference between the two face models' performances is not statistically significant on any level, but that the differences between the models and the random, real and ideal case are significant with a very high level of confidence (more than 99%).

6 REPRODUCING THE BENCHMARK

It is the authors' intent that this benchmark procedure should be easily reproducible by anyone wanting to test their face model and/or face animation system, thus offering a standardized way of measuring and comparing its subjective quality. Thus, all the files necessary for performing the test are available to the public, together with instructions on how to use them. Included in the package are all the video files, the real as well as the synthetic ones, the FAP files (in both ASCII form and binary form) for generating new synthetic sequences and the scripts for generating SMIL files and the instruction screens (shown before and between the videos).

To reproduce the test, the package is available for download on the Web [5]. Then, new synthetic videos should be generated (by means of the FAP files) using the face model to be evaluated, as well as new SMIL files. The SMIL files are generated by a Matlab-script, and input parameters are the number of groups of subjects, the time taken for each test and if a new model (new synthetic sequences) is added to the ones included in the package. Detailed instructions are included in the package.

7 CONCLUSION

From our experiments, it is clear that the face models we have evaluated have a far worse expressive performance than the real sequences, but no significant difference could be measured between the two models. Our main result is the description of a reproducible benchmark that anyone can perform to evaluate his/her face models.

ACKNOWLEDGMENT

The authors thankfully acknowledge the financial support from the national Swedish SSF project VISIT (VISual Information Technology). The authors would also like to thank Telia Research AB and the Natural Language Processing Lab at the Dept. of Computer and Information Science at Linköping University for using their time and tracking equipment, particularly Mustapha Skhiri for his help in acquiring and processing the raw data. We would also like to acknowledge Jacob Ström, formerly at Linköping University, now at Ericsson Media Lab, for his initial involvement in this project and contribution of ideas (in particular the 'small room' sentence).

REFERENCES

[1] I. S. Pandzic, J. Ostermann, D. Millen, "User evaluation: synthetic talking faces for interactive services", *The Visual Computer Journal*, **15**(7-8), 330–340 (1999).

[2] Synchronized Multimedia Integration Language, W3C Recommendation REC-smil-19980615, *www.w3.org/TR/REC-smil/*.

[3] RealPlayer, *www.real.com*.

[4] J. Cohen, *Statistical Power Analysis for the Behavioral Sciences*, 2nd ed., Lawrence Erlbaum Associates, Hillsdale, N.J., 1988.

[5] Image Coding Group, Linköping University, *www.icg.isy.liu.se/fba_benchmark*.

Appendix 2
Web Resources

This is a list of Web resources relevant to this book and to MPEG-4 FA in general.

URL	Description
mpeg.telecomitalialab.com	The MPEG home page
	Data package for the MPEG-4 FBA player benchmark described in Appendix 1
www.icg.isy.liu.se/~igor	Further information and demonstrations of Web-based FA from Chapter 4
www.eptamedia.com	Catalogue of products based on MPEG-4 facial animation technologies. Possibility to download videos and plug-ins.
www.dsp.dist.unige.it/~pok	Reports and references about on-going research on facial animation; contains some standard FAP test sequences.
www.dsp.dist.unige.it/snhc/fba_ce/fba_ce.htm	MPEG-4 FBA Core Experiments official site; contains FAP and FDP test data. Requires MPEG membership password.
www.f2f-inc.com	face2face animation, Inc
www.dis.uniromal.it/~pelachau	Further information on Conversational MPEG-4-based Agent
www.winteractive.fr	W Interactive SARL, a French company producing virtual characters for the Web.
www.vhml.org	Virtual Human Markup Language
playmail.research.att.com	AT&T's talking e-mail service.
www.research.att.com/projects/AnimatedHead/	Further information on facial animation research at AT&T
www.image.ntua.gr/mpeg4book/	Detailed results of experiments described in Chapter 9.
www.visagetechnologies.com	Company providing a range of MPEG-4 FA products.

Index